Problem Solving in Enzyme Biocatalysis

Problem Solving in Enzyme Biocatalysis

ANDRÉS ILLANES, LORENA WILSON AND CARLOS VERA

School of Biochemical Engineering
Universidad Católica de Valparaíso, Chile

WILEY

Library of Congress Cataloging-in-Publication Data

Illanes, Andrés.
  Problem solving in enzyme biocatalysis / Andrés Illanes, Lorena Wilson, Carlos Vera.
    pages cm
  Includes index.
    ISBN 978-1-118-34171-1 (hardback)
  1. Enzymes–Biotechnology. 2. Biocatalysis. 3. Enzyme kinetics. 4. Enzymes–Industrial applications. I. Wilson, Lorena. II. Vera, Carlos. III. Title.
    TP248.65.E59I45 2013
    572′.7–dc23

                                                                                    2013029486

A catalogue record for this book is available from the British Library.

ISBN: 9781118341711

Set in 10/12pt, Times by Thomson Digital, Noida, India
Printed and bound in Malaysia by Vivar Printing Sdn Bhd

1 2014

Contents

Preface

You shall bring forth your work as a mother brings forth her child: out of the blood of your heart. Each act of creation shall leave you humble, for it is never as great as your dream and always inferior to that most marvelous dream which is Nature.
"Decalogue of the Artist," Gabriela Mistral, Chilean Nobel Laureate

This book is primarily intended for chemical and biochemical engineering students, but also for biochemists, chemists, and biologists dealing with biocatalytic processes. It was purposely written in a format that resembles the Schaum's textbooks of my long gone college years. An abridged coverage of the subject is provided in each chapter, followed by a number of sample exercises in the form of solved problems illustrating resolution procedures and the main concepts underlying them, along with supplementary problems for the reader to solve, provided with the corresponding answers. Learning through problem solving is designed to be both challenging and exciting to students. There is no book on enzyme biocatalysis with this format and purpose, so we sincerely hope to have made a contribution to the toolbox of graduate and undergraduate students in applied biology, chemical, and biochemical engineering with formal training in college-level mathematics, organic chemistry, biochemistry, thermodynamics, and chemical reaction kinetics. The book pretends to be a complement of a previous book, *Enzyme Biocatalysis* (Springer, 2008), which I had the privilege and pleasure to edit.

Chapter 1 is an introduction, giving an updated vision of enzyme biocatalysis, its present status, and its potential development. Chapters 2 and 3 refer to enzyme kinetics in homogeneous and heterogeneous systems, respectively, considering both fundamental and practical aspects of the subject. Chapter 4 is devoted to enzyme reactor design and operation under ideal conditions, while Chapters 5 and 6 deal with the main causes of nonideal behavior (mass-transfer limitations and enzyme inactivation, respectively). Chapter 7 presents an overview of the optimization of enzyme reactor operation and different tools for optimization. The book is complemented by two documents. Appendix A refers to mathematical methods for those readers not sufficiently acquainted with the subject, while Epsilon Software Information presents a software program (εpsilon) for solving and representing enzyme reactor performance under different scenarios.

Writing is certainly a most exciting endeavor. Undoubtedly, writing about enzymes is beyond being the subject of our expertise, because enzymes are the catalysts of life, so writing about enzymes lies in the boundaries of writing about life. And life, as the Iraqi poet Abdul Wahab said, is about standing straight, never bent or bow, about remaining glad, never sad or low. We hope our readers will keep straight and glad while getting a little more acquainted with these catalysts of life.

A book is a journey from expectation to consolidation, an act of love from conception to birth. It has been a most rewarding experience to travel and conceive in the company of my co-authors, two brilliant young colleagues of mine, Dr. Carlos Vera and Dr. Lorena Wilson, who were formerly my students and now fly well over my shoulders. My gratitude also to my colleague Dr. Raúl Conejeros, who authored Chapter 7 and beyond that contributed to spice this book throughout with mathematical modeling and elaborated thinking. Do not blame him too much; he is a good mix of wisdom and kindness. Our gratitude also to Dr. Felipe Scott for developing, together with Dr. Carlos Vera, the software Epsilon, presented in Epsilon Software Information to the development of the Epsilon Software.

A significant part of this book was written in Spain, at the Chemical Engineering Department of the Universitat Autònoma de Barcelona, where I was (again) warmly hosted while undertaking a sabbatical (my third one there) during the second semester of 2012. My personal gratitude to my colleagues there, Dr. Josep López and Gregorio Álvaro, who made me feel at home and who shared the ecstasy and the agony of this pregnancy.

My gratitude to the people at my institution, the Pontificia Universidad Católica de Valparaíso, who supported and encouraged this project, particularly to the rector, Professor Claudio Elórtegui, and the Director of my School of Biochemical Engineering, Dr. Paola Poirrier. Special thanks to Dr. Atilio Bustos, Head of the Library System, for his enthusiastic support and advice. My deepest appreciation also to the people at Wiley: Rebecca Stubbs and Sarah Tilley, and Shikha Pahuja at Thomson Digital, who were always helpful, warm, and supportive.

Last but not least, my gratitude to my life partner Dr. Fanny Guzmán, expert in peptide synthesis, loving care, and *savoir-vivre*.

Engineering is about products and processes. A book is both. We hope you will enjoy the product as much as we have enjoyed the process.

Andrés Illanes
Valparaíso, July 2013

Nomenclature

Symbol	Description	Dimensions	Chapter No.
A			
a	molar concentration of substrate A	$[ML^{-3}]$	2, 4
a	cube edge; major axis of oblate ellipsoid	$[L]$	3
a	constant.		Appendix A
A	surface area of catalyst particle	$[L^2]$	3
A	sectional area of reactor	$[L^2]$	5, 7
A	specific activity ratio of intermediate to initial enzyme species		6
A'	specific activity ratio of final (partly active) to initial enzyme species		6
A_S	surface area of catalyst	$[L^2]$	3
a_i	initial molar concentration of substrate A	$[ML^{-3}]$	4, 6
a_{sp}	catalyst-specific activity	$[IUM^{-1}]$	3, 4, 5, 6, 7
B			
b	molar concentration of substrate B	$[ML^{-3}]$	2, 4
b	minor axis of oblate ellipsoid		3
B	vector of parameter		7, Appendix A
B^*	optimum vector parameter		Appendix A
b_0, b_i, b_{ii}	zero-, first-, and second-order parameters for the surface-of-response model		7
B_1^*, B_2^*	vectors of first- and second-order regression coefficients of the surface-of-response model		7
b_i	initial molar concentration of substrate B	$[ML^{-3}]$	4, 6
b_i	parameter i		Appendix A
Bi	Biot number		3, 5
b_{ij}	second-order interaction parameters for the surface-of-response model		7
C			
c	molar concentration of enzyme–substrate complex (ES)	$[ML^{-3}]$	2
C	constant		4, 5, Appendix A
[C]	concentration of analyte C	$[ML^{-3}]$	2, 3, 4, 5, 6, 7, Appendix A

Symbol	Description	Dimensions	Chapter No.
CA_i	annual cost of item i	[USD]	7
ce	coenzyme concentration	$[ML^{-3}]$	2
ce'	modified coenzyme concentration	$[ML^{-3}]$	2
D			
d	molar concentration of enzyme–inhibitor complex (EI)	$[ML^{-3}]$	2
D	differential operator		Appendix A
D_0	diffusion coefficient in water	$[L^2T^{-1}]$	3
D_{eff}	effective diffusion coefficient	$[L^2T^{-1}]$	3, 5, 6, 7
d_P	diameter of catalyst spherical particle	[L]	3, 6
E			
e	concentration of active enzyme	$[IUL^{-3}]$	2, 4, 5, 6, 7, Appendix A
E	enzyme activity	[IU]	4, 5, 6
E	relative error or enzyme species		Appendix A
e_0	initial concentration of active enzyme	$[IUL^{-3}]$	2, 6, 7
E_0	initial enzyme activity	[IU]	6
E_a	energy of activation in Arrhenius equation	$[ML^2T^{-2}]$	2, 5, 7
E_C	contacted enzyme activity	[IU]	3
ee_S	enantiomeric excess of the S-enantiomer		3
E_I	immobilized enzyme activity	[IU]	3
E_{ia}	energy of activation of inactivation in Arrhenius equation	$[ML^2T^{-2}]$	2, 3, 7
$E_{ia, M}$	energy of activation of inactivation in Arrhenius equation in the presence of a modulator	$[ML^2T^{-2}]$	7
E_L	enzyme activity lost by immobilization	[IU]	3
em	molar concentration of enzyme–modulator M complex	$[ML^{-3}]$	6
E_R	enzyme activity remaining in solution after immobilization	[UI]	3
E(t)	enzyme activity over time	[IU]	6
F			
f	molar concentration of enzyme–inhibitor–substrate complex (EIS)	$[ML^{-3}]$	2
f	size distribution frequency		3
f	number of factors assessed by the surface-of-response methodology		7
F	ratio of V_{max} to V'_{max}		2
F	reactor feed flow rate	$[L^3T^{-1}]$	4, 5, 6, 7
F_0	initial reactor feed flow rate	$[L^3T^{-1}]$	6

Symbol	Description	Dimensions	Chapter No.
F_0	Fisher-Snedecor distribution for a null hypothesis		7
$f(x)$	function of vector x		Appendix A
$F(x)$	vector of functions of x		Appendix A
$f(x,y)$	implicit function of vector x and dependent variable y		Appendix A
$F_{\alpha,g,h}$	Fisher–Snedecor distribution for a confidence level of $1 - \alpha$ with g degrees of freedom in the numerator and h degrees of freedom in the denominator		7
H			
h	film volumetric mass-transfer coefficient for substrate	$[L^3T^{-1}]$	3, 6
h	packed-bed reactor length	$[L]$	7
h	integration step size		3,5, Appendix A
H	total reactor length	$[L]$	7
h'	film linear mass-transfer coefficient for substrate	$[LT^{-1}]$	3, 5
h^+	proton (molar) concentration	$[ML^{-3}]$	2, 3
h_0^+	proton (molar) concentration in the bulk medium	$[ML^{-3}]$	3
h_P	film volumetric mass transfer coefficient for product	$[L^3T^{-1}]$	Epsilon Software Information
I			
i	inhibitor molar concentration	$[ML^{-3}]$	2
J			
J	Jacobian matrix		Appendix A
J	substrate flow rate	$[MT^{-1}]$	3
J	objective function		7
K			
k	catalytic rate constant		2, 4, 5, 6, 7, Appendix A
K	equilibrium constant of dissociation of enzyme–substrate complex into enzyme and substrate	$[ML^{-3}]$	2
k_0	pre-exponential term in Arrhenius equation		2, 7
k_B	Boltzmann universal constant		3
k_D	first-order inactivation rate constant	$[T^{-1}]$	2, 3, 6, Appendix A

Symbol	Description	Dimensions	Chapter No.
K_D	dissociation constant of enzyme–substrate complex into enzyme and substrate	$[ML^{-3}]$	2
$k_{D,0}$	pre-exponential term in Arrhenius equation	$[T^{-1}]$	2, 7
k_{D1}	first-order inactivation rate constant in first stage of inactivation	$[T^{-1}]$	6
k_{D2}	first-order inactivation rate constant in second stage of inactivation	$[T^{-1}]$	6
$k_{D,M}$	first-order inactivation rate constant of enzyme–modulator M complex	$[T^{-1}]$	6
$k_{D,M,0}$	pre-exponential term in Arrhenius equation in the presence of modulator M	$[T^{-1}]$	7
K_{eq}	equilibrium constant		2
k_i	Runge–Kutta method coefficients		Appendix A
K_I	inhibition constant	$[ML^{-3}]$	3, 4
K_{IC}	dissociation constant of enzyme–competitive inhibitor complex	$[ML^{-3}]$	2, 6, 7
K_{INC}	dissociation constant of enzyme–noncompetitive inhibitor complex	$[ML^{-3}]$	2, 4, 6
K'_{INC}	dissociation constant of enzyme–substrate–noncompetitive inhibitor tertiary complex	$[ML^{-3}]$	2, 4
K_M	Michaelis–Menten constant	$[ML^{-3}]$	2, 3, 4, 5, 6, 7, Appendix A
K_{MA}	dissociation constant of secondary complex EA into E and A	$[ML^{-3}]$	2, 6
K'_{MA}	dissociation constant of tertiary complex EAB into EB and A	$[ML^{-3}]$	2
K_{MAP}	apparent Michaelis–Menten constant	$[ML^{-3}]$	2, 3
K_{MB}	dissociation constant of secondary complex EB into E and B	$[ML^{-3}]$	2
K'_{MB}	dissociation constant of tertiary complex EAB into EA and B	$[ML^{-3}]$	2, 6
$K_{M,P}$	dissociation constant of EP into E and P (reversible reaction)	$[ML^{-3}]$	2
$K_{M,S}$	dissociation constant of ES into E and S (reversible reaction)	$[ML^{-3}]$	2
K_P	electrostatic partition coefficient at the matrix–medium interface		3
K_P	inhibition constant by product		4, Epsilon Software Information

Symbol	Description	Dimensions	Chapter No.
K_S	dissociation constant of the inactive tertiary complex ESS into ES and S		2, 3, 4, 5, 6, 7, Appendix A
$K_{p,S}$	partition coefficient for substrate		3
K_{p,h^+}	partition coefficient for protons		3
L			
L	optical path in spectrophotometer cell		2
L	catalytic slab width	[L]	3, 5, 7
L_{eq}	equivalent catalyst particle length	[L]	3, 7
L_R	packed-bed reactor length	[L]	5
M			
m_{cat}	concentration of biocatalyst	$[ML^{-3}]$	4, 5, 6
M_{cat}	mass of immobilized enzyme catalyst	[M]	3, 4, 5, 6, 7
m_i	characteristic equation root		Appendix A
MW	molecular weight		7
N			
n	stoichiometric coefficient of product with respect to substrate		4
n	number of observations		7, Appendix A
N	number of sections		Appendix A
N_1	lumped modulation factor in first stage of enzyme inactivation		6
n_{1P}	modulation factor by product in first stage of inactivation		6
n_{1S}	modulation factor by substrate in first stage of inactivation		6
N_2	lumped modulation factor in second stage of enzyme inactivation		6
n_{2P}	modulation factor by product in second stage of inactivation		6
n_{2S}	modulation factor by substrate in second stage of inactivation		6
n_M	modulation factor by modulator M		6, 7
N_R	number of staggered reactors		6
$N_{t1/2}$	number of half-lives of catalyst use		6
O			
OD	optical density		2
P			
p	product molar concentration	$[ML^{-3}]$	2, 3, 4, 7
p	number of parameters		7, Appendix A

Symbol	Description	Dimensions	Chapter No.
p_0	molar concentration of product in the bulk reaction medium	$[ML^{-3}]$	3
P_C	contacted protein	$[M]$	3
P_I	immobilized protein	$[M]$	3
P_R	protein remaining in solution after immobilization	$[M]$	3
p_S	molar concentration of product at the biocatalyst surface	$[ML^{-3}]$	3
p_{sp}	specific protein load		3
Q			
q	coupled analyte concentration	$[ML^{-3}]$	2
q	number of regression variables		7
R			
r	variable radius of spherical particle	$[L]$	3
R	ideal gas constant	$[ML^2T^{-2}\theta^{-1}]$	2, 3, 7
r'	rate of transformation of substrate into product		3
R^2	coefficient of determination		Appendix A
r_a	agitation rate	$[T^{-1}]$	3
R^2_{adj}	adjusted coefficient of determination		Appendix A
R_F	allowable flow-rate fluctuation as a result of downstream operations		6
R_P	radius of catalyst spherical particle	$[L]$	3, 5
S			
s	substrate concentration within the catalyst	$[ML^{-3}]$	2, 3, 4, 5, 7, Appendix A
s_0	molar concentration of substrate in the bulk reaction medium	$[ML^{-3}]$	3
s_{0i}	molar initial (or inlet) substrate concentration in the bulk reaction medium	$[ML^{-3}]$	5
s^2	residual mean square for $n-p$ degrees of freedom		Appendix A
$S(B)$	sum of square error		Appendix A
s_i	initial (or inlet) substrate molar concentration	$[ML^{-3}]$	2, 4, 6, 7
s_S	molar concentration of substrate at the biocatalyst surface	$[ML^{-3}]$	3, 7
SS_E	sum of squares due to residuals (error)		7
SS_R	sum of squares due to regression (model)		7

Symbol	Description	Dimensions	Chapter No.
$SS_{R\,k/,p-k}$	regression sum of squares for k variables, given that p minus k variables are already in the model		7
$SS_{R,p-k}$	regression sum of squares for the reduced model containing p minus k variables		7
T			
t	time	$[T]$	2, 3, 4, 5, 6, 7, Appendix A
T	absolute temperature	$[\theta]$	2, 3, 7
t_0	initial time	$[T]$	7
$t_{1/2}$	half-life of enzyme catalyst	$[T]$	2, 6
t_C	total time of one cycle of reactor operation	$[T]$	6
t_{dless}	dimensionless time of reactor operation		4
t_f	final time	$[T]$	7
T_{lo}	temperature lower bound	$[\theta]$	7
$t(n-p, \alpha/2)$	upper $\alpha/2$ quantile of Student's t-distribution for $n-p$ degrees of freedom		7, Appendix A
t_s	reactor staggering time	$[T]$	6
T_{up}	temperature upper bound	$[\theta]$	7
U			
u	control variable		7
u_{lo}	control variable lower bound		7
u_{up}	control variable upper bound		7
V			
v	initial reaction rate	$[ML^{-3}T^{-1}]$	2, 4, 5, 6, Appendix A
v	model parameter vector or vector whose elements are different from zero		7
V	reactor volume	$[L^3]$	7
v′	enzymatic reaction rate	$[MT^{-1}]$	3
v″	reaction rate per unit volume of catalyst	$[ML^{-3}T^{-1}]$	3
V_{bed}	bed volume of packed-bed reactor	$[L^3]$	4, 5
$V(b_i)$	variance of the parameter b_i		Appendix A
$VC(B)$	variance–covariance matrix of B		Appendix A
V_{eff}	effective volume of reaction	$[L^3]$	4, 5, 6, 7
$v_{effective}$	effective reaction rate	$[ML^{-3}T^{-1}]$	5
V_{max}	maximum reaction rate	$[ML^{-3}T^{-1}]$	2, 4, Appendix A
V_{maxAP}	maximum apparent reaction rate of product formation	$[ML^{-3}T^{-1}]$	2

Symbol	Description	Dimensions	Chapter No.
V_{maxAP0}	pre-exponential term in Arrhenius equation	$[ML^{-3}T^{-1}]$	2
$V_{max,P}$	maximum reaction rate of product to substrate conversion (reversible reaction)	$[ML^{-3}T^{-1}]$	2
$V_{max,S}$	maximum reaction rate of substrate to product conversion (reversible reaction)	$[ML^{-3}T^{-1}]$	2
V'_{max}	maximum reaction rate of product formation from enzyme-substrate-inhibitor tertiary complex	$[ML^{-3}T^{-1}]$	2
V'_{max}	maximum reaction rate	$[MT^{-1}]$	3, Epsilon Software Information
V''_{max}	maximum reaction rate per unit volume of catalyst	$[ML^{-3}T^{-1}]$	3, 7
V_{proc}	total processed volume	$[L^3]$	6
V_R	reaction volume	$[L^3]$	5
v_s	solute molar volume	$[L^3M^{-1}]$	3
X			
x	variable width of catalytic slab	$[L]$	3
x	state variables		7
x	vector of independent variables		Appendix A
X	limiting substrate to product conversion		3, 4, 5, 6
X	matrix of independent variables		7, Appendix A
X_0	initial (steady-state) conversion of limiting substrate into product		6
x_c	coded variables		7
x_g	vector of values for the model variables $(p \times 1)$		7
x_h	highest value for uncoded input variable		7
x_i	input variable i		7
x_l	lowest value for uncoded input variable		7
Y			
y	dependent variable		Appendix A
Y	vector of observed values		7, Appendix A
$\hat{y}$	predicted response by the model		7
$\bar{Y}$	average of the observed values		7, Appendix A
Y_E	enzyme immobilization yield		3
y_i	output variable i		7
y_i	ith observation for the dependent variable		Appendix A

Symbol	Description	Dimensions	Chapter No.
$\hat{y}_i$	predicted response for the ith observation		7
Y_P	protein immobilization yield		3
Z			
z	dimensionless variable width of catalytic slab		3, 7, Appendix A
Z	valence of ionic species		3

Greek Letters

α	ratio of rate constant of formation of B from E'B and transition rate from EA to E' + B (ping-pong mechanism)		2, 4
α	Damkoehler number		3, 5, 6
α	significance level		7, Appendix A
α_S	Damkoehler number for substrate		Epsilon Software Information
α_P	Damkoehler number for product		Epsilon Software Information
β	dimensionless substrate concentration		3, 7, Appendix A
β_0	dimensionless substrate concentration in the bulk reaction medium		3, 5, Appendix A
β_{0i}	initial (or inlet) dimensionless substrate concentration		5
β_c	dimensionless substrate concentration within the catalyst at the center of the slab		7
β_i	initial (or inlet) dimensionless substrate concentration		6
β_S	dimensionless substrate concentration at the biocatalyst surface		3, 5
γ	dimensionless product concentration $(p \cdot K_P^{-1})$		3
γ_0	dimensionless product concentration in bulk reaction medium $(p_0 \cdot K_P^{-1})$		3, 5
γ_{0i}	initial (or inlet) dimensionless product concentration		5
γ_S	dimensionless product concentration at the biocatalyst surface $(p_S \cdot K_P^{-1})$		3
δ	stagnant liquid film width	[L]	3
δ	relative error tolerance		Appendix A
ΔH^0	standard enthalpy change of dissociation of ES into E and S	$[ML^2T^{-2}]$	2

Symbol	Description	Dimensions	Chapter No.
ΔH_I^0	standard enthalpy change of dissociation of EI into E and I or EIS into ES and I	$[ML^2T^{-2}]$	2
ΔS^0	standard entropy change of dissociation of ES into E and S	$[ML^2T^{-2}\theta^{-1}]$	2
ΔS_I^0	standard entropy change of dissociation of EI into E and I, or EIS into ES and I	$[ML^2T^{-2}\theta^{-1}]$	2
ε	error		3
ε	molar extinction coefficient		2
ε	porosity of the catalyst matrix		3
ε	void fraction of catalyst bed		4, 5, 7
ε	error vector		7, Appendix A
E	enantioselectivity		3
ε_i	error for the ith observation		Appendix A
ζ	tortuosity of the catalyst matrix pores		3
η	local (or surface) effectiveness factor		3, 5
η_G	global (mean integral value) effectiveness factor in the catalyst particle		3, 5, 6, 7
η_P	global effectiveness factor under product inhibition		3
κ	ratio of Michaelis to uncompetitive inhibition constants ($K_M \times K_S^{-1}$)		3
λ	optical wavelength	L	2
λ	dimensionless reactor bed length		5
μ	viscosity of the solution	$[ML^{-1}T^{-1}]$	3
ν	dimensionless reaction rate ($v \times V_{max}^{-1}$)		3, 5, 6
π	volumetric productivity of reactor operation	$[ML^{-3}T^{-1}]$	4
π_{sp}	specific productivity of reactor operation	$[T^{-1}]$	4
ρ	dimensionless radius of the spherical catalyst particle		3, 5, Appendix A
ρ_{app}	apparent density of catalyst	$[ML^{-3}]$	4, 5, 6, 7
σ_i	standard deviation of b_i		Appendix A
τ	fluid residence time in the reactor	$[T]$	4, 5
$\varphi(t)$	enzyme decay function		6
Φ	Thièle modulus for substrate		3, 5, 6, Appendix A
Φ_P	Thièle modulus for product		3
Φ^R	Thièle modulus for R enantiomer		3
Φ^S	Thièle modulus for S enantiomer		3
Ψ	electrostatic potential of the support	$[ML^2T^{-2}Q^{-1}]$	3

Other

$\in$	electron charge		3

Epsilon Software Information

Available alongside this book is a copy of the software εpsilon which allows the simulation of the operation of enzymatic reactors. εpsilon has been designed to illustrate the main topics included in this book, offering an additional tool to improve the understanding of the design and operation of enzymatic reactors (specifically Chapters 4 to 6). Guidelines for software installation are given below:

i. Program Installing

Two situations may occur during installation:

- *Matlab® is already installed in user's computer*
  The program was built using Matlab®'s compiler toolbox. Hence, if Matlab® is installed in the user's computer, open the distrib folder and run the file Epsilon_32. exe or Epsilon_64.ex, depending on the system's architecture.
- *Matlab® is not installed in user's computer*
  First, open the Matlab® package (Epsilon_64_pkg or Epsilon_32_pkg, depending on the system's architecture). Two files will be created in the current directory. Second, open MCRInstaller.exe and follow the installer instructions. Once the installation is complete, open the newly created Epsilon_64.exe or Epsilon_32.exe.

ii. Program Description

This software was designed with a simple interface, with the purpose of allowing the comparison of reactor behavior under different scenarios (reaction kinetics and operation modes). The program considers reactor performance for biocatalysts having the most common enzyme kinetics (Michaelis-Menten, competitive and non-competitive inhibition by product and uncompetitive inhibition by substrate) and operating in usual reactor configurations (BSTR, CPBR and CSTR). The program also allows incorporating the effect of external diffusional restrictions (EDR) or catalyst inactivation during reactor operation. Figure 1 shows the interface of εpsilon.

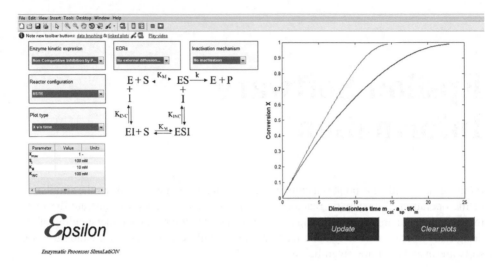

Figure 1 εpsilon interface.

iii. Using Epsilon

The use of εpsilon is straightforward. First, select reaction kinetics; a scheme showing the selected kinetic mechanism should appear. Then input the value of the model parameters into the bottom table. After that, choose the reactor configuration and the plot type desired to show the reactor performance. Furthermore, it is possible to include the effect of EDR and enzyme inactivation in the simulation.

If EDR is selected, introduce the value of the Damkoehler number for substrate, α_s, and product, α_p, (in the case of product inhibition). The former is defined by Equation (3.15); the latter is defined as:

$$\alpha_P = \frac{V'_{max}}{K_P \cdot h_P} \tag{1}$$

where K_P is the inhibition constant by product and h_P is the film volumetric mass transfer coefficient for product. If enzyme inactivation is taken into account, input the value of the corresponding inactivation parameters (see Chapter 6). A scheme of the inactivation mechanism will appear if this option is selected.

To plot reactor performance under the determined conditions, press the "update" button. To compare this condition with another, you will need to repeat the process choosing the new operational conditions, and then press the update button again. Press the "clear plots" button to reset the program.

To simulate the reactor operation under ideal conditions, εpsilon will solve the mathematical expressions presented in Table 4.1 for the corresponding kinetics and reactor mode of operation. The effect of EDR on reactor behavior is simulated using the corresponding Equations (5.7) to (5.9). To calculate the effectiveness factor, it is assumed that, at steady-state, mass transfer rate and reaction rate should be equal at the catalyst surface. So, when product inhibition is considered, the corresponding system of two equations will be solved using Newton's Method (see Appendix A). For the simulation of biocatalyst deactivation, the program considers a two-stage series mechanism (see Equations (6.2) and (6.3)). In the case of continuous reactors, εpsilon will solve Equation (6.16) for a set of given times ranging from zero to final reaction time. Again, the former Equation will be solved using Newton's Method (see Appendix A). For a batch reactor, inactivation kinetics is incorporated into Equation (4.4), similarly as shown by Equation (6.7). The resulting differential equation will be solved by numerical integration.

This software has been developed using Matlab and Matlab Compiler which can be found at the Mathworks website http://www.mathworks.co.uk/products/compiler/mcr/

Acknowledgement

With respect to Dr. Felipe Scott contribution to Epsilon Software Information referred to the Epsilon Software, a special acknowledgement should be done since he made the most significant contribution to the development of that software.

1

Facts and Figures in Enzyme Biocatalysis

1.1 Introduction

1.1.1 Enzyme Properties

Enzymes are the catalysts of life. Each of the chemical reactions that make up the complex metabolic networks found in all forms of life is catalyzed by an enzyme, which is the phenotypic expression of a specific gene. Thus the chemical potential of an organism is dictated by its genomic patrimony and enzymes are the biological entities that convert information into action. They are tightly regulated both by controls at the genomic level and by environmental signals that condition their mode of action once synthesized.

Enzymes are highly evolved complex molecular structures tailored to perform a specific task with efficiency and precision. They can be conjugated with other molecules or not, but their catalytic condition resides in their protein structure. The active site—the molecular niche in which catalysis takes place—represents a very small portion of the enzyme structure (only a few amino acid residues), while the remainder of the molecule acts as a scaffold and provides necessary structural stability. Many enzymes are conjugated proteins associated with other molecules that may or may not play a role in the catalytic process. Those that do are quite important as they will determine to a great extent an enzyme's technological potential.

"Enzyme biocatalysis" refers to the use of enzymes as biological catalysts dissociated from the cell from which they derive; the major challenge in this process is building up robust enzyme catalysts capable of performing under usually very nonphysiological conditions. The goal is to preserve the outstanding properties of enzymes as catalysts (specificity, reactivity under mild conditions) while overcoming their constraints (mostly their poor configurational stability). The pros and cons of enzymes as catalysts are thus the consequence of their complex molecular structure. Enzymes are labile catalysts, with

Problem Solving in Enzyme Biocatalysis, First Edition. Andrés Illanes, Lorena Wilson and Carlos Vera.
© 2014 John Wiley & Sons, Ltd. Published 2014 by John Wiley & Sons, Ltd. Companion Website:
http://www.wiley.com/go/illanes-problem-solving

enzyme stabilization being a major issue in biocatalysis and a prerequisite for most of their applications. Several enzyme stabilization strategies have been proposed, including: searching for new enzymes in the biota and metagenomic pools [1], improving natural enzymes via site-directed mutagenesis [2] and directed evolution [3], catalyst engineering (chemical modification [4], immobilization to solid matrices [5], and auto-aggregation [6]), medium engineering (use of nonconventional reaction media) [7], and, most recently, reactivating enzymes following activity exhaustion [8]. Enzyme immobilization has been a major breakthrough in biocatalysis and has widened its field of application considerably [9].

1.1.2 Enzyme Applications

Enzymes have found a wide spectrum of applications, from a very large number of enzyme-catalyzed industrial processes to use within the toolbox of molecular biology. Besides their use as process catalysts in industrial processes, which will be analyzed in Section 1.2, enzymes have found important applications in:

- Chemical and clinical analysis, due to their high specificity and sensibility, which allow the quantification of various analytes with high precision. Worth mentioning are their uses in flow injection analysis [10], robust electrodes for process control [11], and nanosensors [12]. Enzymes are also widely used in various diagnostic kits and as detectors in immunoassays [13].
- Therapy, due to their high specificity and activity, which allow the precise and efficient removal of unwanted metabolites. Many therapeutic applications have been envisaged, including for such enzymes as asparaginase, billirubin oxidase, carboxypeptidase, α-glucosidase, α-galactosidase, phenylalanine ammonia lyase, streptokinase, urease, uricase, and urokinase. The US Food and Drug Administration (FDA) has already approved applications in cardiovascular disorders, pancreatic insufficiency, several types of cancer, the replacement therapy for genetic deficiencies, the debridement of wounds, and the removal of various toxic metabolites from the bloodstream [14].
- Environmental management (waste treatment and bioremediation). Enzyme specificity allows the removal of particularly recalcitrant pollutants from hard industrial wastes and highly diluted effluents. Enzymes are also used in the final polishing of municipal and industrial effluents following conventional treatment and as enhancers of the hydrolytic potential of the microbial consortia in the first step of anaerobic digestion [15]. In addition, they are increasingly being used in bioremediation of polluted soils and waters with recalcitrant compounds [16], although the high cost of the enzymes is still prohibitive in most such cases [17].
- Biotechnological research and development. Enzymes are fundamental components of the toolbox for biotechnology research, especially in the areas of molecular biology and genetic engineering. Thermostable DNA polymerases are the basis of the polymerase chain reaction (PCR) [18], which is fundamental to gene amplification, while restriction endonucleases and ligases are fundamental tools in recombinant DNA technology [19].

Highly purified enzymes are required for most of these applications, which are sold at very high unitary prices. Therefore, despite their low volume of production, the market size is significant.

1.2 Enzymes as Process Catalysts

Enzymes used as catalysts for industrial processes are generally rather impure preparations sold as commodities at low unitary prices, although this tendency is changing to some extent, with increasing usage in an immobilized form and in organic synthesis. It can be estimated that industrial applications make up roughly 70% of the enzyme market size.

Enzyme transformations can be used in industrial applications to create the desired product (e.g. high-fructose syrup is produced from starch by the sequential operations of liquefaction, saccharification, and isomerization, catalyzed by α-amylase, amyloglucosidase, and glucose (xylose) isomerase, respectively). Enzymes can also be used as additives, in order to confer certain functional properties on the product, as illustrated by many applications in the food sector (amylases and proteases in bread making, phytases and β-glucosidases in animal feed upgrading, pectinases in fruit juice and wine making, β-galactosidases in low-lactose milk and dairy products, and so on).

Enzymes have been used systematically in industry since the beginning of the 20th century. Originally they were mostly crude preparations extracted from plant and animal tissues and fluids, but with the development of fermentation technology and industrial microbiology in the middle of that century, microbial enzymes began to take over. These had the advantages of intensive and reliable production and a wide spectrum of activities. In recent decades, advances in recombinant DNA technology have allowed genes of any origin to be expressed into a suitable microbial host [20], as well as in protein engineering techniques that allow enzyme properties to be improved through progressively rational approaches [21]. While plant and animal enzymes still have some application niches, most of the enzymes used industrially today are produced by microbial strains.

Until recently, most of the enzymes used as industrial catalysts were hydrolases, which are particularly robust and are frequently extracellular proteins that do not require coenzymes, making them well suited to use under harsh process conditions. Most industrial enzymatic processes are catalyzed by carbohydrases, proteases, and lipases, which are mostly used in their hydrolytic capacity to degrade substrates (frequently polymers) into products of lower molecular complexity. Some traditional industrial applications are listed in Table 1.1.

Most of the just-mentioned applications can be considered mature technologies, although significant improvements are ongoing. Major breakthroughs in detersive enzymes have been made rather recently, with the incorporation of highly robust lipases and α-amylases into the already sophisticated proteases currently in use. Phytases have recently been developed and successfully introduced into the fast-growing animal feed market. Major technological breakthroughs are being made in the use of enzymes in biofuel production. α-amylases used in the first-generation production of bioethanol from starchy feedstocks are continuously being improved, although land competition for food production has turned attention to the production of second-generation bioethanol from cellulosic raw materials through the use of cellulase preparations. There are now highly active enzyme cocktails with well-dosed activities that can efficiently hydrolyze cellulose from pretreated woods at very high solids concentrations, delivering high-glucose syrups for ethanol fermentation. High-tonnage plants for second-generation bioethanol are going into operation and their impact on the energy bill is forecasted to be significant by 2020.

Table 1.1   *Enzymes traditionally used as industrial catalysts.*

Enzyme	Source	Application
Carbohydrases		
α-amylase	mold	bakery, confectionery, brewery, bioethanol
α-amylase	bacterium	starch liquefaction, detergent, textile desizing, bioethanol
glucoamylase	mold	glucose syrup
β-amylase	plant, bacterium	glucose syrup, brewery
pectinase	mold	juice and wine manufacture
cellulase	mold	juice extraction and clarification, detergent, denim, digestive-aids, bioethanol
xylanase	mold, bacterium	wood pulp bleaching, bioethanol
β-galactosidase	yeast, mold	low-lactose milk and dairies, whey upgrading
invertase	yeast, mold	confectionery
phytase	bacterium	animal feed
β-glucanase	mold	animal feed, brewery
naringinase	mold	juice debittering
Proteases		
papain	papaya latex	yeast and meat extracts, brewery, protein hydrolysates, meat tenderizer, tanning, animal feed, digestive aids, anti-inflammatory and skin wound-healing preparations
bromelain	pineapple stem	anti-inflammatory and burn-healing preparations, drug absorption
pepsin	animal	cheese manufacture
chymosin	animal, recombinant yeasts	cheese manufacture
neutral protease	mold, bacterium	baking, protein hydrolysates
alkaline protease	bacterium	detergent, stickwater treatment
aminopeptidase	mold, bacterium	protein hydrolysate debittering
Other Hydrolases		
lipases	animal, yeast, mold, bacterium	flavor enhancer, detergent
aminoacylase	mold	food and feed fortification
penicillin acylase	mold, bacterium	β-lactam antibiotics
urease	bacterium	alcoholic beverages, urea removal
Nonhydrolytic enzymes		
glucose isomerase	bacterium, actinomycetes	high-fructose syrups
glucose oxidase	mold	food and beverage preservation
catalase	bacterium	food preservation, peroxide removal in milk
nitrile hydratase	bacterium	acrylamide
aspartate ammonia lyase	bacterium	aspartic acid

1.3 Evolution of Enzyme Biocatalysis: From Hydrolysis to Synthesis

Enzymes were long considered catalysts for the molecular degradation of natural substrates in aqueous media, and most of their traditional industrial uses accord with this, as illustrated in Table 1.1. However, a new paradigm for enzyme biocatalysis that challenges this view has appeared in recent decades and a myriad of technologically relevant reactions of organic synthesis (in which the high selectivity of enzymes is well appreciated), frequently conducted in nonaqueous media and on non-natural substrates, have been developed [22]. Here the goal is molecular construction rather than degradation, which is quite meaningful from an industrial perspective, since the potential added value is certainly much higher than in conventional degradation processes.

Technological breakthroughs in both medium engineering and catalyst engineering are the basis of this change of paradigm. Enzyme catalysis in nonconventional (nonaqueous) media has undergone impressive development since the pioneering work of Klibanov and others in the 1980s [23], so that enzymatic reactions can now be efficiently conducted in several nonaqueous media, such as organic solvents, ionic liquids, supercritical fluids, and semisolid systems. Screening for new enzymes and genetic improvement, along with developments in enzyme immobilization, have been driving forces for novel applications of enzymes, providing robust and finely tuned enzyme catalysts for the performance of organic synthesis reactions. The introduction of catalysts of biological origin in a strictly chemical industry has not been easy, but their enormous potential, stemming from their exquisite selectivity, activity under mild conditions, and compliance with the green chemistry concepts [24], has provided them a firm position [25]. It is not presumptuous to say then that enzyme biocatalysis is there to stay.

Enzymes that perform the metabolic reactions of synthesis are usually complex, unstable coenzyme-requiring proteins, so their use as process biocatalysts involves major technological challenges [26]. However, readily available hydrolases can, under certain conditions, catalyze the reverse reactions of synthesis; that is, formation of a chemical bond instead of its cleavage by water. In this way, robust and readily available carbohydrases can catalyze the formation of glycosidic bonds in order to synthesize oligosaccharides and glycosides; proteases can catalyze the formation of peptide bonds to allow the synthesis of peptides; acylases can catalyze peptide-type bond formation in the synthesis of antibiotics and other bioactive compounds; and lipases can catalyze esterification, transesterification, and interesterification reactions. In order for them to do so, the water activity must be low enough to depress the competing hydrolytic reactions, which means that nonconventional media will be required in most cases [27]. Lipases are particularly well endowed for performance in poorly aqueous media and are therefore the most commonly used class of enzyme in organic synthesis [28].

Examples of the use of hydrolases in industrial synthesis reactions include the following:

- *Carbohydrases*: Several hydrolases are in current use in the synthesis of different types of oligosaccharide and glycoside. Worth mentioning are the industrial production of prebiotic galacto-oligosaccharides with β-glactosidases [29] and fructo-oligosaccharides [30] with β-fructofuranosidases, whose application in the design of functional healthy foods has grown considerably.

- *Proteases*: Several proteases, such as papain, thermolysin, α-carboxypeptidase, alcalase, neutrase, and subtilisin, have been used in the synthesis of functional peptides of industrial interest for therapeutic applications, nutritional supplementation, and flavor enhancement [31,32]. Of particular note is the use of thermolysin in the large-scale synthesis of the leading noncaloric sweetener aspartame (L-α-aspartyl-L-phenylalanine-1-methyl ester) [33].
- *Acylases*: Penicillin acylases, traditionally used for the production of β-lactam nuclei as an intermediate stage in the production of semisynthetic β-lactam antibiotics, are now being used in their synthetic capacity to conduct the side-chain derivatization of such nuclei in order to yield the corresponding antibiotics [34]. The environmentally offensive chemical synthesis is being gradually replaced by enzymatic synthesis with immobilized penicillin acylases, which works on green chemistry principles. Certainly, amoxicillin and cephalexin are currently being produced by biocatalysis at an industrial level. Penicillin acylases are quite versatile enzymes and can be used in the synthesis of a variety of building blocks for the production of bioactive compounds and as catalysts for side reactions in peptide synthesis.
- *Lipases*: Among the hydrolytic enzymes, lipases stand out for their ability to catalyze myriad synthesis reactions, many with clear technological applications. The enantio-selective and stereoselective synthesis reactions conducted using lipases are too numerous to list, being increasingly used in the synthesis of functional compounds for the pharmaceutical and fine chemicals industries [27]. The industrial production of cocoa butter analogues and other functional fats [35], the synthesis of nonionic surfactants for food, cosmetic, and pharmaceutical uses [36], the synthesis of sterol and stanol esters for the production of nutraceuticals [37], and the production of biodiesel by transesterification of different oils and fats [38] are deserving of particular mention.

Enzyme biocatalysis has evolved from the traditional use of enzymes in the molecular degradation of natural compounds, conducted in aqueous media and usually under conditions of homogeneous catalysis, to their more sophisticated use in reactions of organic synthesis from natural or non-natural substrates with hydrolases—and other enzymes—mostly under conditions of heterogeneous catalysis in nonconventional reaction media. Despite the notable specificity of enzymes, they are now considered rather promiscuous catalysts as they are not restricted to their natural substrates, acting instead on a broader spectrum of both natural and non-natural compounds [39]. Their field of application is growing to an extent unimagined a few decades ago.

1.4 The Enzyme Market: Figures and Outlook

Published figures concerning the enzyme market and its projection tend to vary from source to source, mostly because the boundaries of such a market appraisal are ill defined. A demand of USD 5.8 billion estimated for 2010 has been forecasted to increase to 7 billion by 2013 and to 8 billion by 2015, representing an average growth rate of 6.8% for the next quinquennium. This figure includes the so-called "technical enzymes," enzymes for food and feed use, and enzymes for diagnostic and biotechnological use in general. The 2010 market estimate for technical and food and feed enzymes was around USD 3.5 billion

(52% technical enzymes and 48% food and feed enzymes), leaving a figure of around USD 2 billion for other enzymes (if the data are comparable). However, these numbers have to be used with caution, since new partners in the global market (mainly China, but also India and Korea) are gaining increasing importance as enzyme producers and it is not clear to what extent these figures include their production. Further, the figures reflect the demand registered from open trade operations, but many enzyme consumers have their own production facilities or have established joint ventures with enzyme producers for their supply. The economic impact of enzymes may thus be much higher than is suggested by these figures. This is not to contradict the fact that most of the enzymes in use today are supplied by a limited number of big companies, notably Novozymes, with a market share of 47%, and DuPont (including acquired Genencor), with 21%; between them, these two companies represent more than two-thirds of the global enzyme supply market.

Conventional industrial enzyme applications are expected to keep growing, with major opportunities in the detergent and animal feed sectors. Detersive enzymes have experienced an impressive development in recent years, and better and more specialized enzymes are expected to reach the market in the near future. The farm animal and pet feed enzyme market is rapidly expanding and is expected to continue to do so. Major breakthroughs are also expected in the use of enzymes in biofuel production, especially in second-generation bioethanol production, in the coming decades. The market for lignocellulose degrading enzymes may experience an explosive growth as high-tonnage bioethanol plants enter operation. Major opportunities for enzymes will certainly also occur in the organic synthesis industry, where applications will develop at a sustained pace in the pharmaceutical, fine-chemicals, and health-food sectors.

Beyond industrial applications, the market is expected to grow at a higher rate for enzymes involved in human and animal therapy and diagnosis and in the various areas of biotechnology.

The outlook for enzymes is thus certainly promising and their market increase is expected to be sustained in the next few decades. The global evolution of eating habits, health improvement, and environmental sustainability is a megatrend that represents ample room for the development of new enzyme applications.

References

1. Steele, H.L., Jaeger, K.E., Daniel, R., and Streit, W.R. (2009) Advances in recovery of novel biocatalysts from metagenomes. *Journal of Molecular Microbiology and Biotechnology*, **16**, 25–37.
2. Ijima, Y., Matoishi, K., and Terao, Y. (2005) Inversion of enantioselectivity of asymmetric biocatalytic decarboxylation by site-directed mutagenesis based on the reaction mechanism. *Chemical Communications*, **7**, 877–879.
3. Dalby, P.A. (2011) Strategy and success for the directed evolution of enzymes. *Current Opinion in Structural Biology*, **21**, 1–8.
4. Davis, B. (2003) Chemical modification of biocatalysts. *Current Opinion in Biotechnology*, **14**, 379–386.
5. Mateo, C., Palomo, J.M., Fernandez-Lorente, G. *et al.* (2007) Improvement of enzyme activity, stability and selectivity via immobilization techniques. *Enzyme and Microbial Technology*, **40**, 1451–1463.

6. Sheldon, R. (2011) Cross-linked enzyme aggregates as industrial biocatalysts. *Organic. Process Research and Development*, **15**, 213–223.

7. Hari Krishna, S. (2002) Developments and trends in enzyme catalysis in non-conventional media. *Biotechnology Advances*, **20**, 239–267.

8. Romero, O., Guisán, J.M., Illanes, A., and Wilson, L. (2012) Reactivation of penicillin acylase biocatalysts: Effect of the intensity of enzyme-support attachment and enzyme load. *Journal of Molecular Catalysis B: Enzymatic*, **74**, 224–229.

9. Illanes, A. (2011) Immobilized biocatalysts, in *Comprehensive Biotechnology*, 2nd edn, vol. **1** (ed. M. Moo-Young), Elsevier, Boston, USA, pp. 25–39.

10. Yu, J., Liu, S., and Ju, H. (2003) Glucose sensor for flow injection analysis of serum glucose based on immobilization of glucose oxidase in titania sol–gel membrane. *Biosensors and Bioelectronics*, **19**, 401–409.

11. Mulchandani, A. and Bassi, A.S. (1995) Principles and applications of biosensors for bioprocess monitoring and control. *Critical Reviews in Biotechnology*, **15**, 105–124.

12. Kim, J., Grate, J.W., and Wang, P. (2008) Nanobiocatalysis and its potential applications. *Trends in Biotechnology*, **26**, 639–646.

13. Yakovleva, J., Davidsson, R., Lobanova, A. *et al.* (2002) Enzyme immunoassay using silicon microchip with immobilized antibodies and chemiluminescence detection. *Analytical Chemistry*, **74**, 2994–3004.

14. Pandey, A. Webb, C. Soccol, C.R. and Larroche, C. (eds) (2006) *Enzyme Technology*, Springer, London.

15. Karma, J. and Nicell, J.A. (1997) Potential applications of enzymes in waste treatment. *Journal of Chemical Technology and Biotechnology*, **69**, 141–153.

16. Durán, N. and Esposito, E. (2000) Potential applications of oxidative enzymes and phenoloxidase-like compounds in wastewater and soil treatment: a review. *Applied Catalysis B: Environmental*, **28**, 83–99.

17. Alcalde, M., Ferrer, M., Plou, F.J., and Ballesteros, A. (2006) Environmental biocatalysis: from remediation with enzymes to novel green processes. *Trends in Biotechnology*, **24**, 281–287.

18. Weissensteiner, T., Griffin, A.M., and Griffin, H.G. (2003) *PCR Technology: Current Innovations*, CRC Press, Boca Raton.

19. Court, D.L., Sawitzke, J.A., and Thomason, L.C. (2002) Genetic engineering using homologous recombination. *Annual Reviews in Genetics*, **36**, 361–388.

20. Sánchez, S. and Demain, A.L. (2011) Enzymes and bioconversions of industrial, pharmaceutical and biotechnological significance. *Organic Process Research and Development*, **15**, 224–230.

21. Illanes, A., Cauerhff, A., Wilson, L., and Castro, G. (2012) Recent trends in biocatalysis engineering. *Bioresource Technology*, **115**, 48–57.

22. García-Junceda, E., García-García, J.F., Bastida, A., and Fernández-Mayorales, A. (2004) Enzymes in the synthesis of bioactive compounds: the prodigious decades. *Bioorganic and Medicinal Chemistry*, **12**, 1817–1834.

23. Klibanov, A. (2001) Improving enzymes by using them in organic solvents. *Nature*, **409**, 241–246.

24. Ran, N., Zhao, L., Chen, Z., and Tao, J. (2008) Recent applications of biocatalysis in developing green chemistry for chemical synthesis at the industrial scale. *Green Chemistry*, **10**, 361–372.

25. Wandrey, C., Liese, H., and Kihumbu, D. (2000) Industrial biocatalysis: past, present, and future. *Organic Process Research and Development*, **4**, 286–290.

26. Berenguer-Murcia, A. and Fernandez-Lafuente, R. (2010) New trends in the recycling of NADP (H) for the design of sustainable asymmetric reduction catalyzed by dehydrogenases. *Current Organic Chemistry*, **14**, 1000–1021.

27. Bornscheuer, U.T. and Kazlauskas, R.J. (1999) *Hydrolases in Organic Synthesis*, Wiley VCH, Weinheim.

28. Hasan, F., Shah, A.A., and Hameed, A. (2006) Industrial applications of microbial lipases. *Enzyme and Microbial Technology*, **39**, 235–251.
29. Illanes, A. (2011) Whey upgrading by enzyme biocatalysis. *Electronic Journal of Biotechnology*, **14** (6). doi: 10.2225/vol14-issue6-fulltext-11
30. Vaňková, K., Onderková, Z., Antošová, M., and Polakovič, M. (2008) Design and economics of industrial production of fructooligosaccharides. *Chemical Papers*, **62**, 375–381.
31. Kumar, D. and Bhalla, T. (2005) Microbial proteases in peptide synthesis: approaches and applications. *Applied Microbiology and Biotechnology*, **68**, 726–736.
32. Guzmán, F., Barberis, S., and Illanes, A. (2007) Peptide synthesis: chemical or enzymatic. *Electronic Journal of Biotechnology*, **10** (2), 279–314.
33. Schmid, A., Dordick, J., Hauer, B. *et al.* (2001) Industrial biocatalysis today and tomorrow. *Nature*, **409**, 258–268.
34. Illanes, A. and Wilson, L. (2006) Kinetically-controlled synthesis of β-lactam antibiotics. *Chimica Oggi/Chemistry Today*, **24**, 27–30.
35. Aravindam, R., Anbumathi, P., and Viruthagiri, T. (2007) Lipase application in the food industry. *Indian Journal of Biotechnology*, **6**, 141–158.
36. Chang, S.W. and Shaw, J.F. (2009) Biocatalysis for the production of carbohydrate esters. *New Biotechnology*, **26**, 109–116.
37. Olivares, A., Martinez, I., and Illanes, A. (2012) Enzyme assisted fractionation of wood sterols mixture by short path distillation. *Chemical Engineering Journal*, **191**, 557–562.
38. Fjerbaek, L., Christensen, K.V., and Norddahl, B. (2009) A review on the current state of biodiesel production using enzymatic transesterification. *Biotechnology and Bioengineering*, **102**, 1298–1315.
39. Khersonsky, O. and Tawfik, D.S. (2010) Enzyme promiscuity: a mechanistic and evolutionary perspective. *Annual Review of Biochemistry*, **79**, 471–505.

2

Enzyme Kinetics in a Homogeneous System

2.1 Introduction

2.1.1 Concept and Determination of Enzyme Activity

The catalytic potential of an enzyme is denoted as its enzyme activity. Thermodynamically, it refers to the reduction in magnitude of the energy barrier required to convert a substrate into a product. This reduction leads to an increase in the reaction rate, so activity is usually expressed and measured as a rate of reaction, conventionally referred to as the "initial rate of reaction," which represents the maximum catalytic potential of the enzyme.

For the conversion of substrate S into product P:

$$activity = v_{t=0} = -\left(\frac{ds}{dt}\right)_{t=0} \tag{2.1}$$

where s is the molar concentration of species S (the substrate in this case).

When product evolution with time is recorded, a monotonic curve with decreasing slope is observed. The reasons for such a decrease may be enzyme desaturation by the substrate, enzyme inactivation, equilibrium displacement, and, eventually, product inhibition. These conditions will certainly be present in any enzymatic process and will modulate this maximum catalytic potential. "Enzyme kinetics" refers to the quantification of such effects. The determination of enzyme activity is therefore essential to the evaluation of any enzymatic process.

Enzyme activity can be determined by monitoring either substrate (S) consumption or product (P) generation:

$$activity = v_{t=0} = -\left(\frac{ds}{dt}\right)_{t=0} = \left(\frac{dp}{dt}\right)_{t=0} \tag{2.2}$$

Problem Solving in Enzyme Biocatalysis, First Edition. Andrés Illanes, Lorena Wilson and Carlos Vera.
© 2014 John Wiley & Sons, Ltd. Published 2014 by John Wiley & Sons, Ltd. Companion Website:
http://www.wiley.com/go/illanes-problem-solving

Product generation is generally preferred because the margin of error is smaller; measurement of substrate consumption entails calculating a small difference between two rather large numbers (substrate concentration must be high enough to saturate the enzyme).

For enzymes requiring stoichiometric coenzymes, the activity can also be determined by measuring either coenzyme (CE) consumption or generation of the modified coenzyme (CE'):

$$activity = v_{t=0} = -\left(\frac{dce}{dt}\right)_{t=0} = \left(\frac{dce'}{dt}\right)_{t=0} \tag{2.3}$$

This is a sound method for such enzymes because organic CEs are generally quite easy to determine.

Activity can also be determined by coupling the enzymatic reaction to others which transform the product into a more detectable analyte (Q), provided that the transformation of S into P remains the limiting step of the overall reaction:

$$activity = v_{t=0} = \left(\frac{dp}{dt}\right)_{t=0} = \left(\frac{dq}{dt}\right)_{t=0} \tag{2.4}$$

Since the initial rate of reaction is consubstantial to enzyme activity, the conditions used must ensure that it is this rate that is actually being measured. In continuous assay this is straightforward as the analyte concentration will be continuously recorded; in this case, it is necessary only to adjust the enzyme concentration in order to yield a significant response within a reasonable time. In discrete analysis, which is frequently used, samples are taken at time intervals and analyzed for their analyte concentration; in this case, it is necessary to adjust the conditions and verify that the product (generally analyte) concentration varies linearly with time. This requires the taking of several samples or the definition of an enzyme concentration range in which such linearity is observed over a reasonable sampling time—neither too long to produce enzyme inactivation nor too short to make sampling difficult. In practice, sampling for 10–20 minutes is recommended.

A very important feature in any enzyme assay is a control that can account for any non-enzymatic generation of product. Appropriate controls might be the replacement of the enzyme sample by water or buffer, the replacement of the substrate by water or buffer, or the replacement of both. The first option accounts for any product coming from either substrate contamination by product or non-enzymatic chemical conversion of substrate into product; the second for substrate or product present in the enzyme sample. One or the other will be more suitable, depending on the case; for instance, in assaying an enzyme from a commercial concentrate, the former will suffice, since the enzyme preparation must be highly diluted prior to assay; when assaying the activity of an extracellular enzyme during production by fermentation, the latter is adequate, since the enzyme is diluted in the fermentation medium, which is likely to have a high concentration of substrate and/or product. Double control is obviously safer but in most cases is unnecessary.

Although enzyme activity is to be related to the initial rate of reaction, there are certain exceptions in which enzymes act on chemically ill-defined substrates where a departure from this concept may be made. One case worth mentioning is the determination of cellulase activity. Cellulase is actually an enzyme complex that includes endo- and exoglucanase activities. When acting upon a cellulosic (or lignocellulosic) substrate, it will do so preferentially over the amorphous portion of the cellulose complex, which is more amenable for hydrolysis; a measurement of initial rate will therefore give a false picture of its catalytic potential, since the more recalcitrant crystalline portion of cellulose will be acted upon at a later stage of reaction. In such a case the popular filter paper activity assay can be used, which, although not quite orthodox, is accepted as indicative of the whole catalytic potential of the enzyme over its intended substrate.

2.1.2 Definition of a Unit of Activity

The Enzyme Commission (EC) of the International Union of Biochemistry strongly recommends and enforces the expression of enzyme activity, whenever possible, in international units (IU). One IU of enzyme activity has been defined as the amount of enzyme that catalyzes the transformation of one micromole of substrate per minute under standard conditions of temperature, optimal pH, and optimal substrate concentration. Some effort was made to bring this definition into compliance with the International System of Units (SI), according to which reaction rates should be expressed in moles per second. The katal, which represents the number of moles of substrate converted per second, was thus proposed as the proper unit in which to express activity. However, activities expressed in katals are exceedingly small numbers, so the IU has prevailed.

The definition of IU requires some precision. The substrate concentration in the assay should be high enough to saturate the enzyme, so concentrations at least five times higher than the Michaelis constant (K_M) are recommended. This is important not only because it adheres to the idea of activity as maximum catalytic potential but also because during the assay it is convenient to work in a region in which the initial rate of reaction is essentially independent of substrate concentration. The temperature at which enzyme activity should be measured is not a simple decision: "optimum" temperature, regarded as that producing the maximum initial reaction rate, is a rather elusive concept and devoid of practical meaning, since such an optimum is one at which enzyme stability is likely to be very low. A temperature closer to that of operation is thus advised, although this can be a matter of some uncertainty at the time of the assay. The similarly defined pH optimum will be adequate, however, as in most cases no compromise exists between activity and stability with respect to pH.

2.1.3 Measurement of Enzyme Activity

Reaction rates can be determined based on the product, the substrate, a cofactor or a coupled analyte. The choice will be based mostly on economic considerations, accuracy, and simplicity, so the one most amenable as an analyte will be selected. The methods most frequently used for the determination of enzyme activity are listed in Table 2.1.

Spectrophotometry has most of the desired properties of an analytical system and is the most commonly used method of enzyme activity determination. Spectrophotometers, spectrofluorimeters, and refractive index detectors can be coupled to high-performance

Table 2.1 *Methods of analysis of enzyme activity.*

Method/Equipment	Principle	Characteristics
Spectrophotometry/ Spectrophotometer	Light absorption	simple reproducible sensitive accurate versatile commonly available
Fluorometry/ Spectrofluorimeter	Light absorption and irradiation	reproducible sensitive accurate restricted to fluorescent analytes not commonly available
Polarimetry/Refractive index detector	Rotation of polarized light	simple reproducible restricted to optically active analytes commonly available
Viscometry/Viscometers	Reduction in kinematic viscosity	simple sensitive restricted to enzymes acting on polymeric viscous substrates commonly available
Manometry/Warburg manometer (respirometer)	Pressure difference due to gas evolution or consumption	reproducible sensitive accurate restricted to gaseous analytes not commonly available

liquid chromatography (HPLC) delivery systems, allowing the compounds present in the sample to be separated and analyzed. Comprehensive reviews of enzyme activity assays have been provided by Mattiasson and Mosbach [1] and Biswanger [2] and are recommended as supplementary material.

2.2 Theory of Enzyme Kinetics

Enzyme kinetics is the quantitative evaluation of all factors that condition enzyme activity. The most important of these are the concentrations of active enzyme, substrates, and inhibitors, the pH, and the temperature. Theories of enzyme kinetics refer particularly to substrate concentration. All of these factors will be analyzed in Section 2.3.

Enzyme kinetics is a fundamental tool in the determination of the molecular mechanisms of enzyme action, but it is also used to develop models for the design of enzyme reactors and for performance evaluation, where a sound kinetic expression is necessary. This latter aspect will be the subject of this and the following chapters.

Enzyme kinetics is based on the determination of initial rates of reaction, where factors affecting the maximum catalytic potential of the enzyme do not come into play. The quantitative determination of the impact of such factors is precisely the subject of enzyme kinetics.

It is generally assumed that enzyme activity is proportional to the concentration of active enzyme protein, although some deviations may occur in particular cases. This assumption is fundamental to enzyme kinetics.

Hypotheses concerning enzyme kinetics have been developed based on studies of homogeneous catalysis; that is, systems in which the enzyme, substrates, and products of reaction are dissolved in the reaction medium. Enzyme kinetics under conditions of heterogeneous catalysis will be analyzed in Chapter 3. The two main theories of enzyme kinetics are the rapid-equilibrium and steady-state hypotheses. Both are based on a molecular sequential mechanism, as proposed by Henri [3], in which the substrate is first linked to the enzyme at its active site and then processed by the catalytic amino acids at this site, which convert it into the product. The first step was considered reversible, while the second was assumed to be irreversible. This mechanism can be schematically represented as:

$$E + S \underset{k_2}{\overset{k_1}{\rightleftarrows}} ES \overset{k}{\longrightarrow} E + P$$

According to the rapid-equilibrium hypothesis, proposed by Michaelis and Menten [4], the binding step is much faster than the conversion step, so that the former can be properly considered in equilibrium. In this way, the dissociation of ES into E and S can be described by:

$$K = \frac{k_2}{k_1} = \frac{(e - c)(s - c)}{c} \tag{2.5}$$

but $c < e \lll s$, so that Equation 2.5 simplifies to:

$$K = \frac{k_2}{k_1} = \frac{(e - c)s}{c} \tag{2.6}$$

and:

$$v = k \cdot c = \frac{k \cdot e \cdot s}{K + s} \tag{2.7}$$

Defining $V_{max} = k \cdot e$, Equation 2.7 is now written in parametric form as:

$$v = \frac{V_{max} \cdot s}{K + s} \tag{2.8}$$

Later, Briggs and Haldane [5] put forward the steady-state hypothesis. In this theory, no equilibrium is assumed, but instead it is postulated that following a very short transient state the enzyme–substrate complex reaches steady state, so that its concentration remains constant throughout the reaction when substrate is being transformed into product. According to this hypothesis:

$$\frac{dc}{dt} = k_1(e - c)s - (k_2 + k)c = 0 \tag{2.9}$$

so that:

$$v = k \cdot c = \frac{k \cdot e \cdot s}{\dfrac{k_2 + k}{k_1} + s} \tag{2.10}$$

The dissociation constant of the ES complex into E and S is:

$$K_D = \frac{k_2 + k}{k_1} \tag{2.11}$$

so that Equation 2.10 is now rewritten as:

$$v = k \cdot c = \frac{V_{max} \cdot s}{K_D + s} \tag{2.12}$$

Both hypotheses are usually referred to by the names of their proponents (i.e. as the Michaelis–Menten hypothesis and the Briggs–Haldane hypothesis).

In parametric terms, both equations are similar; however, the dissociation constant of the enzyme–substrate complex is an equilibrium constant according to the Michaelis–Menten hypothesis but is not in the case of the Briggs–Haldane hypothesis. The first hypothesis can simply be considered a particular case of the latter ($K = K_D$, if k is significantly smaller than k_1 and k_2) that does not rely on the equilibrium assumption. Despite this, the Michaelis–Menten hypothesis has been widely used to describe enzyme kinetics and the assumption it makes is valid in most cases. It has the advantage over the Briggs–Haldane hypothesis of simpler mathematics and parametrization of the rate equations. For more complex kinetics (see Sections 2.3 and 2.4), steady-state expressions are quite complex to develop, requiring mathematical algorithms like that of King and Altmann [6]; even with these, parametrization is also a complex task. Computer programs have been proposed to aid with this, and the reader is referred to the one proposed by Myers and Palmer [7].

Whatever the hypothesis, the rate equation is expressed in terms of two parameters, K_M and V_{max}:

$$v = \frac{V_{max} \cdot s}{K_M + s} \tag{2.13}$$

K_M, usually termed the "Michaelis constant," is a constant in the sense of not being dependent on either enzyme or substrate concentration, while V_{max} is a lumped parameter containing the enzyme concentration (e) and the catalytic rate constant (k). The latter is a fundamental property of the enzyme and its dimension will be determined by that in which the enzyme concentration is expressed. When enzyme concentration is expressed in moles per unit volume for example, k has the dimension of reciprocal time (turnover number). This is the most appropriate way of expressing k, but in order to do so the molecular weight, specific activity, and number of active sites of the enzyme must be known, so for practical purposes k is frequently expressed in dimensions of mass of substrate converted per unit time and unit of enzyme activity. If the latter is expressed in IU, k reduces to a dimensionless value of 1, making it equivalent to V_{max}.

2.3 Single-Substrate Reactions

Equation 2.13 described the kinetics of irreversible single-substrate reactions. When reversibility is important (as in the case of reactions of isomerization catalyzed by isomerases or reverse reactions of synthesis by hydrolases in nonaqueous media) the steady-state hypothesis should be used to develop kinetic expressions. In this case, the simplest reaction scheme is:

$$E + S \underset{k_2}{\overset{k_1}{\rightleftharpoons}} ES \underset{k_4}{\overset{k_3}{\rightleftharpoons}} E + P$$

so that:

$$\frac{dc}{dt} = k_1(e - c)s + k_4(e - c)p - (k_2 + k_3)c = 0 \tag{2.14}$$

$$v = k_3 \cdot c - k_4(e - c)p \tag{2.15}$$

Solving Equations 2.14 and 2.15 and rearranging:

$$v = \frac{\dfrac{V_{max,S} \cdot s}{K_{M,S}} - \dfrac{V_{max,P} \cdot p}{K_{M,P}}}{\dfrac{s}{K_{M,S}} + \dfrac{p}{K_{M,P}} + 1} \tag{2.16}$$

where: $K_{M,S} = \dfrac{k_2 + k_3}{k_1}$; $K_{M,P} = \dfrac{k_2 + k_3}{k_4}$; $V_{max,S} = k_3 \cdot e$; $V_{max,P} = k_2 \cdot e$

From Equation 2.16, the equilibrium constant of the reaction ($p_{eq} \cdot s_{eq}^{-1}$) is:

$$K_{eq} = \frac{V_{max,S}}{V_{max,P}} \cdot \frac{K_{M,P}}{K_{M,S}} \tag{2.17}$$

Equations 2.13 and 2.16 represent the rate expressions for single-substrate irreversible and reversible reactions, respectively. The former applies to many reactions of practical interest, including the reactions of hydrolysis catalyzed by hydrolases in aqueous media (most of the enzyme technology still relies on such types of reaction). Although these reactions imply a second substrate (water), water also acts as the solvent, since its molar concentration is high enough not to affect enzyme kinetics. Such reactions in aqueous media can thus properly be considered single-substrate reactions.

2.3.1 Kinetics of Enzyme Inhibition

Enzymes are prone to inhibition, which is undesirable for use as process catalysts. Inhibitors should be kept out of the reaction medium, but this is not feasible when the reaction products or even the substrates themselves are inhibitors—a not uncommon situation. Inhibition is a reversible phenomenon that will last as long as the inhibitor is present.

Inhibitors are classified according to the step in the catalytic mechanism in which they act. Competitive inhibitors are defined as those acting at the substrate binding step, noncompetitive inhibitors are those acting at the enzyme-bound substrate transformation step, while mixed-type inhibitors are those acting at both steps. The terms "competitive" and "noncompetitive" stem from the inhibitor's competing with the substrate for binding to the active site in the former case and its binding to another site and not interfering with substrate binding, but instead hindering chemical processing of the substrate at the active site, in the latter. A special case of mixed-type inhibition is so-called "uncompetitive inhibition," in which the enzyme has no preformed site for inhibitor binding, the site being formed only after the substrate has bound to the enzyme; a relevant case of uncompetitive inhibition is inhibition by high substrate concentration.

A convenient way of expressing rate equations of inhibition is to use the parametric form of the Michaelis–Menten equation:

$$v = \frac{V_{maxAP} \cdot s}{K_{MAP} + s} \tag{2.18}$$

in which both kinetic parameters (K_M and V_{max}) are apparent, in the sense of being dependent on the inhibitor concentration. Expressed in parametric form, competitive inhibitors will affect K_{MAP}, noncompetitive inhibitors will affect V_{maxAP} and mixed-type inhibitors will affect both, as illustrated in Table 2.2.

With mixed-type inhibition I and II, the net effect on the reaction rate ought to be negative, otherwise it will be an activation. If $V_{maxAP} > V_{max}$ and $K_{AP} < K_M$, it will always be an activation.

Proceeding according to the Michaelis–Menten hypothesis, rate equations can be derived as for simple Michaelis–Menten kinetics. Table 2.3 summarizes the values of the apparent kinetic parameters for each possible mechanism of enzyme inhibition. Apparent parameters are expressed in terms of the corresponding inhibition constants (K_I) and inhibitor molar concentration (i).

Table 2.2 *Effects of inhibition on apparent kinetic parameters.*

Mechanism	V_{maxAP}	K_{MAP}
Simple Michaelis–Menten	V_{max}	K_M
Competitive inhibition	V_{max}	$>K_M$
Noncompetitive inhibition	$<V_{max}$	K_M
Mixed-type inhibition I	$<V_{max}$	$>K_M$
Mixed-type inhibition II	$<V_{max}$	$<K_M$
Mixed-type inhibition III	$>V_{max}$	$>K_M$

Table 2.3 *Apparent parameters of enzyme inhibition kinetics.*

Mechanism	V_{maxAP}	K_{MAP}	Number of Kinetic Parameters
Simple Michaelis–Menten	V_{max}	K_M	2
Competitive inhibition	V_{max}	$K_M\left(1 + \frac{i}{K_{IC}}\right)$	3
Total noncompetitive inhibition	$\dfrac{V_{max}}{1 + \frac{i}{K_{INC}}}$	K_M	3
Partial noncompetitive inhibition	$\dfrac{V_{max} + V'_{max}\frac{i}{K_{INC}}}{1 + \frac{i}{K_{INC}}}$	K_M	4
Total mixed-type inhibition	$\dfrac{V_{max}}{1 + \frac{i}{K'_{INC}}}$	$K_M \dfrac{1 + \frac{i}{K_{INC}}}{1 + \frac{i}{K'_{INC}}}$	4
Partial mixed-type inhibition	$\dfrac{V_{max} + V'_{max}\frac{i}{K'_{INC}}}{1 + \frac{i}{K'_{INC}}}$	$K_M \dfrac{1 + \frac{i}{K_{INC}}}{1 + \frac{i}{K'_{INC}}}$	5
Total uncompetitive inhibition	$\dfrac{V_{max}}{1 + \frac{i}{K_S}}$	$\dfrac{K_M}{1 + \frac{i}{K_S}}$	3
Partial uncompetitive inhibition	$\dfrac{V_{max} + \frac{V'_{max}\cdot i}{K_S}}{1 + \frac{i}{K_S}}$	$\dfrac{K_M}{1 + \frac{i}{K_S}}$	4
Double inhibition (competitive by I_1 and noncompetitive by I_2)	$\dfrac{V_{max}}{1 + \frac{i_2}{K_{INC}}}$	$K_M \dfrac{1 + \frac{i_1}{K_{IC}} + \frac{i_2}{K_{INC}}}{1 + \frac{i_2}{K_{INC}}}$	4

Noncompetitive and mixed-type inhibition can be either total or partial, depending on whether the tertiary enzyme complex formed (ESI) is fully inactive or partially active.

2.4 Multiple-Substrate Reactions

Enzymes are increasingly being used as catalysts for reactions of organic synthesis, in which more than one substrate is usually involved. The enzyme kinetics of such reactions is thus a very important subject in enzyme biocatalysis today.

2.4.1 Reaction Mechanisms

The reaction mechanisms for multiple-substrate reactions can be broadly divided into those in which all substrates must be bound to the enzyme (sequential mechanism) and those in which the product can be formed before all substrates have been bound (oscillatory or ping-pong mechanism). Sequential mechanisms can be subdivided into those in which substrates are bound to the enzyme in a predetermined order (ordered) and those in which they are not (random). The name "ping-pong mechanism" implies that the enzyme exists in two alternative catalytic configurations, each of them binding and transforming one substrate into one product. Considering the bi–bi reaction:

$$A + B \xrightarrow{\text{E}} P + Q$$

and using Cleland's nomenclature [8], one gets:

- *Ordered sequential*:

- *Random sequential*:

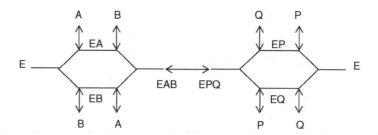

- *Ping-pong*:

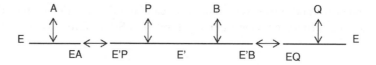

2.4.2 Kinetics of Enzyme Reactions with Two Substrates

Rate expressions for sequential mechanisms according to the Michaelis–Menten hypothesis can be derived as for simple Michaelis–Menten kinetics. As in the case of enzyme inhibition, it is convenient to express the rate equations in terms of apparent parameters. For the previously given bi–bi reaction:

$$v = \frac{V_{maxAP(B)} \cdot a}{K_{MAP(B)} + a} = \frac{V_{maxAP(A)} \cdot b}{K_{MAP(A)} + b} \tag{2.19}$$

where a and b are the molar concentrations of substrates A and B, respectively.

This is merely an artifact, since when expressing the rate with respect to substrate A, parameters are a function of the concentration of B and *vice versa*. However, this is quite convenient for the determination of kinetic parameters (Section 2.6).

Table 2.4 *Apparent parameters of two-substrate enzyme kinetics.*

Mechanism	$V_{maxAP(B)}$	$K_{MAP(B)}$	$V_{maxAP(A)}$	$K_{MAP(A)}$
Ordered sequential	$\dfrac{V_{max} \cdot b}{b + K'_{MB}}$	$\dfrac{K_{MA} \cdot K'_{MB}}{b + K'_{MB}}$	V_{max}	$\dfrac{a \cdot K'_{MB} + K_{MA} \cdot K'_{MB}}{a}$
Random sequential	$\dfrac{V_{max} \cdot b}{b + K'_{MB}}$	$\dfrac{b \cdot K'_{MA} + K_{MA} \cdot K'_{MB}}{b + K'_{MB}}$	$\dfrac{V_{max} \cdot a}{a + K'_{MA}}$	$\dfrac{a \cdot K'_{MB} + K_{MA} \cdot K'_{MB}}{a + K'_{MA}}$
Ping-pong	$\dfrac{V_{max} \cdot b}{b + K''_{MB}}$	$\dfrac{\alpha \cdot K_{MA} \cdot b}{b + K''_{MB}}$	$\dfrac{V_{max} \cdot a}{a + \alpha \cdot K_{MA}}$	$\dfrac{K''_{MB} \cdot a}{a + \alpha \cdot K_{MA}}$

Rate equations for ping-pong mechanisms are fairly complex to derive and several assumptions are usually required in order to obtain sound kinetic expressions. A simplified rate expression has been proposed by Marangoni based on the fact that the rate-limiting step is the transition from the secondary complex $E'B$ to EQ [9].

Table 2.4 summarizes the values of the apparent kinetic parameters for sequential and simplified ping-pong mechanisms.

2.5 Multiple-Enzyme Reactions

Multiple-enzyme catalyzed processes, in which the product of one enzyme is acted upon by another, and so on until the desired product is formed, are becoming increasingly popular, particularly for organic synthesis reactions [10]. The use of two or more enzymes in a single reactor is known as a "one-pot reaction." The main advantage of this system is that intermediates do not require isolation or purification, so downstream operation costs are reduced significantly. This must be balanced against working under a set of conditions that are not optimal for any of the enzymes but are the result of a compromise between them all. In such reactions, enzymes can act concertedly under three basic configurations: cascade, parallel, or network [11]. More complex configurations can be obtained by combinations of these three. In (a) the cascade scheme, a product of the first enzymatic reaction acts as substrate to the next one and so on; in (b) parallel reactions, two or more enzymes compete for the same substrate to yield different products; and in (c) the network pattern, two or more reactions are connected by one or more compound to drive the reaction network in a particular direction. These three schemes are illustrated for a two-enzyme system:

As with single-enzyme reactions, the modeling of multiple-enzyme reactions requires a mass balance and a rate equation for each of the species involved. It is usually not possible to find analytical expressions that describe the course of the reaction, because enzyme rate equations are nonlinear. The profiles of the substrate and products must therefore be determined using numerical methods [12,13]. However, if first-order kinetics is assumed for the enzymatic reactions, analytical expressions for substrate and product profiles can be found by solving the resulting system of linear ordinary differential equations [14,15]. On the other hand, when the intermediate compounds of a reaction pathway are in steady state, the system can be described easily using stoichiometric modeling. Let the following multiple-enzyme system be considered by way of an example, where the intermediate compounds A, B, C, and D are in steady state:

Writing the molar balance equations for the intermediate compounds, one gets:

$$\frac{da}{dt} = 0 = v_1 - v_2 \tag{2.20}$$

$$\frac{db}{dt} = 0 = v_2 - v_3 - v_4 \tag{2.21}$$

$$\frac{dc}{dt} = 0 = v_3 - v_5 \tag{2.22}$$

$$\frac{dd}{dt} = 0 = v_4 - v_6 \tag{2.23}$$

$$\begin{bmatrix} 1 & -1 & 0 & 0 & 0 & 0 \\ 0 & 1 & -1 & -1 & 0 & 0 \\ 0 & 0 & 1 & 0 & -1 & 0 \\ 0 & 0 & 0 & 1 & 0 & -1 \end{bmatrix} \cdot \begin{bmatrix} v_1 \\ v_2 \\ v_3 \\ v_4 \\ v_5 \\ v_6 \end{bmatrix} = \begin{bmatrix} 0 \\ 0 \\ 0 \\ 0 \end{bmatrix} \tag{2.24}$$

Note that the balance equations yield a linear system of equations, which can be represented concisely in matrix form (see Equation 2.24), where the left-hand-side matrix in Equation 2.24 is called the stoichiometric matrix (T) and the vector of v_i is designated the flux vector (v). Under steady state, $T \cdot v = 0$. Since Equation 2.24 comprises four independent equations and six unknowns, the system is under-determined. Thus, two fluxes have to be quantified in order to calculate the remaining four.

2.6   Determination of Kinetic Parameters

In order for any rate equation to be meaningful, sound values must be determined for the kinetic parameters. Kinetic parameters are determined experimentally, with most methods

Table 2.5 *Methods of linearization for the determination of kinetic parameters.*

Method	y-axis	x-axis	Intercept in y-axis	Intercept in x-axis	Slope
Lineweaver–Burk (double-reciprocal)	$\frac{1}{v}$	$\frac{1}{s}$	$\frac{1}{V_{maxAP}}$	$-\frac{1}{K_{MAP}}$	$\frac{K_{MAP}}{V_{maxAP}}$
Eaddie–Hofstee	v	$\frac{v}{s}$	V_{maxAP}	$\frac{V_{maxAP}}{K_{MAP}}$	$-K_{MAP}$
Hanes	$\frac{s}{v}$	s	$\frac{K_{MAP}}{V_{maxAP}}$	$-K_{MAP}$	$\frac{1}{V_{maxAP}}$

based on measuring initial reaction rates at varying concentrations of substrate(s) and effectors. They can also be determined from the evolution of the reaction with time, although precautions must be taken to leave out factors affecting enzyme activity other than substrate depletion.

Experimental design for the determination of kinetic parameters consists in a series of measurements of initial reaction rates at different concentrations of substrates and effectors. This leads to a matrix of data, which is then processed to yield the corresponding values of the parameters. It is quite convenient to express the kinetic rate equations in a Michaelis–Menten format, as represented by Equation 2.18 for single-substrate reactions and by Equation 2.19 for two-substrate reactions. Parameter determination can be done either by linearization of the rate data or by nonlinear regression to the rate equation. Several methods have been proposed for the linearization of Equation 2.18, with the most common shown in Table 2.5.

In the case of simple Michaelis–Menten kinetics, V_{maxAP} and K_{MAP} are just V_{max} and K_M, so they can be determined from a series of data on initial reaction rates at different initial substrate concentrations. Eaddie–Hofstee and Hanes have the advantage of a more evenly distributed error compared to Lineweaver–Burk, but the Lineweaver-Burk method is the only one in which the variables are separated, which, aside from being more orthodox, allows for a better appreciation of the results. The accuracy of the results obtained will very much rely upon the experimental design adopted. Some guidelines should be followed:

- Use a significant number of data points (no less than eight).
- Distribute the data evenly (a geometric progression of substrate concentration, as can be easily obtained using serial substrate dilution, is better than an arithmetic progression).
- Use a substrate concentration range that contains the value of the dissociation constants. Ideally, the value of the determined parameter (K_M) should be in the midpoint of that range; if the calculated value is outside the range, it is advisable to repeat the experimental design with this primary calculated value as the midpoint.

With a reversible reaction, the same procedure is applicable, except that now initial reaction rates of substrate conversion into product at different substrate concentrations and initial rates of product conversion into substrate at different product concentrations will have to be determined. According to Equation 2.16, the former data will yield the values of $V_{max,S}$ and $K_{M,S}$ and the latter the values of $V_{max,P}$ and $K_{M,P}$.

In the case of enzyme inhibition, the experimental design will consist in generating a matrix of data on the initial reaction rates at varying substrate and inhibitor concentrations.

Firstly, the mechanism of inhibition must be determined; this can properly be done by representing the data in a linearized correlation. The double-reciprocal plot is particularly useful in this respect: as easily deduced from Table 2.3, when using varying inhibitor concentrations, competitive inhibition will yield a common intercept in the y-axis, noncompetitive inhibition will yield a common intercept in the x-axis, mixed-type inhibition will yield lines intersecting at some point in quadrant II, while uncompetitive inhibition will yield parallel lines. Secondly, the values of apparent parameters (V_{maxAP} and K_{MAP}) can be determined as before, followed by the corresponding inhibition parameters—according to the mechanism of inhibition (see Table 2.3)—again by linearization with respect to the concentration of the inhibitor. The same guidelines apply as in the case of simple Michaelis–Menten kinetics, although obviously the number of experiments will increase. Additionally, in this case the inhibitor concentration range used must contain the values of the inhibition dissociation constants.

In the case of a two-substrate reaction, the experimental design will consist in generating a matrix of data on initial reaction rates at varying concentrations of each substrate. Firstly, the mechanism of reaction must be determined, which can be done by representing the data in a linearized correlation. The double-reciprocal plot is also particularly useful in this respect. The ordered sequential mechanism can easily be distinguished from the other mechanisms because $V_{maxAP(A)}$ is not a function of a, being simply V_{max}. Secondly, apparent parameters can easily be determined from Equation 2.19 by plotting v^{-1} (reciprocal of initial reaction rates) versus a^{-1} (values of $V_{maxAP(B)}$ and $K_{MAP(B)}$ are obtained as a function of b) and versus b^{-1} (values of $V_{maxAP(A)}$ and $K_{MAP(A)}$ are obtained as a function of a). The corresponding kinetic parameters can be determined according to the mechanism of reaction (see Table 2.4), again by linearization of the apparent parameters with respect to the concentration of the other substrate. The same guidelines apply as in the case of simple Michaelis–Menten kinetics, except the number of experiments required for the same level of accuracy will be squared. In this case, the substrate concentration ranges for substrates A and B must contain the values of K_{MA} and K_{MB}, respectively.

With the precautions indicated before, the integration of the Michaelis–Menten Equation 2.13:

$$v = -\frac{ds}{dt} = \frac{V_{max} \cdot s}{K_M + s} \tag{2.25}$$

allows the kinetic parameters to be determined by linearization of the integrated equation:

$$\frac{1}{t} \cdot ln\frac{s_i}{s} = \frac{V_{max}}{K_M} - \frac{s_i - s}{K_M} \cdot \frac{1}{t} \tag{2.26}$$

2.7 Effects of Operational Variables on Enzyme Kinetics

The catalytic potential of an enzyme is strongly affected by environmental variables, with pH and temperature being the most important. These variables affect both enzyme activity and enzyme stability, although not necessarily in the same direction nor with the same magnitude.

The effect of pH stems from the fact that enzymes are polyionic polymers, so that changes in pH produce changes in the ionization stage of the active site and in the charge distribution at the protein surface, according to the pK values of the ionizable amino acid residues. This effect is quite relevant because most enzymatic processes are conducted in aqueous media; for synthesis reactions performed in nonaqueous media, the pH has no meaning, but even so enzymes tend to express their activity with respect to the pH of the aqueous milieu from which they were obtained, a phenomenon referred to as "pH memory."

Temperature is the most important variable in any enzymatic process and will affect both enzyme activity and enzyme stability, although in opposite directions. Temperature will increase the rate of any chemical reaction, leading to a positive effect in terms of enzyme kinetics, but will also increase the rate of enzyme inactivation caused by structural alterations, so stability will be impaired. This means that the optimal temperature for any enzymatic process results from a balance of both effects, making it a key aspect in biocatalysis.

2.7.1 Effects of pH

The effect of pH on enzyme kinetics is quite complex and approaches to quantifying it are necessarily an oversimplification. A well accepted approach is the one proposed by Michaelis and Davidsohn [16], which is based on two main assumptions: (1) the active site of the enzyme exists in three ionic configurations of successive numbers of charges and (2) of these three, only the intermediate ionic species is active, being able to bind and transform the substrate. These assumptions, although arguable, are supported by experimental evidence in many cases and are the basis for making the effect of pH on kinetic parameters explicit. From the scheme presented in this section, where n denotes the number of charges at the active site (arbitrarily set as negative) and K_i are the ionic equilibrium constants of the free enzyme (E) and the enzyme–substrate complex (ES):

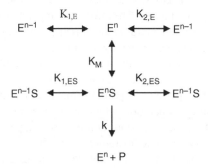

explicit functions of pH can be obtained for the apparent (with respect to pH) kinetic parameters in the rate equation:

$$v = \frac{V_{maxAP} \cdot s}{K_{MAP} + s}$$

being:

$$V_{maxAP} = \frac{V_{max}}{\dfrac{h^+}{K_{1,ES}} + 1 + \dfrac{K_{2,ES}}{h^+}} \tag{2.27}$$

$$K_{MAP} = K_M \frac{\dfrac{h^+}{K_{1,E}} + 1 + \dfrac{K_{2,E}}{h^+}}{\dfrac{h^+}{K_{1,ES}} + 1 + \dfrac{K_{2,ES}}{h^+}} \tag{2.28}$$

where   is the proton (molar) concentration.

An experimental design, in which initial rates of reaction are determined at varying pHs, allows a determination of the apparent kinetic parameters V_{maxAP} and K_{MAP}. Plotting log V_{maxAP} versus pH $(-\log[h^+])$ and $\log(K_{MAP} \cdot V_{maxAP}^{-1})$ versus pH, all kinetic parameters in Equations 2.27 and 2.28 can be determined, allowing a pH-explicit rate equation to be evaluated:

$$v = \frac{V_{maxAP} \cdot s}{K_{MAP} + s} = \frac{\dfrac{V_{max}}{\dfrac{h^+}{K_{1,ES}} + 1 + \dfrac{K_{2,ES}}{h^+}} \cdot s}{K_M \dfrac{\dfrac{h^+}{K_{1,E}} + 1 + \dfrac{K_{2,E}}{h^+}}{\dfrac{h^+}{K_{1,ES}} + 1 + \dfrac{K_{2,ES}}{h^+}} + s} \tag{2.29}$$

Equation 2.29 allows the optimum pH in terms of reaction rate to be determined. Since enzymes tend to be more stable at the pH at which they are most active, this represents valuable information by which to determine the optimum operational pH in an enzymatic process.

2.7.2   Effects of Temperature

Enzymes are labile protein molecules, so temperature has a strong effect on their configuration, affecting both activity and stability. The effects of temperature run in opposite directions: activity tends to increase with temperature as a consequence of the increase in reaction rate, while stability tends to decrease, since temperature increases the rate of enzyme inactivation. Unlike with pH, the temperature that maximizes activity (initial reaction rate) is usually detrimental to preserving it. There is thus a clear compromise between enzyme activity and stability. The value usually reported as optimum (maximum activity) may have very little meaning with respect to process operation, where the enzyme must be active for an extended period of time. As a consequence, the optimum process temperature may be significantly lower than the reported optimum temperature. The determination of optimum process temperature is a complex task that will be analyzed in Section 7.1.3. It requires temperature-explicit functions to be derived for kinetic and inactivation parameters.

2.7.2.1   Effect on Kinetic Parameters

Assuming that affinity parameters (Michaelis constant and inhibition constants) can be considered equilibrium constants, the Gibbs–Helmholz thermodynamic correlation can be applied:

$$\Delta H^0 - T \cdot \Delta S^0 = R \cdot T \cdot \ln K_{MAP} \tag{2.30}$$

$$\Delta H_I^0 - T \cdot \Delta S_I^0 = R \cdot T \cdot \ln K_{IAP} \tag{2.31}$$

From Equations 2.30 and 2.31, K_{MAP} and K_{IAP} can be obtained as temperature-explicit functions:

$$K_{MAP} = K_{MAP,0} \cdot \exp\left(\frac{\Delta H^0}{R \cdot T}\right) \tag{2.32}$$

$$K_{IAP} = K_{IAP,0} \cdot \exp\left(\frac{\Delta H_I^0}{R \cdot T}\right) \tag{2.33}$$

An experimental design in which initial rates of reaction are determined at varying temperatures allows the determination of temperature-explicit functions for K_{MAP} and K_{IAP} through plotting of $\ln K_{MAP}$ (or $\ln K_{IAP}$) versus T^{-1}.

The effect of temperature on the catalytic rate constant, k, can be determined from the semi-empirical Arrhenius equation:

$$k = k_0 \cdot \exp\left(\frac{-E_a}{R \cdot T}\right) \tag{2.34}$$

which can be rewritten as:

$$V_{max\,AP} = V_{max\,AP,0} \cdot \exp\left(\frac{-E_a}{R \cdot T}\right) \tag{2.35}$$

From the same experimental design used before, a temperature-explicit function can be obtained for V_{maxAP} by plotting $\ln V_{maxAP}$ versus T^{-1}.

For simple Michaelis–Menten kinetics, a temperature-explicit rate equation is thus:

$$v = \frac{V_{maxAP} \cdot s}{K_{MAP} + s} = \frac{V_{maxAP,0} \cdot \exp\left(\frac{-E_a}{R \cdot T}\right) \cdot s}{K_{MAP,0} \cdot \exp\left(\frac{\Delta H^0}{R \cdot T}\right) + s} \tag{2.36}$$

and for competitive inhibition:

$$v = \frac{V_{maxAP} \cdot s}{K_{MAP} + s} = \frac{V_{maxAP,0} \cdot \exp\left(\frac{-E_a}{R \cdot T}\right) \cdot s}{K_{MAP,0} \cdot \exp\left(\frac{\Delta H^0}{R \cdot T}\right)\left(1 + \frac{i}{K_{IAP,0} \cdot \exp\left(\frac{\Delta H_I^0}{R \cdot T}\right)}\right) + s} \tag{2.37}$$

2.7.2.2 Effect on Stability Parameters

Enzyme stability under process conditions is a major concern in biocatalysis. Enzymes tend to be poorly stable under operating conditions, so biocatalyst stabilization is a central issue in enzyme technology. Several strategies have been proposed, from molecular redesign to immobilization, but no matter how stable the enzyme biocatalyst, its efficient use will force it to the point of inactivation; knowledge of enzyme inactivation kinetics is thus necessary in order to properly evaluate biocatalyst performance [17]. This aspect will be thoroughly reviewed in Chapter 6. For now, a very simple (but often useful) mechanism of inactivation will be proposed, in which the enzyme is assumed to suffer a conformational transition from a native (active) structure to an unfolded, completely inactive form. This simple mechanism leads to a first-order kinetic model:

$$-\frac{de}{dt} = k_D \cdot e \tag{2.38}$$

Integrating Equation 2.38 leads to:

$$\frac{e}{e_0} = exp(-k_D \cdot t) \tag{2.39}$$

In this simple model, the parameter describing enzyme inactivation is the first-order transition rate constant (inactivation rate constant), k_D. A very simple experimental design in which the enzyme is incubated at the corresponding temperature and samples are taken at intervals of time in order to analyze residual activity allows k_D to be determined as the slope of $ln(e \cdot e_0^{-1})$ versus t plot. If k_D is considered as a rate constant, an Arrhenius correlation can be made to describe the effect of temperature on enzyme inactivation:

$$k_D = k_{D,0} \cdot exp\left(\frac{-E_{ia}}{R \cdot T}\right) \tag{2.40}$$

From data on k_D at a range of temperatures, the parameters of Equation 2.33 can be determined by plotting $ln\, k_D$ versus T^{-1}.

A common parameter used to evaluate enzyme stability is the half-life ($t_{1/2}$) of the catalyst, defined as the time after which enzyme activity has been reduced to one-half of its original value:

$$t_{1/2} = t\Big|_{e=\frac{e_0}{2}}$$

From Equation 2.39:

$$t_{1/2} = \frac{ln\, 2}{k_D} \tag{2.41}$$

In the case of simple first-order inactivation kinetics—but only in this case—$t_{1/2}$ is tightly related to the inactivation rate constant. However, enzyme inactivation is a very complex process, so Equation 2.39 can be considered only as a first approximation to describing it. More complex mechanisms of enzyme inactivation and the effects of other process variables (such as substrate and product concentration) on enzyme inactivation kinetics have been proposed. In such more complex mechanisms, several inactivation parameters may arise, such that $t_{1/2}$ is not related to any of them individually, being only an operational parameter that roughly describes enzyme inactivation. These aspects will be analyzed in Sections 6.1 and 6.2 respectively.

Solved Problems

2.1 β-galactosidase (β-D-galactoside galactohydrolase, EC 3.2.1.23) catalyzes the hydrolysis of lactose into glucose and galactose and is a widely used enzyme in the dairy industry. A commercial liquid β-galactosidase preparation from *Kluyveromyces marxianus var lactis* is being used in the production of low-lactose milk. It is known that 500 IU of β-galactosidase per liter of milk is required to reduce the lactose content in milk by 80% during storage (considered satisfactory) and you must calculate the enzyme dosage required (mL of enzyme preparation to be added per L of milk).

To do so, the β-galactosidase activity of the enzyme preparation must be determined. The enzyme assay is based on determining glucose in an HPLC system with a refractive index detector. The analytical procedure consists in adding 0.3 mL of the enzyme preparation, diluted three times, to 1.2 mL of $150\,g \cdot L^{-1}$ lactose solution and incubating the mixture at $35\,°C$ and pH 6.4. Samples are taken at different reaction times and subjected to enzyme inactivation by boiling and then filtering prior to the HPLC assay. The following results were thus obtained:

t (mins)	0	2	4	6	8	12	18	26	36	48
g (mM)	0	150	307	461	597	912	942	976	999	1015

t: time of incubation
g: glucose concentration

Resolution: Enzyme activity is determined within the linear zone of product concentration versus reaction time. From the data gathered, the cumulative product formation rate (Δ) is calculated at each incubation time:

t (mins)	0	2	4	6	8	12	18	26	36	48
Δ (mM · min^{-1})	0	75.0	76.7	76.8	74.6	76.0	52.3	37.5	27.8	21.2

As can be seen, linearity extends up to 12 minutes, so enzyme activity can be determined in that range. From a linear correlation of g versus t data in the time range from 0 to 12 minutes, $\Delta = 75.6\,mM \cdot min^{-1}$ is obtained with a correlation coefficient close to 1. The activity of the enzyme preparation, considering the dilution factor and the equimolar conversion of lactose into glucose, is thus:

$$activity = 75.6 \left[\frac{\mu moles\ glucose}{mL \cdot min}\right] \cdot \frac{1.5}{0.3} \cdot 3 = 1134 \left[\frac{\mu moles\ glucose}{mL\ enzyme \cdot min}\right]$$

$$= 1134 \left[\frac{IU}{mL\ enzyme}\right]$$

The initial concentration of lactose in the reaction assay was $150 \cdot 1.2 \cdot 1.5^{-1} = 120\,g \cdot L^{-1}$; that is, 350 mM, which is very high and is likely to be saturating. The initial reaction rate measured can thus properly be considered as activity. This should be checked to prove that this concentration is at least five times the K_M value of the enzyme.

Answer: The commercial liquid β-galactosidase preparation of *Kluyveromyces marxianus var lactis* has a volumetric activity of $1134 \, IU \cdot mL^{-1}$. The dose required per liter of milk is thus:

$$dose = \frac{500 \left[\dfrac{IU}{L \, milk} \right]}{1134 \left[\dfrac{IU}{mL \, enzyme} \right]} = 0.44 \left[\frac{mL \, enzyme}{L \, milk} \right]$$

2.2 A crude extract of glucose isomerase (EC 5.3.1.5) has been obtained after cultivation of *Streptomyces flavogriseus* and cell disruption in a homogenizer. The enzyme catalyzes the reversible isomerization of glucose into fructose. A 1 mL sample of this extract, diluted as indicated in the following table, was put in contact with 4 mL of 0.1 M glucose ($C_6H_{12}O_6$) solution. Glucose concentration was monitored during the reaction and the following results were obtained:

Reaction Time (minutes)	Glucose Concentration ($g \cdot L^{-1}$)	
	1 : 50 dilution	1 : 100 dilution
0	14.40	14.40
2	13.91	14.15
4	13.58	13.91
6	13.27	13.64
8	13.04	13.46
10	13.03	13.34

Calculate the activity of glucose isomerase in the crude extract, expressing it in $IU \cdot mL^{-1}$, and discuss the result.

Resolution: From the data obtained, the glucose consumed ($-\Delta S$) at the two dilutions is:

Reaction Time (minutes)	Glucose Consumed ($-\Delta s$) ($g \cdot L^{-1}$)	
	1 : 50 dilution	1 : 100 dilution
0	—	—
2	0.49	0.25
4	0.82	0.49
6	1.13	0.76
8	1.36	0.94
10	1.37	1.06

Plotting $(-\Delta S)$ versus time, linearity is obtained at $1 : 100$ dilution for up to 6 minutes. Enzyme activity can thus be determined as the slope of the straight line extending to that time:

$$-\frac{\Delta s}{t} = \frac{0.76}{6} = 0.1267 \left[\frac{g}{L \cdot min}\right] = 0.1267 \frac{10^6}{10^3 \cdot 180} \left[\frac{\mu moles}{mL \cdot min}\right] = 0.704 \left[\frac{\mu moles}{mL \cdot min}\right]$$

$$activity = 0.704 \cdot 5 \cdot 100 = 352 \left[\frac{\mu moles}{mL \cdot min}\right] = 352 \left[\frac{IU}{mL}\right]$$

Answer: The crude extract has $352\ IU \cdot mL^{-1}$ glucose isomerase activity. Data obtained at $1:50$ dilution cannot be properly used to determine activity as the enzyme is still too concentrated, meaning that linearity is not revealed even at a short reaction time (2 minutes). Glucose concentration at the end of the assay is still 95% of its initial value, so the enzyme can be assumed to still be saturated (it is reasonable to assume that the assay was designed to have a saturating initial glucose concentration, but this assumption should be checked by determining the K_M value; substrate concentration should be no less than five times K_M. The reversibility of the reaction is not yet revealed and the time of assay is too short to produce enzyme inactivation, so the value obtained can be properly considered the enzyme activity.

2.3 A glucoamylase producer claims that its commercial preparation has an activity of $0.033\ katals \cdot L^{-1}$ and you have to check the accuracy of such information. To do so, you will follow the procedure recommended by the enzyme producer, which determines the glucoamylase activity using maltose as a substrate (maltose $+ H_2O \rightarrow 2$ glucose) and an enzymatic kit (Glucostat) that quantifies glucose according to:

Glucose + ATP $\xrightarrow{\text{hexokinase}}$ Glucose 6-Pi + ADP

Glucose 6-Pi + NAD$^+$ $\xrightarrow{\text{glucose 6-Pi dehydrogenase}}$ 6-Pi gluconate + NADH + H$^+$

The NADH formed can conveniently be measured spectrophotometrically by recording optical density (OD) at 340 nm. The extinction coefficient of NADH is $90\ mL \cdot cm^{-1} \cdot mmol^{-1}$. According to the stoichiometry, 1 mol of glucose reacted yields 1 mol of NADH.

The following results have been obtained with the properly diluted glucoamylase preparation by putting it in contact with a saturated maltose solution under an excess of ATP, NAD$^+$, and coupling enzyme activities at the conditions recommended by the enzyme producer:

Reaction Time (minutes)	(OD)$_{340}$	
	1:1500 dilution	1:7500 dilution
0	0.030	0.005
1	0.259	0.0563
2	0.542	0.107
3	0.800	0.159
4	0.905	0210
5	0.943	0.261

Resolution

According to the Beer–Lambert law:

$$(OD)_\lambda = \varepsilon \cdot L \cdot [C] \tag{2.42}$$

So NADH concentration can be calculated as:

$$[NADH] = \frac{(OD)_{340}}{90 \cdot 1} \cdot 1000 \quad [mM]$$

The data obtained thus correspond to:

Reaction Time (minutes)	[NADH] (mM)	
	1 : 1500 dilution	1 : 7500 dilution
0	0.333	0.056
1	2.878	0.626
2	6.022	1.189
3	8.889	1.767
4	10.06	2.333
5	10.478	2.900

Linearity is obtained only at 1 : 7500 dilution for up to 5 minutes of reaction, so enzyme activity can be determined as the slope of the straight line extending to that time:

$$activity = \frac{2.9}{5} \cdot 7500 = 4350 \left[\frac{\mu moles\, glucose\, reacted}{mL \cdot min} \right]$$

One mole of hydrolyzed maltose yields two moles of glucose, so:

$$activity = \frac{4350}{2} = 2175 \left[\frac{\mu moles\, maltose\, reacted}{mL \cdot min} \right] = 2175 \frac{IU}{mL}$$

$$activity = \frac{2175}{10^6 \cdot 60} \left[\frac{moles\, maltose\, reacted}{mL \cdot sec} \right] = 3.625 \cdot 10^{-5} \frac{katals}{mL}$$

Answer: The declared activity ($3.3 \cdot 10^{-5}$ katals $\cdot$ mL^{-1}) is almost 10% lower than that measured. This is not surprising, as some enzyme producers market their products on the safe side so as to guarantee to their costumers that under no circumstances will a product with less than the declared activity be received.

2.4 The activity of a recombinant esterase from *Pseudomonas fluorescens* (ethyl acetate + H$_2$O → acetic acid + ethanol) is determined by measuring ethanol using gas chromatography. 1 µL of purified esterase sample is incubated with 10 mL of saturating ethyl acetate solution (500 mM) at pH 7.5 and 25 °C and ethanol evolution is measured over time. The purified enzyme sample is diluted to 1 : 4 and 1 : 8 prior to assay and the following results are obtained:

Incubation Time (minutes)	Ethanol Produced (mM)			
	Undiluted	Diluted 1 : 4	Diluted 1 : 8	Diluted 1 : 8[*]
2	1.01	0.28	0.14	0.125
5	2.18	0.71	0.35	0.31
8	2.22	1.04	0.57	0.50
13	2,25	1.38	0.90	0.81

[*]Values obtained at 100 mM ethyl acetate.

Calculate the volumetric activity of the purified enzyme, its maximum reaction rate, and its Michaelis constant for ethyl acetate. Make all assumptions explicit.

Resolution: From the data obtained, the cumulative product formation rate is calculated at each incubation time:

Incubation Time (minutes)	Cumulative Ethanol Formation Rate (mM · min^{-1})			
	Undiluted	Diluted 1 : 4	Diluted 1 : 8	Diluted 1 : 8[*]
2	0.505	0.14	0.07	0.063
5	0.436	0.142	0.07	0.062
8	0.278	0.130	0.071	0.063
13	0.173	0.106	0.069	0.062

[*]Values obtained at 100 mM ethyl acetate.

As can be seen from these data, the reaction is linear up to 13 minutes at a 1 : 8 dilution. From a linear correlation of ethanol produced versus t data in the time range from 0 to 13 minutes at 1 : 8 dilution, the cumulative ethanol formation rates obtained at 500 and 100 mM ethyl acetate are 0.0695 and 0.0623 mM · min^{-1}, respectively, with correlation coefficients close to 1.

At 500 mM ethyl acetate it is assumed that the enzyme is saturated with substrate, so that the maximum reaction rate is being measured:

$$v_{500\,nM} = \frac{0.0695\left[\dfrac{mmoles\ ethanol}{L \cdot min}\right]}{0.0125\left[\dfrac{mL\ enzyme}{L}\right]} = 5.56\left[\frac{mmoles\ ethanol}{mL\ enzyme \cdot min}\right] = 5560\left[\frac{\mu moles\ ethanol}{mL\ enzyme \cdot min}\right]$$

Answer: Since one mole of ethanol is produced from one mole of ethyl acetate, the activity of the purified enzyme is:

$$activity = 5560\,IU \cdot mL^{-1}$$

With 100 mM ethyl acetate, from the data in the last column of the table, the initial reaction rate is:

$$v_{100\,nM} = \frac{0.0623\left[\dfrac{mmoles\ ethanol}{L \cdot min}\right]}{0.0125\left[\dfrac{mL\ enzyme}{L}\right]} = 4.98\left[\frac{mmoles\ ethanol}{mL\ enzyme \cdot min}\right] = 4980\left[\frac{\mu moles\ ethanol}{mL\ enzyme \cdot min}\right]$$

Given that Michaelis–Menten kinetics applies, from Equation 2.13:

$$4980 = \frac{5560 \cdot 100}{K_M + 100}$$

$$K_M = 11.7\,mM$$

This is just a rough estimate of K_M, since only one substrate concentration has been used in its determination.

2.5 A certain enzyme has no preformed active site and the binding of a modulator molecule is required to build it up so that the substrate can be bound to it and acted upon by it.

Develop a rate equation for such a mechanism and describe the experimental procedure used to determine the corresponding kinetic parameters.

Resolution: Let S, X, and Pi represent the substrate, the modulator, and the products of reaction, respectively (lower-case letters represent the corresponding molar concentrations):

$$E + X \rightleftharpoons EX \qquad K_X = \frac{(e - d - f)x}{d} \qquad (2.43)$$

$$EX + S \rightleftharpoons EXS \qquad K'_M = \frac{d \cdot s}{f} \qquad (2.44)$$

$$EX \xrightarrow{k} E + Pi \qquad v = k \cdot f \qquad (2.45)$$

where e, d, and f are the molar concentrations of the free enzyme and the enzyme complexes EX and EXS, respectively.

Answer: From Equations 2.43 to 2.45:

$$v = \frac{k \cdot e \cdot s}{\dfrac{K'_M \cdot K_X}{x} + K'_M + s} = \frac{V_{max} \cdot s}{\dfrac{K'_M \cdot K_X}{x} + K'_M + s} \qquad (2.46)$$

Equation 2.41 can be represented in the form of Equation 2.18, where:

$$V_{maxAP} = V_{max}$$

$$K_{MAP} = K'_M\left(\frac{K_X}{x} + 1\right)$$

Experimental design consists in a matrix of data on initial rates or reaction at different initial concentrations of S and X. Parameters can be determined by linearization of Equation 2.46:

$$\frac{1}{v} = \frac{K_{MAP}}{V_{maxAP} \cdot s} + \frac{1}{V_{maxAP}}$$

A family of straight lines at constant x will be obtained which have a common intercept in the y-axis (V_{max}^{-1}) and variable intercept in the x-axis, corresponding to $-K_{MAP}^{-1}$.

Plotting the obtained values of K_{MAP} versus x^{-1}, a straight line will be obtained with their intercepts in the y- and x-axis at K_M' and $-K_X^{-1}$, respectively.

2.6 Phenylalanine ammonia lyase (EC 4.3.1.5) from *Rhodotorula glutinis* catalyzes the conversion of phenylalanine (X) into trans-cinnamic acid (Y) and ammonia (Z), where Y is a competitive inhibitor and Z a partial uncompetitive inhibitor.

Develop a parametric kinetic expression of the form of Equation 2.18 and evaluate the apparent kinetic parameters in terms of the corresponding Michaelis and inhibition constants.

***Resolution*:**

$$E + X \rightleftharpoons EX \qquad K_M = \frac{(e - c - d - f)x}{c} \tag{2.47}$$

$$E + Y \rightleftharpoons EY \qquad K_Y = \frac{(e - c - d - f)y}{d} \tag{2.48}$$

$$EX + Z \rightleftharpoons EXZ \qquad K_Z' = \frac{c \cdot z}{f} \tag{2.49}$$

$$EX \xrightarrow{k} E + Y + Z$$

$$EXZ \xrightarrow{k'} EZ + Y + Z$$

$$v = k \cdot c + k' \cdot f \tag{2.50}$$

where x, y, and z are the molar concentration of the corresponding species and e, c, d, and f are the molar concentration of the free enzyme and the EX, EY, and EXZ enzyme complexes, respectively.

Values of c and f are obtained from Equations 2.47 to 2.49 and replaced in Equation 2.50:

$$v = \frac{e \cdot x \left(k + k' \frac{z}{K_Z'}\right)}{K_M \frac{1 + \frac{y}{K_Y}}{1 + \frac{z}{K_Z'}} + x} = \frac{\left(V_{max} + V_{max}' \frac{z}{K_Z'}\right) \cdot x}{K_M \frac{1 + \frac{y}{K_Y}}{1 + \frac{z}{K_Z'}} + x} \tag{2.51}$$

***Answer*:** From Equations 2.18 and 2.51:

$$V_{maxAP} = \left(V_{max} + V_{max}' \frac{z}{K_Z'}\right)$$

$$K_{MAP} = K_M \frac{1 + \frac{y}{K_Y}}{1 + \frac{z}{K_Z'}}$$

2.7 Draw the profile of sucrose concentration across time when invertase (β-fructofur-anosidase, EC 3.2.1.26) is added at a level of $1 \, g \cdot L^{-1}$ to a 50 mM sucrose solution. The kinetic parameters of the enzyme under the conditions of reaction are: $K_M = 10 \, mM$ and $V_{max} = 1000 \, \mu moles \cdot min^{-1} \cdot g_{enzyme}^{-1}$. Justify any assumption required.

Resolution: Assuming that sucrose hydrolysis can be represented by simple Michaelis–Menten kinetics:

$$v = -\frac{ds}{dt} = \frac{V_{max} \cdot s}{K_M + s} = \frac{k \cdot e \cdot s}{K_M + s} \tag{2.52}$$

Integrating Equation 2.52:

$$\int_{s}^{s_i} \frac{K_M + s}{s} ds = \int_{0}^{t} k \cdot e \cdot dt \tag{2.53}$$

$$K_M \cdot ln\frac{s_i}{s} + (s_i - s) = k \cdot e \cdot t \tag{2.54}$$

where s is the variable concentration of substrate and s_i is its initial value:

$$s_i = 50 \, mM$$

$$V_{max} = k \cdot e = 1000 \left[\frac{\mu moles}{min \cdot g}\right] \cdot 1 \left[\frac{g}{L}\right] = 1 \left[\frac{mM}{min}\right]$$

Equation 2.54 is obtained only if e is considered constant; that is, if it is assumed that during the reaction time the enzyme does not suffer inactivation. Then Equation 2.54 becomes:

$$10 \cdot ln\frac{50}{s} + (50 - s) = t \tag{2.55}$$

which can be solved using modified secant methods (see Section A.1) for t = 3 minutes. Figure 2.1 shows the reaction course for times between 0 and 120 minutes. Each point in the figure was determined by solving Equation 2.55.

Answer: The sucrose time profile is represented in Figure 2.1.

The integrated form of the Michaelis–Menten equation can be used to determine kinetic parameters from the time course of the enzymatic reaction, provided that within that time the enzyme remains fully active and saturated with substrate.

The problem can then be solved backwards. From the data on the reaction profile with time, calculate the kinetic parameters of the enzyme.

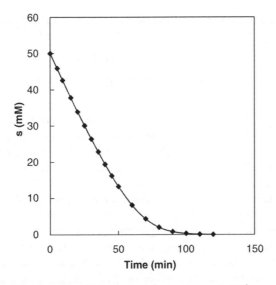

Figure 2.1   *Time profile of sucrose (S) hydrolysis by invertase. Initial concentration of sucrose = 50 mM, $V_{max} = 1\,mM \cdot min^{-1}$, $Km = 10\,mM$.*

Resolution: From Equation 2.26:

$$\frac{1}{t} \cdot ln\frac{50}{s} = \frac{V_{max}}{K_M} - \frac{(50-s)}{K_M \cdot t} \qquad (2.56)$$

so V_{max} and K_M can be determined from the intercept and slope of the $t^{-1} \cdot ln(50 \cdot s^{-1})$ versus $(50\text{-}s) \cdot t^{-1}$ plot. The following set of data can be obtained from Equation 2.55 (columns 1 and 2), and from them the values in columns 3 to 5 can be obtained:

1 t (minutes)	2 s (mM)	3 $ln(50 \cdot s^{-1})$	4 $t^{-1} \cdot ln(50 \cdot s^{-1})$	5 $(50\text{-}s) \cdot t^{-1}$
3	47.51	0.05 108	0.017 027	0.830
5	45.86	0.08 643	0.017 286	0.828
10	41.79	0.1794	0.017 937	0.821
15	37.79	0.2800	0.01 867	0.814
20	33.88	0.3892	0.01 946	0.806
25	30.08	0.5082	0.02 033	0.797
30	26.39	0.6390	0.02 130	0.787
35	22.83	0.7840	0.02 240	0.7763
40	19.44	0.9447	0.02 362	0.764
45	16.24	1.1246	0.02 499	0.750
50	13.27	1.3265	0.02 653	0.735
60	8.15	1.8140	0.03 023	0.698

Answer: From a linear correlation of the data in columns 4 and 5 and Equation 2.56:

$$V_{max} = 1.001 \, mM \cdot min^{-1}$$
$$K_M = 10.04 \, mM$$

These values are very close to the ones given in the first part of the exercise. This can be considered a good method for determining kinetic parameters as they can be determined from a single kinetic experiment, which may save time. However, the experiment has to be planned very carefully so as to ensure that the results are the consequence of substrate desaturation only, and that neither enzyme inactivation nor inhibition of any kind occurs during the experiment. This may be troublesome, especially for labile or product inhibition-sensitive enzymes. In such cases, measurement will have to be made during very short reaction times and this might produce significant error, because small differences between large numbers will have to be calculated. Besides this, if the final point is taken at a substrate concentration close to the K_M value, or higher, error will also be produced, because the initial portion of the v versus s correlation will not be represented.

2.8 A preliminary experiment with invertase (β-fructofuranosidase) from *Saccharomyces cerevisiae* at pH 4.5 and 25 °C gave the following results:

Sucrose Initial Concentration (mM)	Initial Reaction Rate (μmoles glucose $\cdot$ min^{-1} $\cdot$ g$_{enzyme}$$^{-1}$)
1	26.06
5	95.21
10	139.05
25	179.18
50	174.56
100	141.35

From such data it is apparent that sucrose exerts inhibition at high concentrations. Therefore, further experiments were conducted to obtain data in the low (below K_M value) and high (much higher than K_M and up to saturation) range of sucrose concentration. Sucrose solubility at 25 °C is 850 g $\cdot$ L^{-1}. The following data were gathered in both ranges:

Sucrose Initial Concentration (mM)	Initial Reaction Rate (μmoles glucose $\cdot$ min^{-1} $\cdot$ g$_{enzyme}$$^{-1}$)
0.1	2.830
0.25	6.976
0.5	13.633
0.75	19.990
1	26.063
1.25	31.869

Sucrose Initial Concentration (mM)	Initial Reaction Rate (μmoles glucose $\cdot$ min^{-1} $\cdot$ g$_{enzyme}^{-1}$)
900	28.066
1200	21.564
1500	17.507
1800	14.734
2100	12.720
2400	11.190

Determine the mechanism of inhibition and the values of the corresponding kinetic parameters

Resolution: The enzyme is subjected to inhibition by high sucrose concentration. It is reasonable to assume uncompetitive inhibition. According to such a mechanism:

$$E + S \rightleftharpoons ES \qquad K_M = \frac{(e - c - f)s}{c}$$

$$ES + S \rightleftharpoons ESS \qquad K_S = \frac{c \cdot s}{f}$$

$$ES \xrightarrow{k} E + G + Fr \qquad v = k \cdot c = \frac{k \cdot e \cdot s}{K_M + s + \dfrac{s^2}{K_S}} = \frac{V_{max} \cdot s}{K_M + s + \dfrac{s^2}{K_S}} \tag{2.57}$$

where S stands for sucrose, G for glucose and Fr for fructose.

Equation 2.57 is formally similar to the one in Table 2.3 for uncompetitive inhibition, but essentially different because in this case the substrate itself is the inhibitor, so it cannot be represented by Equation 2.18. Being non-Michaelian, Equation 2.57 cannot be linearized as before; however, inverting it produces:

$$\frac{1}{v} = \frac{K_M}{V_{max} \cdot s} + \frac{1}{V_{max}} + \frac{s}{V_{max} \cdot K_S} \tag{2.58}$$

so linear regions will be obtained at very low and very high values of s, as represented by Equations 2.59 and 2.60, respectively.

For: $s \lll K_M$:

$$\frac{1}{v} = \frac{K_M}{V_{max} \cdot s} + \frac{1}{V_{max}} \tag{2.59}$$

For: $s \ggg K_M$:

$$\frac{1}{v} = \frac{1}{V_{max}} + \frac{s}{V_{max} \cdot K_S} \tag{2.60}$$

From the data in the first zone ($s \ll K_M$):

s (mM)	v ($\mu mol \cdot min^{-1} \cdot g^{-1}$)	s^{-1} $(mM)^{-1}$	v^{-1} ($\mu mol^{-1} \cdot min \cdot g$)
0.1	2.830	10	0.3534
0.25	6.976	4	0.1434
0.5	13.633	2	0.07 335
0.75	19.990	1.333	0.05 003
1	26.063	1	0.03 837
1.25	31.869	0.8	0.03 138

From a linear correlation of data in the third and fourth columns (Equation 2.59):

$V_{max} = 296.9 \, \mu mol \cdot min^{-1} \cdot g^{-1}$
$K_M = 10.4 \, mM$
$r \sim 1$

However, these values are still approximate. If one takes only the four lower values:

$V_{max} = 297.3 \, \mu mol \cdot min^{-1} \cdot g^{-1}$
$K_M = 10.4 \, mM$
$r \sim 1$

and if only the first three:

$V_{max} = 297.8 \, \mu mol \cdot min^{-1} \cdot g^{-1}$
$K_M = 10.4 \, mM$
$r \sim 1$

One can then assume that values should be close to:

$V_{max} = 298 \, mol \cdot min^{-1} \cdot g^{-1}$
$K_M = 10.4 \, mM$

These values can be better approximated if more data in the very low s range are gathered. However, experimental error will certainly increase in that zone.

From the data in the second zone ($s \gg K_M$):

s (mM)	v ($\mu mol \cdot min^{-1} \cdot g^{-1}$)	v^{-1} ($\mu mol^{-1} \cdot min \cdot g$)
900	28.066	0.03 563
1200	21.564	0.04 637
1500	17.507	0.05 712
1800	14.734	0.06 787
2100	12.720	0.07 862
2400	11.190	0.08 937

From a linear correlation of data in the first and third columns (Equation 2.60):

$K_S = 94.3 \, mM$
$r \sim 1$

Again, these values are still approximate. If one takes only the four higher values:

$K_S = 94.1\,mM$

$r \sim 1$

and if only the three higher:

$K_S = 94.0\,mM$

$r \sim 1$

Answer: The results are consistent with a mechanism of uncompetitive inhibition and, in view of the value of K_M obtained (10.4 mM), the s ranges used are adequate

$V_{max} = 298.0\,\mu mol \cdot min^{-1} \cdot g^{-1}$

$K_M = 10.4\,mM$

$K_S = 94.0\,mM$

The resulting rate equation is then:

$$v = \frac{298 \cdot s}{10.4 + s + \dfrac{s^2}{94}} \tag{2.61}$$

Predicted values from Equation 2.61 can now be compared with actual experimental data obtained across the whole s range:

s (mM)	v (experimental) ($\mu mol \cdot min^{-1} \cdot g^{-1}$)	v (calculated by Equation 2.61) ($\mu mol \cdot min^{-1} \cdot g^{-1}$)	Error (%)
0.1	2.830	2.838	0.28
0.5	13.633	13.666	0.24
1	26.063	26.116	0.20
5	95.21	95.11	0.11
10	139.05	138.84	0.15
25	179.18	177.18	1.12
50	174.56	171.27	1.88
100	141.35	137.47	2.74
900	28.066	28.150	0.30
1800	14.734	14.786	0.35
2400	11.190	11.230	0.36

Error can be considered acceptable and within the margin of analytical error. However, error tends to be higher in the middle portion of the s range. No experimental data were gathered in the range from 100 to 900 mM and it is important that these be obtained, since error is expected to be higher in that zone, which is certainly quite important from a technological perspective. In fact, suppose that a batch reactor is fed with a 1 M sucrose concentration (342 g · L) and that the reaction is conducted up to 90% conversion (quite usual for an industrial operation of sucrose inversion) so that the final sucrose concentration is 100 mM. The reaction will normally be in this "twilight zone," where the error is probably higher than the values reported here. This analysis highlights the complexity in accurately determining kinetic parameters of substrate inhibition kinetics by linearization of initial rate data (the case presented here allows a wide range of substrate concentration to be covered, which is quite convenient but seldom possible in practice).

In this case, it is recommended that once the mechanism of uncompetitive inhibition has been identified, nonlinear regression to the rate curve (Equation 2.57) be used for parameter determination. The methodology for the estimation of parameters by nonlinear regression is given in Section A.2.2. Taking a confidence interval of 95% for the estimated value of each kinetic parameter, the following values are obtained:

$V_{max} = 305 \pm 4.005 \, \mu mol \cdot min^{-1} \cdot g^{-1}$
$K_M = 10.81 \pm 0.27 \, mM$
$K_S = 93.94 \pm 2.70 \, mM$

with an R^2 of 0.999.

In this particular example, minor differences are observed among the values estimated by linear and nonlinear regression. This is because in this didactic example the two limit zones for s (s $\ll K_M$ and s $\gg K_M$) were both expanded with a considerable number of data points. In practice, this is seldom possible. A more realistic example is included in Section A.2.2 in order to illustrate this point.

2.9 α-L-fucosidase (EC 3.2.1.51) catalyzes the hydrolysis of α-L-fucosides into L-fucose and an alcohol. Enzyme activity is determined by measuring initial rates of hydrolysis of the synthetic substrate *p*-nitrophenyl α-l-fucopyranoside [18]. The following results have been obtained with a purified preparation of α-L-fucosidase from the marine mollusk *Pecten maximus* at pH 4.5 and 50 °C:

s	1	2	3	4	5	10	15	20
v	51.52	64.15	69.86	73.12	75.22	79.81	81.47	82.34

s: initial concentration of *p*-nitrophenyl α-l-fucopyranoside (mM)
v: initial reaction rate (μmoles *p*-nitrophenol $\cdot$ min$^{-1} \cdot$ mg^{-1})

Determine the kinetic parameters of the enzyme using the linearization methods of Lineweaver–Burk, Eaddie–Hofstee, and Hanes and the nonlinear regression to the Michaelis–Menten equation. Discuss the results.

Resolution: From the results just quoted, the following data are obtained:

C1	C2	C3	C4	C5	C6
s	v	s^{-1}	v^{-1}	$v \cdot s^{-1}$	$s \cdot v^{-1}$
1	51.52	1	0.0194	51.520	0.0194
2	64.15	0.5	0.0156	32.075	0.0312
3	69.86	0.333	0.0143	23.287	0.0429
4	73.12	0.25	0.0137	18.280	0.0547
5	75.22	0.2	0.0133	15.044	0.0665
10	79.81	0.1	0.0125	7.981	0.1253
15	81.47	0.667	0.0123	5.431	0.1841
20	82.34	0.05	0.0121	4.117	0.2429

From Equation 2.13:

$$\text{Lineweaver–Burk}: \quad \frac{1}{v} = \frac{K_M}{V_{max} \cdot s} + \frac{1}{V_{max}} \tag{2.62}$$

$$\text{Eaddie–Hofstee}: \quad v = V_{max} - \frac{v}{s} \cdot K_M \tag{2.63}$$

$$\text{Hanes}: \quad \frac{s}{v} = \frac{K_M}{V_{max}} + \frac{s}{V_{max}} \tag{2.64}$$

Answer: According to Lineweaver–Burk (Equation 2.62) and data in columns 3 and 4:

$$V_{max} = 85.05\ \mu mol \cdot min^{-1} \cdot mg^{-1}$$
$$K_M = 0.651\ mM$$

According to Eaddie–Hofstee (Equation 2.63) and data in columns 2 and 5:

$$V_{max} = 85.00\ \mu mol \cdot min^{-1} \cdot mg^{-1}$$
$$K_M = 0.650\ mM$$

According to Hanes (Equation 2.64) and data in columns 1 and 6:

$$V_{max} = 85.00\ \mu mol \cdot min^{-1} \cdot mg^{-1}$$
$$K_M = 0.650\ mM$$

By nonlinear regression to Equation 2.13 (see Section A.2.2):

$$V_{max} = 85.00\ \mu moles \cdot min^{-1} \cdot mg^{-1}$$
$$K_M = 0.650\ mM$$

As can be seen, differences between methods are negligible in this case, which reflects the accuracy of the reported data. However, note that the value of K_M lies outside the substrate range considered, so it is advisable to explore data at lower substrate concentrations.

A different situation will arise if more experimental error is involved. Assume for such purpose the following data obtained with α-L-fucosidase from abalone liver in which enzyme kinetics is determined by measuring initial rates of hydrolysis of the synthetic substrate 2-chloro-4-nitrophenyl-α-L-fucopyranoside into α-L-fucoside and 2-chloro-4-nitrophenol, which is kinetically quantified by measuring the absorbance at 405 nm (Diazyme kit):

s	0.5	1	1.5	2	4	8	16	20
v	2.2	4.8	5.73	7.19	9.68	12.04	13.66	14

s: initial concentration of 2-chloro-4-nitrophenyl-α-L-fucopyranoside (mM)
v: initial reaction rate (μmoles 2-chloro-4-nitrophenol · min^{-1} · mg^{-1})

Resolution: From the results in the table above the following data are obtained:

C1	C2	C3	C4	C5	C6
s	v	s^{-1}	v^{-1}	$v \cdot s^{-1}$	$s \cdot v^{-1}$
0.5	2.2	2.000	0.455	4.400	0.227
1	4.8	1.000	0.208	4.800	0.208
1.5	5.73	0.667	0.175	3.820	0.262
2	7.19	0.500	0.139	3.595	0.278
4	9.68	0.250	0.103	2.420	0.413
8	12.04	0.125	0.083	1.505	0.664
16	13.66	0.063	0.073	0.854	1.171
20	14	0.050	0.071	0.700	1.429

Answer: According to Lineweaver-Burk (Equation (2.62)) and data in columns C3 and C4:

$$V_{max} = 19.16 \ \mu moles \cdot min^{-1} \cdot mg^{-1}$$
$$K_M = 3.66 \ mM$$
$$R^2 = 0.9828$$

According to Eaddie-Hofstee (Equation (2.63)) and data in columns C2 and C5:

$$V_{max} = 15.92 \ \mu moles \cdot min^{-1} \cdot mg^{-1}$$
$$K_M = 2.62 \ mM$$
$$R^2 = 0.944$$

According to Hanes (Equation (2.64)) and data in columns C6 and C1:

$$V_{max} = 15.87 \ \mu moles \cdot min^{-1} \cdot mg^{-1}$$
$$K_M = 2.60 \ mM$$
$$R^2 = 0.999$$

By non-linear regression to Equation (2.13):

$$V_{max} = 15.8 \ \mu moles \cdot min^{-1} \cdot mg^{-1}$$
$$K_M = 2.51 \ mM$$
$$R^2 = 0.999$$

As can been seen the Lineweaver-Burk method results in the overestimation of V_{max} and K_M, since the higher error at low s values produce a "leverage effect" in the regression. Eaddie-Hofstee method results in a low value of R^2 (0.944) indicating a poor fitting to the experimental data, so the estimated values for the parameters are not reliable. Hanes method offers a better fit, explaining 99.9% of the experimental data variability ($R^2 = 0.999$), being this the consequence of a more even distribution of error in such plot. Therefore, among these three methods, the Hanes allowed the best estimation of the enzyme kinetic parameters based on the experimental data provided. However, the use of non-linear regression represents the best fit, being nowadays a sound method for parameter estimation.

2.10 Inosine nucleosidase (EC 3.2.2.2) catalyzes the hydrolysis of inosine monophosphate (IMP) to hypoxanthine (Hx) and ribose phosphate (R5Pi). An enzyme preparation from *Azotobacter vinelandii* has shown partial noncompetitive inhibition by adenine (A). The following results for the initial rate of hydrolysis of IMP have been obtained without the addition of adenine:

[IMP] (mM)	v (μmoles Hx $\cdot$ min^{-1} $\cdot$ mg$_{enzyme\ protein}^{-1}$)
0.23	40
0.5	75
1	120
1.33	141
1.73	161
2.1	175

The maximum apparent reaction rate at saturating adenine concentration is reduced by 70%. At an adenine concentration of 35 mM it is 125 μmol $\cdot$ min^{-1} $\cdot$ mg^{-1}.

Determine the values of the kinetic parameters of the enzyme and the kinetic rate expression for the hydrolysis of IMP.

Resolution:

[IMP] (mM)	v (μmol $\cdot$ min^{-1} $\cdot$ mg^{-1})	[IMP]$^{-1}$ (mM^{-1})	v^{-1} (μmol $\cdot$ min^{-1} $\cdot$ mg^{-1})
0.23	40	4.348	0.025
0.5	75	2.00	0.0133
1	120	1.00	0.00 833
1.33	141	0.752	0.00 709
1.73	161	0.578	0.00 621
2.1	175	0.476	0.00 571

Answer: According to Lineweaver–Burk (Equation 2.62) and data in the third and fourth columns:

$V_{max} = 300\ \mu$moles $\cdot$ min^{-1} $\cdot$ mg^{-1}

$K_M = 1.5$ mM

$r \sim 1$

$V'_{max} = 0.3 \cdot V_{max} = 90\ \mu$moles $\cdot$ min^{-1} $\cdot$ mg^{-1}

For partial noncompetitive inhibition (see Table 2.3):

$$V_{maxAP} = \frac{V_{max} + V'_{max} \dfrac{[A]}{K_{INC}}}{1 + \dfrac{[A]}{K_{INC}}}$$

So:

$$125 = \frac{300 + 90 \dfrac{35}{K_{INC}}}{1 + \dfrac{35}{K_{INC}}}$$

$$K_{INC} = 7 \, mM$$

Then:

$$v = \frac{\dfrac{V_{max} + V'_{max} \dfrac{[A]}{K_{INC}}}{1 + \dfrac{[A]}{K_{INC}}} \cdot [IMP]}{K_M + [IMP]} = \frac{300 + 12.86 \cdot [A]}{1 + 0.143 \cdot [A]}$$

2.11 The following results were obtained with a glucose 6 phosphatase (EC 3.1.3.9) purified from an extract of rat liver. The enzyme catalyzes the hydrolysis of glucose 6 phosphate (G6P) to glucose (G) and inorganic phosphate (Pi).

[G6P] (mM)	v (μmoles glucose $\cdot$ min^{-1} $\cdot$ mg enzyme^{-1})			
	[G] = 0 mM	[G] = 20 mM	[G] = 40 mM	[G] = 80 mM
10	57.1	42.6	33.9	24.1
20	88.9	67.8	54.8	39.6
40	123.1	96.4	79.2	58.4
80	152.4	122.1	101.9	76.6

Determine the type of inhibition exerted by glucose. Propose a plausible rate expression for the hydrolysis of G6P with glucose 6 phosphatase.

Resolution: From the data in the table:

[G6P]$^{-1}$ (mM^{-1})	v^{-1} (min $\cdot$ mg $\cdot$ μmol^{-1})			
	[G] = 0 mM	[G] = 20 mM	[G] = 40 mM	[G] = 80 mM
0.1	0.0175	0.0235	0.0295	0.0415
0.05	0.0112	0.0148	0.0183	0.0253
0.025	0.00 810	0.0104	0.0126	0.0171
0.0125	0.0066	0.0082	0.0098	0.0131
V_{maxAP}^{*} (μmol $\cdot$ min^{-1} $\cdot$ mg^{-1})	200	165.83	143.03	110.84
K_{MAP}^{*} (mM)	25.0	30.0	32.22	36.0

* Values of V_{maxAP} and K_{MAP} were determined by linear regression of the data.

Answer: From data at [G] = 0: $V_{max} = 200$ (μmol $\cdot$ min^{-1} $\cdot$ mg^{-1}) and $K_M = 25$ mM. Both apparent parameters are a function of glucose concentration, so glucose exerts mixed-type inhibition on glucose 6 phosphatase. From Table 2.3:

$$V_{maxAP} = \frac{V_{max}}{1 + \frac{[G]}{K'_{INC}}} \qquad K_{MAP} = K_M \frac{1 + \frac{[G]}{K_{INC}}}{1 + \frac{[G]}{K'_{INC}}}$$

So:

$$\frac{1}{V_{maxAP}} = \frac{1}{V_{max}} + \frac{[G]}{V_{max} \cdot K'_{INC}} \qquad \frac{K_{MAP}}{V_{maxAP}} = \frac{K_M}{V_{max}} + \frac{K_M}{V_{max} \cdot K_{INC}}[G]$$

Both expressions are linear with respect to [G], so from the data:

[G] (mM)	V_{maxAP}^{-1} $(min \cdot mg \cdot \mu mol^{-1})$	$K_{MAP} \cdot V_{maxAP}^{-1}$ $(mM \cdot min \cdot mg \cdot \mu mol^{-1})$
0	0.005	0.125
20	0.00 603	0.181
40	0.00 699	0.225
80	0.00 902	0.325

By linear regression, $K'_{INC} = 100\,mM$ and $K_{INC} = 51.5\,mM$.

The rate expression is:

$$v = \frac{\frac{200}{1 + \frac{[G]}{100}} \cdot [G6P]}{25 + \frac{1 + \frac{[G]}{51.5}}{1 + \frac{[G]}{100}} + [G6P]}$$

It was assumed that inhibition was total. The data are insufficient to sustain this assumption and data over a wider range of [G] will be required to assess whether it is total or partial.

2.12 A recombinant *Pseudomonas* sp. cephalosporin C acylase (CCA) expressed in *Escherichia coli* was evaluated in the hydrolysis of cephalosporin C (CC) into 7-aminocephalosporanic acid (7ACA) and glutaric acid (GA). It has been claimed that 7ACA is a competitive inhibitor of cephalosporin C acylase from *Pseudomonas diminuta* [19]. You have recently isolated and purified a CCA from an unknown soil bacterium and have obtained the following results at saturating CC and with varying initial 7ACA concentrations:

[7ACA] (mM)	0	1	2	5	10	20	30	40
v ($\mu mol \cdot min^{-1} \cdot mL^{-1}$)	10	9.808	9.630	9.167	8.571	7.778	7.273	6.923

v: initial rate of CC hydrolysis at saturating CC concentration

Determine the mechanism of inhibition exerted by 7ACA on CCA and evaluate their kinetic parameters. Leave aside the eventual inhibition by GA.

Resolution: The issue is to define whether noncompetitive inhibition is total or partial, which cannot be ascertained by the usual linearization method. However, Whiteley [20] has proposed an elegant method for discerning this (the method applies to both noncompetitive and mixed-type inhibition). From Equation 2.18 and Table 2.3, rate equations for total and partial inhibition are:

Total	Partial
$$v = \dfrac{\dfrac{V_{max}}{1+\frac{i}{K_{INC}}} \cdot s}{K+s}$$	$$v = \dfrac{\dfrac{V_{max}+V'_{max}\frac{i}{K'_{INC}}}{1+\frac{i}{K'_{INC}}} \cdot s}{K+s}$$

At saturating substrate concentration, these equations are simply:

Total	Partial
$$v = \dfrac{V_{max}}{1+\frac{i}{K'_{INC}}}$$	$$v = \dfrac{V_{max}+V'_{max}\frac{i}{K'_{INC}}}{1+\frac{i}{K'_{INC}}}$$

At zero initial i, both mechanisms lead to $v_0 = V_{max}$, so:

Total	Partial
$$\dfrac{v_0}{v} = 1+\dfrac{i}{K'_{INC}}$$	$$\dfrac{v_0}{v} = \dfrac{1+\dfrac{i}{K'_{INC}}}{1+F\dfrac{i}{K'_{INC}}}$$

where:

$$F = \frac{V'_{max}}{V_{max}}$$

These equations can be linearized as proposed by Whiteley:

Total	Partial
$$\dfrac{v}{v_0-v} = K'_{INC}\cdot\dfrac{1}{i}$$	$$\dfrac{v}{v_0-v} = \dfrac{K'_{INC}}{1-F}\cdot\dfrac{1}{i}+\dfrac{F}{1-F}$$

so that both mechanisms can be easily discerned by the magnitude of the intercept in the y-axis (0 in the case of total and $F\cdot(1-F)^{-1}$ in the case of partial inhibition).

The data obtained with CCA are now tabulated:

[7ACA] (mM)	$[7ACA]^{-1}$ (mM^{-1})	v $(\mu mol \cdot min^{-1} \cdot mL^{-1})$	$v \cdot (v_0 - v)^{-1}$
0	—	10	—
1	1	9.808	51.083
2	0.5	9.630	26.027
5	0.2	9.167	11.005
10	0.1	8.571	5.998
20	0.05	7.778	3.500
30	0.033	7.273	2.667
40	0.025	6.923	2.250

Linearizing $v \cdot (v_0 - v)^{-1}$ vs $[7ACA]^{-1}$:

$$\frac{F}{1-F} = 0.993 \qquad \frac{K'_{INC}}{1-F} = 50.08$$

Answer: The inhibition is partial:

$$V_{max} = 10 \ \mu mol \cdot min^{-1} \cdot mL^{-1}$$

$$F = 0.5$$

$$V'_{max} = 5 \ \mu mol \cdot min^{-1} \cdot mL^{-1}$$

$$K'_{INC} = 25.04 \ mM$$

2.13 A certain arylesterase (EC 3.1.1.29) catalyzes the hydrolysis of phenyl acetate (S) into phenol (P) and acetic acid (A). Phenol is a (total) competitive inhibitor but also exerts (total) uncompetitive inhibition. The following results have been obtained with a preparation of arylesterase, where v is the initial rate of phenyl acetate hydrolysis:

s mM	v (μmoles S hydrolyzed $\cdot min^{-1} \cdot g^{-1}$)			
	p = 0 mM	p = 2 mM	p = 5 mM	p = 10 mM
1	6.667	4.403	2.917	1.867
2.5	9.091	6.446	4.487	2.979
5	10.345	7.625	5.469	3.717
7.5	10.843	8.121	5.899	4.051
10	11.111	8.393	6.140	4.242
15	11.392	8.685	6.402	4.452

Determine the kinetic expression for the initial rate of phenylacetate hydrolysis, calculate the corresponding kinetic parameters and discuss the accuracy of the results obtained.

Resolution: Phenol exerts both competitive and uncompetitive inhibition on arylesterase. According to this mechanism:

$$E + S \rightleftharpoons ES \qquad K_M = \frac{(e - c - d - f) \cdot s}{c}$$

$$E + P \rightleftharpoons EP \qquad K_I = \frac{(e - c - d - f) \cdot p}{d}$$

$$ES + P \rightleftharpoons EPS \qquad K_I' = \frac{c \cdot p}{f}$$

$$ES \xrightarrow{k} E + P + A$$

where c, d, and f are the molar concentrations of the ES, EP, and EPS complexes, respectively, and s and p are the molar concentrations of phenyl acetate and phenol, respectively. Solving:

$$v = k \cdot c = \frac{V_{max} \cdot s}{K_M \left(1 + \frac{p}{K_I}\right) + s\left(1 + \frac{p}{K_I'}\right)} = \frac{\dfrac{V_{max} \cdot s}{1 + \frac{p}{K_I'}}}{K_M \dfrac{1 + \frac{p}{K_I}}{1 + \frac{p}{K_I'}} + s}$$

the above rate equation can be expressed as:

$$v = \frac{V_{maxAP} \cdot s}{K_{MAP} + s}$$

where:

$$V_{maxAP} = \frac{V_{max}}{1 + \frac{p}{K_I'}} \qquad K_{MAP} = K_M \frac{1 + \frac{p}{K_I}}{1 + \frac{p}{K_I'}}$$

From the initial rate data provided:

$s^{-1}\,mM^{-1}$	$v^{-1}\,min \cdot g \cdot \mu mol^{-1}$			
	p = 0 mM	p = 2 mM	p = 5 mM	p = 10 mM
1	0.150	0.227	0.343	0.536
0.4	0.110	0.155	0.223	0.336
0.2	0.0967	0.131	0.183	0.269
0.133	0.0922	0.123	0.170	0.247
0.1	0.090	0.119	0.163	0.236
0.0667	0.0878	0.115	0.156	0.225
V_{maxAP}	12	9.346	6.994	4.936
K_{MAP}	0.8	1.122	1.399	1.646
$K_{MAP} \cdot V_{maxAP}^{-1}$	0.0667	0.120	0.200	0.334

$$\frac{1}{V_{maxAP}} = \frac{1}{V_{max}} + \frac{p}{V_{max} \cdot K'_I}$$

Plotting V_{maxAP}^{-1} versus p: $V_{max} = 12 \ \mu mol \cdot min^{-1} g^{-1}$; $K'_I = 6.95$ mM:

$$\frac{K_{MAP}}{V_{maxAP}} = \frac{K_M}{V_{max}} \cdot \left(1 + \frac{p}{K_I}\right)$$

Plotting $K_{MAP} \cdot V_{maxAP}^{-1}$ versus p: $K_I = 2.5$ mM.

Answer:

$$v = \frac{V_{max} \cdot s}{K_M \left(1 + \dfrac{p}{K_I}\right) + s\left(1 + \dfrac{p}{K'_I}\right)}$$

$V_{max} = 12 \ \mu mol \cdot min^{-1} g^{-1}$; $K_M = 0.8$ mM; $K'_I = 6.95$ mM; $K_I = 2.5$ mM.

Inhibition constants (K_I and K'_I) are within the range of inhibitor concentrations used for their determination, but the value of K_M is outside the substrate concentration range. Therefore, in order to improve accuracy, initial rate data at concentrations below 1 mM (at least two) should be determined, which will also increase the number of data points, which is also advisable in this case.

2.14 The enzyme amino acid ester hydrolase (EC 3.1.1.43) catalyzes the hydrolysis of cephalosporin C (CC) into 7-aminocephalosporanic acid (7ACA) and α-aminoadipic acid (AA). A certain amino acid ester hydrolase is subjected to mixed-type partial inhibition by AA but is not inhibited by 7ACA. The following results have been obtained with a preparation of such an enzyme at very high AA concentrations:

CC (mM)	200	300	400	500	600
v (μmoles 7ACA $\cdot$ min^{-1} $\cdot$ mg$_{protein}^{-1}$)	1.3	1.7	2.0	2.2	2.4

v: initial rate of 7ACA synthesis

Determine the value of the maximum rate of 7ACA formation from the enzyme–CC–AA tertiary complex and the dissociation constant of that tertiary complex into enzyme–AA secondary complex and CC.

Resolution: Let S, I, and P represent CC, AA, and 7ACA, respectively, with s, i, and p the corresponding molar concentrations. The following pseudo-equilibrium reactions represent the mixed-type partial inhibition:

Equation	Pseudo-equilibrium constant
$E + S \rightleftarrows ES$	K_M
$E + I \rightleftarrows ES$	K_{INC}
$EI + S \rightleftarrows EIS$	K'_M
$ES + I \rightleftarrows EIS$	K'_{INC}
$ES \xrightarrow{k} E + P + I$	
$EIS \xrightarrow{k'} EI + P + I$	

The corresponding rate equation is:

$$v = \frac{\dfrac{V_{max} + V'_{max}\frac{i}{K'_{INC}}}{1 + \frac{i}{K'_{INC}}} \cdot s}{K_M \dfrac{1 + \frac{i}{K_{INC}}}{1 + \frac{i}{K'_{INC}}} + s}$$

where $V_{max} = k \cdot e$ and $V'_{max} = k' \cdot e$.

This rate equation above can be rewritten as:

$$v = \frac{\dfrac{\frac{V_{max}}{i} + \frac{V'_{max}}{K'_{INC}}}{\frac{1}{i} + \frac{1}{K'_{INC}}} \cdot s}{K_M \dfrac{\frac{1}{i} + \frac{1}{K_{INC}}}{\frac{1}{i} + \frac{1}{K'_{INC}}} + s}$$

so that at $i \ggg K'_{INC}$ (and K_{INC}):

$$v = \frac{V'_{max} \cdot s}{K_M \frac{K'_{INC}}{K_{INC}} + s} = \frac{V'_{max} \cdot s}{K'_M + s}$$

Now, by linearization of the previous data:

$s^{-1} \cdot 10^3$ $(mM)^{-1}$	v^{-1} $\mu mol^{-1} \cdot min \cdot mg_{protein}$
5.0	0.769
3.33	0.588
2.5	0.50
2.0	0.455
1.67	0.417

$V_{max} = 4.16\,\mu mol \cdot min^{-1} \cdot mg_{protein}^{-1}$
$K_M = 438.7\,mM$

Answer: The maximum rate of 7ACA formation from the enzyme–CC–AA tertiary complex is $4.16\,\mu mol \cdot min^{-1} \cdot mg_{protein}^{-1}$; the dissociation constant of that tertiary complex enzyme–CC–AA into an enzyme–AA secondary complex and CC is 438.7 mM.

2.15 Describe how to discriminate between ordered sequential and random sequential mechanisms in the enzyme-catalyzed reaction:

$$A + B \rightarrow P$$

by using Eaddie–Hofstee plots (initial reaction rate versus ratio of initial reaction rate to substrate concentration). In the ordered sequential mechanism case, A represents the first substrate to bind to the enzyme.

Resolution: From Equation 2.19:

$$v = \frac{V_{maxAP(A)} \cdot b}{K_{MAP(A)} + b}$$

According to Eaddie–Hofstee:

$$v = V_{maxAP(A)} - \frac{v}{b} \cdot K_{MAP(A)}$$

So, plotting v versus $v \cdot b^{-1}$ at varying a, the intercept in the y-axis corresponds to $V_{maxAP(A)}$ and the slope corresponds to $K_{MAP(A)}$.

From Table 2.4:

Ordered Sequential	$V_{maxAP(A)} = V_{max}$	$K_{MAP(A)} = \dfrac{a \cdot K'_{MB} + K_{MA} \cdot K'_{MB}}{a}$
Random Sequential	$V_{maxAP(A)} = \dfrac{V_{max} \cdot a}{a + K'_{MA}}$	$K_{MAP(A)} = \dfrac{a \cdot K'_{MB} + K_{MA} \cdot K'_{MB}}{a + K'_{MA}}$

Answer: For an ordered sequential, the intercept in the y-axis ($V_{maxAP(A)}$) is independent of a, while for a random sequential it will vary with a, providing a simple way of determining between the two (see Figure 2.2). In the case of an ordered sequential, it will also allow determination of which of the substrates is first to bind to the enzyme.

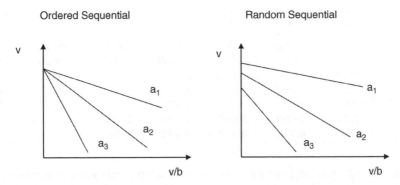

Figure 2.2 *Eaddie–Hofstee plots for ordered-sequential and random-sequential mechanisms of the reaction $A + B \rightarrow P$. a and b are the molar concentrations of A and B respectively.*

2.16 The following results have been obtained in the synthesis of ascorbyl oleate (ascorbic acid + oleic acid → ascorbyl oleate + water) with *Candida rugosa* lipase (EC 3.1.1.3) immobilized on γ-Fe_2O_3 magnetic nanoparticles:

			v		
[A]	[B] = 0.1	[B] = 0.25	[B] = 0.5	[B] = 0.75	[B] = 1
0.5	2.432	5.562	9.742	12.997	15.605
1	3.626	7.961	13.234	16.984	19.788
1.5	4.336	9.298	15.031	18.919	21.73
3	5.392	11.175	17.391	21.351	24.093
6	6.139	12.429	18.873	22.817	25.479

v: initial reaction rate of ascorbyl oleate synthesis ($\mu mol \cdot min^{-1} \cdot mg_{catalyst}^{-1}$)
[B]: oleic acid concentration (mM)
[A]: ascorbic acid concentration (mM)

Identify the mechanism of reaction, determine the values of the corresponding kinetic parameters and discuss the accuracy of the values reported.

Resolution: The generalized rate expression for a two-substrate reaction can be represented by Equation 2.19. Linearizing this:

$$\frac{1}{v} = \frac{K_{MAP(B)}}{V_{maxAP(B)}} \cdot \frac{1}{[A]} + \frac{1}{V_{maxAP(B)}} = \frac{K_{MAP(A)}}{V_{maxAP(A)}} \cdot \frac{1}{[B]} + \frac{1}{V_{maxAP(A)}}$$

where the values of the apparent kinetic parameters are those in Table 2.4 for the different kinetic mechanisms.

From the data obtained, the following table is constructed:

	v^{-1} ($\mu mol^{-1} \cdot min \cdot mg_{catalyst}$)					$V_{maxAP(A)}$	$K_{MAP(A)}$
$[A]^{-1}$			$[B]^{-1}$				
	10	4	2	1.333	1		
2	0.4111	0.1800	0.1027	0.0769	0.0641	39.10	1.507
1	0.2758	0.1256	0.0756	0.0589	0.0505	39.20	0.981
0.667	0.2306	0.1077	0.0665	0.0529	0.0460	39.16	0.803
0.333	0.1855	0.0895	0.0575	0.0468	0.0415	39.23	0.628
0.167	0.1629	0.0805	0.0530	0.0438	0.0392	39.23	0.539
$V_{maxAP(B)}$	7.23	14.00	20.64	24.50	27.06		
$K_{MAP(B)}$	0.984	0.759	0.560	0.442	0.367		

The values of the apparent kinetic parameters were determined by linear regression of the double reciprocal plot (see the linearized version of Equation 2.19) and are given in the last two columns and last two rows of the table.

As can be seen, $V_{maxAP(A)}$ is constant and is independent of [A], so according to Table 2.4, the mechanism corresponds to an ordered sequential, being the only mechanism in which $V_{maxAP(A)}$ is not a function of [A] and equal to V_{max}. So:

$$K_{MAP(A)} = K'_{MB} + \frac{K_{MA} \cdot K'_{MB}}{[A]}$$

Plotting $K_{MAP(A)}$ versus $[A]^{-1}$, a straight line should be obtained whose intercepts in the y- and x-axes are K'_{MB} and $-K_{MA}^{-1}$, respectively. By linear regression of the data:

$$K'_{MB} = 0.45 \, mM$$

$$K_{MA} = 1.17 \, mM$$

These values can be checked against the values obtained for $V_{maxAP(B)}$. According to Table 2.4:

$$V_{maxAP(B)} = \frac{V_{max} \cdot [B]}{[B] + K'_{MB}}$$

so that:

$$\frac{1}{V_{max}} = \frac{1}{V_{max}} + \frac{K'_{MB}}{V_{max}+} \cdot \frac{1}{[B]}$$

Plotting $V_{maxAP(B)}^{-1}$ versus $[B]^{-1}$, a straight line should be obtained whose intercepts in the y- and x-axes are V_{max}^{-1} and $-K'_{MB}^{-1}$, respectively. By linear regression of the data in the table:

$$V_{max} = 39.2 \, \mu moles \cdot min^{-1} \cdot mg^{-1}$$

$$K'_{MB} = 0.45 \, mM$$

Answer: The mechanism of synthesis of ascorbyl oleate with *C. rugosa* lipase is ordered sequential, with ascorbic acid bound to the active site of the enzyme first and then oleic acid.

The kinetic parameters are:

$$V_{max} = 39.2 \, \mu moles \cdot min^{-1} \cdot mg^{-1}$$

$$K_{MA} = 1.17 \, mM$$

$$K'_{MB} = 0.45 \, mM$$

The values of the dissociation constants are within the range of the corresponding substrate concentrations, so the results can be considered accurate. However, more data points are advisable.

Nanoparticles are essentially free of mass-transfer limitations, so the parameters obtained can be considered intrinsic (see Chapter 3).

2.17 The following rate data have been obtained with a glucoamylase from *Aspergillus awamori*. Kinetic parameters assuming simple Michaelis–Menten kinetics were obtained using maltose as substrate:

pH	V_{maxAP} $\mu mol \cdot min^{-1} \cdot mL^{-1}$	K_{MAP} mM	pH	V_{maxAP} $\mu mol \cdot min^{-1} \cdot mL^{-1}$	K_{MAP} mM
0.5	0.063	31.62	5.8	185.4	11.58
1	0.20	31.60	6.8	122.5	18.37
1.5	0.63	31.58	7.8	27.36	28.66
2	1.98	31.41	8.8	3.12	31.29
3	18.18	29.66	9.3	1.0	31.52
4	100.0	20.82	9.8	0.316	31.59
5	180.0	12.14	10.3	0.10	31.61
5.5	188.0	11.29	10.8	0.032	31.62

Based on the Michaelis–Davidsohn hypothesis, determine the ionic equilibrium constants corresponding to the enzyme–substrate complex and the free enzyme, and obtain the pH-explicit rate expression for the hydrolysis of maltose.

Resolution: Data are presented in terms of the decimal logarithms of V_{maxAP} and $K_{MAP} \cdot V_{maxAP}^{-1}$ (Δ_{AP}):

pH	$\log V_{maxAP}$	$\log \Delta_{AP}$
0.5	−1.2	2.70
1	−0.7	2.199
1.5	−0.2	1.70
2	0.3	1.20
3	1.26	0.213
4	2.0	−0.682
5	2.255	−1.171
5.5	2.274	−1.222
5.8	2.268	−1.204
6.8	2.088	−0.824
7.8	1.437	0.02
8.8	0.494	1.00
9.3	−0.0013	1.50
9.8	−0.5	2.00
10.3	−1.0	2.50
10.8	−1.5	3.00

According to the Michaelis–Davidsohn hypothesis (Equations 2.27 and 2.28):

$$log V_{maxAP} = log V_{max} - log\left(\frac{h^+}{K_{1ES}} + 1 + \frac{K_{2ES}}{h^+}\right)$$

$$log\Delta_{AP} = log\Delta_{max} + log\left(\frac{h^+}{K_{1E}} + 1 + \frac{K_{2E}}{h^+}\right)$$

At low pHs (significantly lower than pK_{1ES} and pK_{1E}), these equations can be simplified to:

$$log V_{maxAP} = log V_{max} + pH - pK_{1ES}$$
$$log \Delta_{AP} = log \Delta - pH + pK_{1E}$$

so that straight lines (slope $= 1$ and -1, respectively) are obtained in such a region (from the data obtained, the range is below pH 2). The corresponding equations are:

$$V_{maxAP} = pH - 1.7$$
$$log \Delta_{AP} = 3.2 - pH$$

At high pHs (significantly higher than pK_{2ES} and pK_{2E}), the equations can be simplified to:

$$V_{maxAP} = log V_{max} + pK_{2ES} - pH$$
$$log \Delta_{AP} = log \Delta + pH - pK_{2E}$$

so that straight lines (slope $= -1$ and 1, respectively) are obtained in such a region (from the data obtained, the range is above pH 7.8). The corresponding equations are:

$$V_{maxAP} = 9.27 - pH$$
$$log \Delta_{AP} = pH - 7.8$$

At intermediate values of pH (significantly higher than pK_{1ES} and pK_{1E} but significantly lower than pK_{2ES} and pK_{2E}), the equations can be simplified to:

$$log V_{maxAP} = log V_{max}$$
$$log \Delta_{AP} = log \Delta$$

This zone may actually be quite small and corresponds to the optimum pH of the enzyme. The corresponding equations are:

$$log V_{maxAP} = log V_{max} = 2.274$$
$$V_{max} = 188 \ \mu moles \cdot min^{-1} \cdot mL^{-1}$$
$$log \Delta_{AP} = log \Delta = -1.222$$
$$\Delta = 0.06 \ mM \cdot min \cdot mL \cdot \mu moles^{-1}$$
$$K_M = 11.28 \ mM$$

The respective intercepts of the straight lines in the low and intermediate pH ranges in the log V_{maxAP} versus pH and log Δ_{AP} versus pH plots are:

$$pH = pK_{1ES}$$
$$pH = pK_{1E}$$

The respective intercepts of the straight lines in the high and intermediate pH ranges in the log V_{maxAP} versus pH and log Δ_{AP} versus pH plots are:

$$pH = pK_{2ES}$$
$$pH = pK_{2E}$$

Answer: From the experimental data:

$pK_{1ES} = 3.974$	$K_{1ES} = 1.06 \cdot 10^{-4}$ M
$pK_{2ES} = 7.0$	$K_{2ES} = 10^{-7}$ M
$pK_{1E} = 4.422$	$K_{1E} = 3.78 \cdot 10^{-5}$ M
$pK_{2E} = 6.578$	$K_{2E} = 2.64 \cdot 10^{-7}$ M

The pH-explicit rate equation is:

$$v = \frac{V_{maxAP} \cdot s}{K_{MAP} + s} = \frac{\frac{188}{\frac{[h^+]}{1.06\cdot10^{-4}} + 1 + \frac{10^{-7}}{[h^+]}} \cdot s}{11.28 \cdot \frac{\frac{[h^+]}{3.78\cdot10^{-5}} + 1 + \frac{2.64\cdot10^{-7}}{[h^+]}}{\frac{[h^+]}{1.06\cdot10^{-4}} + 1 + \frac{10^{-7}}{[h^+]}} + s} \quad [\mu moles \cdot min^{-1} \cdot mL^{-1}]$$

2.18 The pKs of the equilibrium of the three successive ionic forms of an enzyme–substrate complex are 5 and 7.5. According to the Michaelis–Davidsohn hypothesis, (a) what is the optimum pH of the enzyme and (b) at what pH will the maximum reaction rate be one-quarter of the value obtained at optimum pH?

Resolution (a): From Equation 2.27:

$$\frac{dV_{maxAP}}{d[h^+]} = -V_{max}\left[\frac{[h^+]}{K_{1,ES}} + 1 + \frac{K_{2,ES}}{[h^+]}\right]^{-2} \cdot \left[\frac{1}{K_{1,ES}} - \frac{K_{2,ES}}{[h^+]^2}\right]$$

Setting this equation to zero, optimum $[H^+]$ (asterisk denotes the optimum) is obtained. So:

$$\frac{1}{K_{1,ES}} - \frac{K_{2,ES}}{h^{+2*}} = 0$$

$$h^{+*} = \sqrt{K_{1,ES} \cdot K_{2,ES}}$$

$$log h^{+*} = \frac{1}{2}\left(log K_{1,ES} + K_{2,ES}\right)$$

$$pH^* = \frac{1}{2}\left(pK_{1,ES} + pK_{2,ES}\right)$$

Answer (a):

$$pH^* = \frac{1}{2}(5 + 7.5) = 6.25$$

The optimum pH for this enzyme is 6.25.

Resolution (b): From Equation 2.27:

$$\frac{V_{maxAP}}{V_{max}} = \frac{1}{\dfrac{h^+}{K_{1,ES}} + 1 + \dfrac{K_{2,ES}}{h^+}} = \frac{1}{4}$$

$$\frac{h^+}{K_{1,ES}} + 1 + \frac{K_{2,ES}}{h^+} = 4 \qquad h^{+2} - 3h^+ \cdot K_{1,ES} + K_{1,ES} \cdot K_{2,ES} = 0$$

$$h^+_1 = 3 \cdot 10^{-5} \qquad h^+_2 = 1.05 \cdot 10^{-8}$$
$$pH_1 = 4.52 \qquad pH_2 = 7.98$$

Answer (b): The maximum reaction rate is one-quarter of its optimum value at pH 4.52 and 7.98

2.19 Succinate dehydrogenase (succinate–ubiquinone oxidoreductase, EC 1.6.99.3) from *Escherichia coli* catalyzes the oxidation of succinate into fumarate. Initial reaction rates of fumarate formation have been determined with such enzymes at two different temperatures:

[succinate] (mM)	$v_{35\,°C} \cdot 10^2$ ($\mu mol \cdot min^{-1} \cdot mg_{protein}^{-1}$)	$v_{25\,°C} \cdot 10^2$ ($\mu mol \cdot min^{-1} \cdot mg_{protein}^{-1}$)
2.5	7.5	3.7
5	9.2	4.8
10	10.4	5.7
15	10.9	6.1
20	11.2	6.3
25	11.3	6.4

$v_{35\,°C}$: initial reaction rates of fumarate formation at 35 °C
$v_{25\,°C}$: initial reaction rates of fumarate formation at 25 °C

Determine the energy of activation and the enthalpy change of the formation of fumarate from the enzyme–succinate complex. Assume a value for the universal gas constant R of 1.98 cal · g-mole^{-1} · K^{-1}. Discuss the validity of the results obtained. Make any assumption explicit.

Resolution: Assuming that the reaction is described by simple Michaelis–Menten kinetics, from Equation 2.13:

$$\frac{1}{v} = \frac{K_M}{V_{max} \cdot s} + \frac{1}{V_{max}} \tag{2.65}$$

where s in this case is the molar concentration of succinate.

From the table:

$s^{-1} (mM)^{-1}$	$V_{35 °C}^{-1}$ $(\mu mol^{-1} \cdot min \cdot mg_{protein})$	$V_{25 °C}^{-1}$ $(\mu mol^{-1} \cdot min \cdot mg_{protein})$
0.4	13.33	27.03
0.2	10.87	20.83
0.1	9.62	17.54
0.667	9.17	16.39
0.05	8.93	15.87
0.04	8.85	15.63

According to Equation 2.65 and the s^{-1} versus v^{-1} data given:

$V_{max,35 °C} = 0.12 \, \mu mol \cdot min^{-1} \cdot mg_{protein}^{-1}$	$V_{max,25 °C} = 0.07 \, \mu mol \cdot min^{-1} \cdot mg_{protein}^{-1}$
$K_{M,35 °C} = 1.5 \, mM$	$K_{M,25 °C} = 2.2 \, mM$

From Equation 2.35:

$$\frac{V_{max,35°C}}{V_{max,25°C}} = exp\left[-\frac{E_a}{R}\left(\frac{1}{308} - \frac{1}{298}\right)\right]$$

$$1.714 = exp\left[5.5 \cdot 10^{-5} \cdot E_a\right]$$

$$E_a = 9797 \, cal \cdot mole^{-1}$$

From Equation 2.32:

$$\frac{K_{M,35°C}}{K_{M,25°C}} = exp\left[-\frac{\Delta H^0}{R}\left(\frac{1}{308} - \frac{1}{298}\right)\right]$$

$$0.682 = exp\left[5.5 \cdot 10^{-5} \cdot \Delta H^0\right]$$

$$\Delta H^0 = -6959 \, cal \cdot mole^{-1}$$

Answer: The energy of activation and the enthalpy change of the formation of fumarate from the enzyme–succinate complex are 9797 and −6959 $cal \cdot mole^{-1}$, respectively. Data at more temperatures are required to validate Equations 2.35 and 2.32. The results are only valid in the 25–35 °C temperature range and should not be extrapolated outside that range. Values of E_a obtained are within the usual range for enzyme-catalyzed reactions (from 2 to 40 $kcal \cdot mole^{-1}$).

2.20 Xylose isomerase (D-xylose ketol-isomerase EC 5.3.1.5) catalyzes the reversible conversion of xylose (Xo) into xylulose (Xu). With a preparation of xylose isomerase from *Actinoplanes missouriensis* immobilized in agarose gel particles, the following initial rates of isomerization of Xu into Xo (v) were obtained at pH 8.0 and 27 °C:

[Xu] (mM)	20	30	40	50	60	70	80
v ($\mu mol \cdot min^{-1} \cdot mL_{gel}^{-1}$)	80.0	93.3	101.8	107.7	112.0	115.3	117.9

v: initial rate of Xo formation from Xu

Under these conditions, the equilibrium constant of the reaction is 1.25 and the affinity of the enzyme can be considered the same for both sugars and essentially independent of temperature.

Evaluate the rate equation for the enzymatic isomerization of Xo into Xu and discuss the accuracy of the values obtained. If the values of the energy of activation of the forward and reverse reactions are 21 and 25 kcal · mole^{-1}, respectively, determine the value of the equilibrium constant of the isomerization of Xo into Xu in the temperature range from 20 to 70 °C.

Resolution: The rate equation (Equation 2.16) for the reversible conversion of Xo into Xu can be expressed as:

$$v = \frac{\pm \dfrac{V_{maxXo} \cdot [Xo]}{K_{MXo}} \mp \dfrac{V_{maxXu} \cdot [Xu]}{K_{M,Xu}}}{\dfrac{[Xo]}{K_{MXo}} + \dfrac{[Xu]}{K_{MXu}} + 1}$$

The signs in the numerator are positive or negative depending on the intended direction of the reaction, so the initial rate of xylose formation from xylulose is:

$$v = \frac{V_{maxXu} \cdot [Xu]}{K_{MXu} + [Xu]}$$

From the data provided:

$[Xu]^{-1}$ (mM^{-1})	0.05	0.033	0.025	0.02	0.0167	0.0143	0.0125
v^{-1} ($\mu oles^{-1} \cdot min \cdot mL_{gel}$)	0.0125	0.0107	0.0098	0.0093	0.0089	0.0087	0.0085

V_{maxXu} and K_{MXu} are determined by linear regression in the double reciprocal plot:

$$V_{max}Xu = 139.8 \; \mu moles \cdot min^{-1} \cdot mL_{gel}^{-1}$$

$$K_{MXu} = 14.9 \; mM$$

$$K_{MXo} = 14.9 \; mM \; (enzyme \; has \; the \; same \; affinity \; for \; Xo \; than \; for \; Xu)$$

From Equation 2.17:

$$K_{eq} = \frac{V_{maxXo}}{V_{maxXu}} \cdot \frac{K_{MXu}}{K_{MXo}} = 1.25$$

$$V_{max,Xo} = 1.25 \cdot V_{max,Xu} = 174.8 \; (\mu oles \cdot min^{-1} mL_{gel}^{-1})$$

From Equation 2.35:

$$V_{maxXo} = V_{maxXo,0} \cdot \exp\left(\frac{-E_{af}}{R \cdot T}\right)$$

$$V_{maxXu} = V_{maxXu,0} \cdot \exp\left(\frac{-E_{ar}}{R \cdot T}\right)$$

where E_{af} and E_{ar} are the energies of activation of the forward and the reverse reaction (21 and 25 kcal $\cdot$ mole^{-1}, respectively).

Since K_{MXo} and K_{MXu} are equal and are not dependent on temperature:

$$K_{eq} = \frac{V_{maxXo}}{V_{maxXu}}$$

$$K_{eq} = \frac{V_{maxXo,0} \cdot \exp\left(\frac{-E_{af}}{R \cdot T}\right)}{V_{maxXu,0} \cdot \exp\left(\frac{-E_{ar}}{R \cdot T}\right)} = \frac{V_{maxXo,0}}{V_{maxXu,0}} \cdot \exp\left[\frac{E_{ar} - E_{af}}{R \cdot T}\right]$$

At 27 °C:

$$1.25 = \frac{V_{maxXo,0}}{V_{maxXu,0}} \cdot \exp\left[\frac{4000}{1.98 \cdot 300}\right]$$

$$\frac{V_{maxXo,0}}{V_{maxXu,0}} = 1.49 \cdot 10^{-3}$$

Then:

$$K_{eq} = 1.49 \cdot 10^{-3} \cdot \exp\left[\frac{4000}{1.98 \cdot T}\right]$$

which allows evaluation of the temperature dependence of the equilibrium constant of the reaction (K_{eq}).

Answer: The rate equation for the enzymatic isomerization of Xo into Xu at 27 °C is:

$$v = \frac{\dfrac{174.8 \cdot [Xo]}{14.9} - \dfrac{139.8 \cdot [Xu]}{14.9}}{\dfrac{[Xo]}{14.9} + \dfrac{[Xu]}{14.9} + 1} = \frac{174.8 \cdot [Xo] - 139.8 \cdot [Xu]}{[Xo] + [Xu] + 14.9}$$

The value of K_{MXu} (and K_{MXo}) is outside the range of [Xu] used for its determination, so the accuracy of the results will be increased by measuring v at a lower [Xu].

The values of K_{eq} in the range from 20 to 70 °C are:

T (°C)	20	27	30	40	50	60	70
K_{eq}	1.47	1.25	1.17	0.95	0.78	0.64	0.54

2.21 Nitrile hydratase (nitrile hydro-lyase, EC 4.2.1.84) is a very important industrial enzyme used in the bulk production of acrylamide (AA) from acrylonitrile (AN). *Rhodococcus rhodochrous* NCIMB 11216 cells containing nitrile hydratase activity at a level of $1000 \, IU \cdot g_{cell}^{-1}$ were used to study the effect of acrylonitrile and acrylamide on enzyme kinetics. The following results were obtained at 15 and 25 °C with an amount of cells corresponding to 40 mg of protein:

[AN]			v					
	T = 15 °C				T = 25 °C			
	[AA] = 0	[AA] = 10	[AA] = 20	[AA] = 30	[AA] = 0	[AA] = 10	[AA] = 20	[AA] = 30
0.5	22.5	15.7	12.0	9.7	32.0	23.8	18.9	15.7
1	40.0	28.8	22.5	18.5	56.4	43.3	35.1	29.5
2	65.5	49.7	40.0	33.5	91.2	73.1	61.1	52.4
4	96.0	77.8	65.5	56.5	131.7	111.8	97.1	85.9
8	125.2	108.7	96.0	86.0	169.3	151.9	137.8	126.1
16	147.7	135.5	125.2	116.4	197.5	185.2	174.3	164.6

[AN] = initial millimolar concentration of acrylonitrile
[AA] = initial millimolar concentration of acrylamide
v = initial reaction rate of acrylamide synthesis ($\mu mol \cdot min^{-1} \cdot g_{cell}^{-1}$)

Determine the mechanism of the enzymatic reaction, evaluate the kinetic parameters at both temperatures, estimate the values of the energy of activation of acrylamide synthesis and the enthalpy changes in the dissociation of the enzymatic complexes and provide the temperature-explicit rate equation for the synthesis of acrylamide from acrylonitrile. Discuss the results.

Resolution: From the table, the reciprocals of substrate concentration and initial reaction rate are:

$[AN]^{-1}$	$v^{-1} \cdot 10^{-3}$							
	$T = 15\,°C$				$T = 25\,°C$			
	$[AA] = 0$	$[AA] = 10$	$[AA] = 20$	$[AA] = 30$	$[AA] = 0$	$[AA] = 10$	$[AA] = 20$	$[AA] = 30$
2	44.44	63.69	83.33	103.09	31.25	42.02	52.91	63.69
1	25.00	34.72	44.44	54.05	17.73	23.09	28.49	33.90
0.5	15.27	20.12	25.00	29.85	10.97	13.68	16.37	19.08
0.25	10.42	12.85	15.27	17.70	7.59	8.94	10.30	11.64
0.125	7.99	9.20	10.42	11.63	5.91	6.58	7.26	7.93
0.0625	6.77	7.38	7.99	8.59	5.06	5.40	5.74	6.08
V_{maxAP}	180.0	179.1	179.9	182.1	237.3	237.1	237.7	237.6
K_{MAP}	3.50	5.21	6.99	8.88	3.21	4.48	5.79	7.07

By linear regression of the double reciprocal data, the values of V_{maxAP} ($\mu mol \cdot min^{-1} \cdot g_{cell}^{-1}$) and K_{MAP} (mM) are obtained at different [AA]. The calculated values are in the last two rows of the table.

K_{MAP} varies with [AA], while V_{maxAP} does not. So inhibition of nitrile hydratase by AA is competitive. Then, according to Table 2.3:

$$K_{MAP} = K_M \left(1 + \frac{[AA]}{K_{IAA}} \right)$$

where K_{IAA} is the competitive inhibition constant of nitrile hydratase by AA.

[AA]	$T = 15\,°C$				$T = 25\,°C$			
	0	10	20	30	0	10	20	30
K_{MAP}	3.50	5.21	6.99	8.88	3.21	4.48	5.79	7.07

By linear regression of K_{MAP} versus [AA], the values of K_M and K_{IAA} correspond to the intercepts in the y-axis and the x-axis (negative value), respectively. V_{max} is essentially independent of [AA]. So:

$K_{M15} = 3.46\ mM$	$K_{M25} = 3.20\ mM$
$K_{IAA15} = 19.3\ mM$	$K_{IAA25} = 24.9\ mM$
$V_{max15} = 180.3\ \mu moles \cdot min^{-1} \cdot g_{cell}^{-1}$	$V_{max25} = 237.4\ \mu moles \cdot min^{-1} \cdot g_{cell}^{-1}$

This means that the affinity for the substrate is higher at 25 than at 15 °C, while the affinity for the inhibitor is higher at 15 °C. This indicates that 25 °C is a better temperature at which to conduct the reaction.

From Equation 2.35:

$$\frac{V_{max25}}{V_{max15}} = \exp\left[\frac{-E_a}{R}\left(\frac{1}{298} - \frac{1}{288}\right)\right]$$

$$\frac{237.4}{180.3} = \exp\left[\frac{-E_a}{1.98}\left(\frac{1}{298} - \frac{1}{288}\right)\right]$$

$$E_a = 4675\ cal \cdot mole^{-1}$$

$$V_{maxAP,0} = \frac{237.4}{\exp\left(\frac{-4675}{1.98 \cdot 298}\right)} = \frac{180.3}{\exp\left(\frac{-4675}{1.98 \cdot 288}\right)}$$

$$V_{max,0} = 6.55 \cdot 10^{-5}$$

From Equation 2.32:

$$\frac{K_{M25}}{K_{M15}} = \exp\left[\frac{\Delta H^0}{R}\left(\frac{1}{298} - \frac{1}{288}\right)\right]$$

$$\frac{3.20}{3.46} = \exp\left[\frac{\Delta H^0}{1.98}\left(\frac{1}{298} - \frac{1}{288}\right)\right]$$

$$\Delta H^0 = 1327\ cal \cdot mole^{-1}$$

$$K_{MAP,0} = \frac{3.2}{\exp\left(\frac{1327}{1.98 \cdot 298}\right)} = \frac{3.46}{\exp\left(\frac{1327}{1.98 \cdot 288}\right)}$$

$$K_{MAP,0} = 0.338\ mM$$

From Equation 2.33:

$$\frac{K_{IAA25}}{K_{IAA15}} = \exp\left[\frac{\Delta H_I^0}{R}\left(\frac{1}{298} - \frac{1}{288}\right)\right]$$

$$\frac{24.9}{19.3} = \exp\left[\frac{\Delta H_I^0}{1.98}\left(\frac{1}{298} - \frac{1}{288}\right)\right]$$

$$\Delta H_I^0 = -4329\ cal \cdot mole^{-1}$$

$$K_{IAA,0} = \frac{24.9}{\exp\left(\frac{-4329}{1.98 \cdot 298}\right)} = \frac{19.3}{\exp\left(\frac{-4329}{1.98 \cdot 288}\right)}$$

$$K_{IAA,0} = 38241\ mM$$

Answer: Competitive inhibition by acrylamide:

- $K_{M15} = 3.46$ mM	$K_{M25} = 3.20$ mM
$K_{IAA15} = 19.3$ mM	$K_{IAA25} = 24.9$ mM
$V_{max15} = 180.3$ µmoles · min^{-1} · g_{cell}^{-1}	$V_{max25} = 237.4$ µmoles · min^{-1} · g_{cell}^{-1}

The energy of activation of acrylamide synthesis is 4675 cal · mole^{-1}. The change in enthalpy of the reaction of dissociation of the enzyme–acrylonitrile complex is 1327 cal · mole^{-1} and that of the enzyme–acrylamide complex is -4329 cal · mole^{-1}.

From Equation 2.37:

$$v = \frac{6.55 \cdot 10^{-5} \cdot exp\left(\frac{-2361.1}{T}\right) \cdot [AN]}{0.338 \cdot exp\left(\frac{670.2}{T}\right)\left(1 + \frac{[AA]}{38241 \cdot exp\left(\frac{-2186.4}{T}\right)}\right) + [AN]}$$

2.22 A neutral lipase (EC 3.1.13 glycerol ester hydrolase) from *Rhizopus delemar* was subjected to inactivation in 20% (v/v) aqueous dioxane at different temperatures and apparent pHs, with the following results obtained:

Temperature (°C)	Half-life (h)		
	pH 6	pH 7	pH 8
30	12.01	14.98	8.02
40	6.50	7.92	4.11
50	3.66	4.35	2.20
60	2.13	2.48	1.23

Determine the variation in the energy of activation of the inactivation of lipase with respect to (apparent) pH in a 20% (v/v) dioxane aqueous solution. Make any assumptions explicit.

Resolution: Assuming that lipase inactivation can be described by one-stage first-order kinetics, from the data obtained and Equation 2.41:

Temperature (K)	$k_D \cdot 10^2$ (h^{-1})		
	pH 6	pH 7	pH 8
303	5.77	4.63	8.64
313	10.67	8.75	16.87
323	18.94	15.93	31.51
333	32.54	27.95	56.35

Assuming that the temperature dependence of the first-order inactivation rate constant is of Arrhenius type, from Equation 2.40:

$$ln\, k_D = ln\, k_{D,0} - \frac{E_{ia}}{R \cdot T}$$

Then, by linear regression of $\ln k_D$ versus T^{-1}, the values of E_{ia} presented in the last row of the following table are obtained:

$T^{-1} \cdot 10^4$ (K)$^{-1}$	$\ln k_D$ (h^{-1})		
	pH 6	pH 7	pH 8
33.00	-2.852	-3.073	-2.449
31.95	-2.238	-2.436	-1.780
30.96	-1.664	-1.837	-1.155
30.03	-1.123	-1.275	-0.574
$E_{ia} \cdot R^{-1}$ (K)	5819	6054	6314
E_{ia} (cal $\cdot$ mole^{-1})	11 522	11 986	12 502

A graphical representation of the Arrhenius equations at the three pHs is presented in Figure 2.3.

The correlation coefficients were close to 1 in all cases, so the Arrhenius equation provided a good description of the effect of temperature on the inactivation rate constant within the temperature range studied.

Answer:

pH	E_{ia} (kcal $\cdot$ mole^{-1})
6	11.52
7	11.99
8	12.50

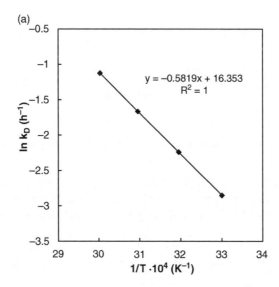

Figure 2.3 *Arrhenius plots at pH (a) 6, (b) 7 and (c) 8 of lipase from R. delemar.*

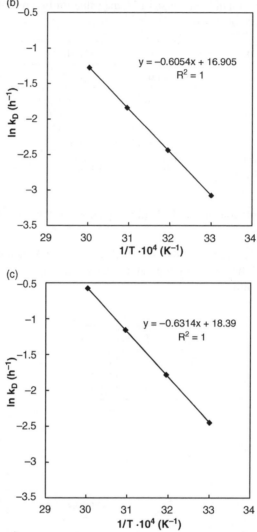

Figure 2.3 *(Continued)*

The energy of activation of inactivation increased almost linearly with pH within the range studied.

2.23 A mutant lipase from the psychrophilic bacterium *Psychrobacter articus* of potential application in room-temperature laundry was evaluated for its thermal stability in the range from 10 to 40 °C. The following values of half-life ($t_{1/2}$) were obtained:

T (°C)	10	15	20	25	30	35	40	45
$t_{1/2}$ (h)	2888	704.4	183.4	49.9	20.7	9.18	4.19	1.96

Determine the energy of activation of inactivation of the enzyme.

Resolution: Assuming that the enzyme is inactivated according to one-stage first-order kinetics, from Equation 2.41:

T (°C)	10	15	20	25	30	35	40	45
$k_D \cdot 10^4$ (h)$^{-1}$	2.40	9.84	37.80	138.9	334.9	755.0	1654	3537

Assuming an Arrhenius-type correlation for k_D, from Equation 2.40:

$$\ln k_D = \ln k_{D,0} - \frac{E_{ia}}{R \cdot T}$$

So, plotting $\ln k_D$ versus T^{-1}, a straight line should be obtained whose slope corresponds to $E_{ia} \cdot R^{-1}$. From the data in the table:

$T^{-1} \cdot 10^3$ (K)$^{-1}$	3.533	3.472	3.413	3.356	3.300	3.247	3.195	3.145
$\ln k_D$ (h)$^{-1}$	−8.335	−6.924	−5.578	−4.277	−3.397	−2.584	−1.799	−1.039

These data are represented in Figure 2.4a, from which:

$$E_{ia} = 18,642 \cdot 1.98 = 36,911 \; cal \cdot mole^{-1}$$

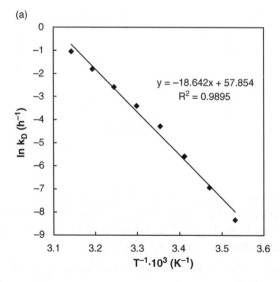

Figure 2.4 *Arrhenius plots of lipase from* Psychrobacter articus: *(a) for the whole temperature range from 10 to 45°C; (b) for the temperature range from 10 to 25°C; (c) for the temperature range from 30 to 45°C.*

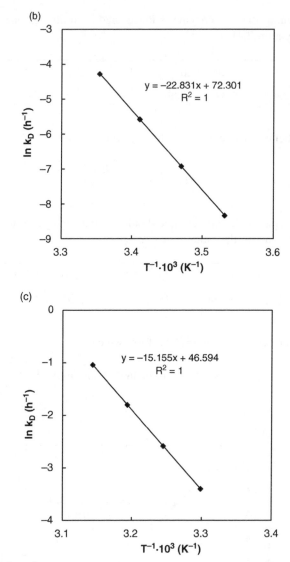

Figure 2.4 (*Continued*).

However, as can be seen, correlation is poor and two regions of different slopes can be distinguished, suggesting that at certain temperature the enzyme suffers some conformational change. If data from 10–25 °C and from 30–45 °C are now plotted separately, Figure 2.4b and c are obtained, according to which:

$E_{ia} = 22,831 \cdot 1.98 = 45,206 \; cal \cdot mole^{-1}$ $10 \,°C < T < 25 \,°C$

$E_{ia} = 15,155 \cdot 1.98 = 30,007 \; cal \cdot mole^{-1}$ $30 \,°C < T < 45 \,°C$

Answer: The energy of activation of inactivation of the lipase from *Psychrobacter articus* is 45,206 cal · mole^{-1} in the temperature range from 10 to 25 °C and 30,007 cal · mole^{-1} in the temperature range from 30 to 45 °C.

2.24 Maltose hydrolysis can be catalyzed either with mineral acids or enzymatically with glucoamylase. Acid hydrolysis can be described by second-order kinetics with respect to maltose concentration [M], and enzymatic hydrolysis by first-order kinetics with respect to the same. The following results have been obtained in the time course of both processes:

t (minutes)	0	0.1	0.5	1	2	4	8	16
[M]$_C$ (mM)	100	99.2	96.2	92.6	86.2	75.8	61.0	43.9
[M]$_E$ (mM)	150	148.2	141.3	133.1	118.1	92.9	57.5	22.1

The suffixes C and E refer to the chemical and enzymatic reactions, respectively. (a) Determine the kinetic constants of the chemical and enzymatic reaction of hydrolysis. (b) At what time is the residual maltose concentration the same in both processes?

Resolution (a):

	Rate equation	Integrated Rate Equation
Chemical	$-\dfrac{d[M]}{dt} = k_Q \cdot [M]^2$	$[M] = \dfrac{100}{1 + 100 \cdot k_Q \cdot t}$
Enzymatic	$-\dfrac{d[M]}{dt} = k_E \cdot [M]$	$[M] = 150 \cdot exp(-k_E \cdot t)$

Answer (a): By linear regression of the data to the corresponding integrated rate equations:

$$k_Q = 8 \cdot 10^{-4} \, min^{-1} \times mM^{-1}$$

$$k_E = 0.12 \, min^{-1}$$

Resolution (b):

$$\frac{100}{1 + 0.08 \cdot t^*} = 150 \cdot exp(-0.12 \cdot t^*)$$

Answer (b): Solving for t*:

$$t^* = 7.15 \, min$$

$$[M]^* = 63.6 \, mM$$

where the asterisk denotes the conditions under which residual maltose concentration is the same in both processes.

Supplementary Problems

2.25 Cholesterol hydroxylation to 7-α-hydroxycholesterol is the first step in the degradation of cholesterol into bile acids, catalyzed by cholesterol 7-α-hydroxylase (cytochrome P450 7A1). Cholesterol can be measured by flurometric assay and cholesterol concentration can be determined using a proper calibration curve. The following results have been obtained by adding 1 mL of a sample containing cholesterol 7-α-hydroxylase activity to 9 mL of a $50 \, g \cdot L^{-1}$ solution of cholesterol in oil (the molecular mass of cholesterol is 386):

Reaction time (seconds)	0	30	50	100	180	240	320	360	420
Cholesterol ($mg \cdot mL^{-1}$)	0.45	0.432	0.420	0.390	0.342	0.306	0.285	0.271	0.263

Determine the activity in the 7-α-hydroxylase sample.

Answer: $0.94 \, IU \cdot mL^{-1}$ (reaction is linear up to 4 minutes).

2.26 The enzyme formate dehydrogenase (EC 1.2.1.2) catalyzes the NAD^+-dependent oxidation of formic acid into CO_2, being a key enzyme in cofactor regeneration in synthesis reactions. The assay for the enzyme consisted in incubating 2.5 mg of enzyme into 10 mL of $26 \, g \cdot L^{-1}$ formic acid solution containing $0.1 \, M \, NAD^+$ at 37 °C and pH 7.0. At 1-minute intervals, 0.1 mL samples were taken, and following enzyme inactivation, the optical density was read at 340 nm in a spectrophotometer; this was then converted into NADH concentration using a suitable calibration curve. One mole of NADH is produced per mole of formic acid oxidized to CO_2. The results obtained are presented in the first column of the following table. Since the enzyme was too concentrated, no linearity between product concentration and time was observed. The enzyme was then serially diluted to one-half, one-quarter, and one-eighth of its concentration and the new results are presented in columns 2, 3, and 4, respectively.

Time (minutes)	mmoles NADPH			
	Column 1	Column 2	Column 3	Column 4
1	0.0095	0.0070	0.0030	0.0014
2	0.0094	0.0085	0.0059	0.0029
3	0.0098	0.0091	0.0088	0.0043
4	0.0095	0.0095	0.0099	0.0057
5	0.0098	0.0098	0.0069	0.0069

Determine the specific activity of the formate dehydrogenase in $IU \cdot g^{-1}$ of enzyme preparation.

Answer: $4416 \, IU \cdot g^{-1}$.

2.27 The enzyme malate dehydrogenase (decarboxylating; NADP; EC 1.1.1.40) catalyzes the reductive carboxylation of pyruvate:

$$Pyruvate + NADPH + CO_2 \rightarrow (S)\text{-malate} + NADP^+$$

The activity of the enzyme is determined by measuring the decrease in optical density at 340 nm (OD_{340}), which corresponds to the decrease in NADPH concentration. 10 μL of enzyme solution containing 10 mM NADPH was added to 2 mL of a 0.1 M pyruvate solution under excess CO_2 and the decrease in OD_{340} with time was recorded, with the following results being obtained:

Reaction time (seconds)	0	10	25	40	70	100	125	150
OD_{340}		0.310	0.307	0.303	0.298	0.288	0.279	0.276	0.274

Determine the molar extinction coefficient of NADPH and calculate the activity of the enzyme solution.

Answer: $\varepsilon = 6.2$ $[\text{mM}]^{-1}[\text{cm}]^{-1}$; activity $= 0.606$ $\text{IU} \cdot \text{mL}_{\text{enzyme}}^{-1}$.

2.28 The enzyme succinate dehydrogenase (EC 1.3.99.1) catalyzes the conversion of succinate (S) into fumarate (F). The following data were obtained with a purified succinate dehydrogenase from a mammalian tissue:

[S]	0.1	0.25	0.5	0.75	1	1.5	2	3	4
$v_{S \to F}$	3.85	9.09	16.67	23.08	28.57	37.50	44.44	54.55	61.53

[S]: concentration of succinate (mM)
[F]: concentration of fumarate (mM)
$v_{S \to F}$: initial rate of fumarate synthesis (μmoles fumarate $\cdot$ min^{-1} $\cdot$ mg$_{\text{protein}}^{-1}$)

A 100 mM succinate solution was contacted with a certain amount of the purified preparation of succinate dehydrogenase under the same conditions and the time course of reaction gave the following results:

Time (minutes)	[S] (mM)	[F] (mM)
0	100	0
20	74.7	25.2
40	59.3	40.6
60	49.7	50.3
80	40.5	59.6
100	37.6	62.3
120	37.6	62.4
150	37.7	62.3

The affinity of the enzyme was about the same for succinate and for fumarate. Determine the equilibrium constant (K_{eq}) of the reaction of conversion of succinate into fumarate. Evaluate the rate equation for the reaction.

Answer: $K_{eq} = 1.66$;

$$v = \frac{40 \cdot [S] - 24.1 \cdot [F]}{0.4 \cdot [S] - 0.4 \cdot [F] + 1}.$$

2.29 The following kinetic expression has been obtained for the hydrolysis of stachyose (a galactosyl–galactosyl–glucosyl–fructose tetrasaccharide) with α-D-galactosidase (EC 3.2.1.22) from *Aspergillus niger*:

$$v = \frac{15 \cdot [S]]}{45 + [S] + 0.2 \cdot [S]^2}$$

where [S] is the millimolar concentration of stachyose and v is the initial rate of stachyose hydrolysis ($\mu mol \cdot sec^{-1} \cdot mL^{-1}$).

Identify the kinetic mechanism that corresponds to such a rate equation and determine the optimum concentration of stachyose (the one maximizing the initial reaction rate of stachyose hydrolysis) and the maximum initial rate of reaction.

Answer: Total uncompetitive inhibition by substrate. Optimum stachyose concentration: 15 mM; maximum initial reaction rate: $2.14 \, \mu mol \cdot sec^{-1} \cdot mL^{-1}$.

2.30 In the following table there are four rate equations for enzyme-inhibited reactions. Identify the mechanism of inhibition and evaluate the corresponding parameters (K_M is the Michaelis constant and K_I, K_I', and K' are the dissociation constants of the enzyme inhibitor–secondary complex, the enzyme–inhibitor–substrate tertiary complex, and the enzyme–substrate–substrate tertiary complex, respectively; s and i are the molar concentrations of substrate and inhibitor, respectively):

(1)	$v = \dfrac{5 \cdot s}{0.3 + s + 0.06 \cdot i + 0.2 \cdot i \cdot s}$
(2)	$v = \dfrac{10.72 \cdot s}{5.19 + s + 0.19 \cdot i}$
(3)	$v = \dfrac{36 \cdot s}{17 + s + 0.02 \cdot s \cdot i}$
(4)	$v = \dfrac{14.7}{\dfrac{2.55}{s} + 1 + 1.2 \cdot s}$

Answer:

Kinetic Model	Kinetic Mechanism of Inhibition	K_M	K_I	K_I'	K'
(1)	Noncompetitive	0.3	∞	5	∞
(2)	Competitive	5.19	27.32	∞	∞
(3)	Uncompetitive	17	∞	50	—
(4)	Uncompetitive	14.7	2.55	—	0.833

2.31 The following results were obtained in the hydrolysis reaction of raffinose (30 mM) with $100 \, mg \cdot L^{-1}$ of α-galactosidase (α-D-galactoside galactohydrolase, EC 3.2.1.22) from *Gibberella fujikuroi* at 55 °C and pH 5.8:

t	3	4.5	6	7.5	9	12	15	20	25
[R]	28.718	28.081	27.445	26.812	26.181	24.296	24.351	23.958	23.762

t: reaction time (minutes)
[R]: raffinose concentration (mM)

From these data, determine the kinetic parameters of the enzyme: K_M (mM) and V_{max} ($\mu mol \cdot min^{-1} \cdot g_{enzyme}^{-1}$). Specify the information used and discuss the validity of the result obtained.

Answer: $K_M = 10.1 \, mM$; $V_{max} = 5 \, \mu mol \cdot min^{-1} \cdot g_{enzyme}^{-1}$. The linearity of Equation 2.26 is lost after 12 minutes, probably due to enzyme inactivation at such a high temperature; therefore, only data up to 12 minutes can be used for the determination of kinetic parameters. However, after 12 minutes, substrate concentration is still higher than the calculated value of K_M, which affects the quality of the result. This is a limitation that is frequently faced when using this method of kinetic parameter determination.

2.32 Xylose isomerase (D-xylose ketol isomerase, EC 5.3.1.5) catalyzes the reversible isomerization of D-xylose (S) into D-xylulose (P). A crude enzyme preparation from *Actinoplanes missouriensis* was tested in the isomerization at pH 8.0 and 27 °C of a birch xylan hydrolysate containing $24 \, g \cdot L^{-1}$ of D-xylose ($C_5H_{10}O_5$). Serial dilutions of the hydrolysate were carried out in buffer and the following results were obtained:

Dilution	None	2×	4×	8×	16×	32×
v	127.9	117.9	101.8	80.0	56.0	35.0

v: initial rate of xylose isomerization ($\mu mol \cdot min^{-1} \cdot mL^{-1}$)

The equilibrium constant of the reaction at 27 °C and pH 8.0 was 0.25 and the affinity of the enzyme for D-xylulose was 75% higher than that for D-xylose.
Evaluate the kinetic parameters and propose a rate expression for such reaction.

Answer: $V_{max,S} = 140 \, \mu mol \cdot min^{-1} \cdot mL^{-1}$; $K_{M,S} = 15 \, mM$.

$V_{max,P} = 320 \, \mu mol \cdot min^{-1} \cdot mL^{-1}$; $K_{M,P} = 8.57 \, mM$.

$$v = \frac{9.33 \cdot [S] - 37.34[P]}{0.067 \cdot [S] + 0.1167 \cdot [P] + 1}$$

2.33 Lipases are used to allow the separation of racemic mixtures by the selective hydrolysis of one of the esters of the racemic substrate. A lipase from *Candida cylindracea* is used for the hydrolysis of racemic R-S ibuprofen trifluoroethyl ester (TFEE). The enzyme acts selectively on the S-enantioner so that S-ibuprofen is produced with trifluoroethanol (TFE) as a product of hydrolysis. The following initial reaction rate data were obtained with such an enzyme:

[TFEE] (mM)	v (μmoles TFE $\cdot$ min^{-1} $\cdot$ g$_{enzyme}$$^{-1}$)			
	[TFE] = 0 mM	[TFE] = 20 mM	[TFE] = 40 mM	[TFE] = 90 mM
4	33.3	32.4	31.6	29.6
8	57.1	54.6	52.2	47.1
16	88.9	82.8	77.4	66.7
24	109.1	100.0	92.3	77.4
32	123.1	111.6	102.1	84.2

Identify the mechanism of inhibition exerted by TFE on *Candida cylindracea* lipase, calculate the corresponding kinetic parameters and determine the rate equation. Discuss the accuracy of the kinetic parameters determined.

Answer: Trifluoroethanol exerts uncompetitive inhibition. $V_{max} = 200\ \mu mol \cdot min^{-1} \cdot g_{enzyme}^{-1}$; $K_M = 20\ mM$; $K_S = 120\ mM$. More data points are required. The range of substrate concentration is adequate but the range of inhibitor concentration is well below the K_S value, so data at higher initial concentrations of TFE are required.

2.34 The enzyme naringinase (an α-L-rhamnosidase, EC 3.2.1.40) is inhibited by 5-*epi*-deoxyrhamnojirimycin (LRJ) [21]. The following results were obtained with a naringinase preparation from *Penicillium decumbens* using *p*-nitrophenyl-α-L-rhamnopyranoside (pNR) as an artificial substrate:

[pNR] (mM)	v (μmoles *p*-nitrophenol $\cdot$ min^{-1} $\cdot$ mL^{-1})				
	[LRJ] = 0 mM	[LRJ] = 3 mM	[LRJ] = 6 mM	[LRJ] = 9 mM	[LRJ] = 15 mM
0.1	3.125	1.820	1.284	0.992	0.681
0.25	6.579	3.710	2.584	1.982	1.352
0.5	10.417	5.676	3.900	2.971	2.012
0.75	12.931	6.893	4.699	3.564	2.404
1	14.706	7.721	5.234	3.959	2.662
1.25	17.046	8.774	5.908	4.453	2.984

Determine the type of inhibition of LRJ on naringinase (competitive, non-competitive, or mixed-type; total or partial) and evaluate the kinetic parameters.

Answer: Total mixed-type inhibition (both V_{maxAP} and K_{MAP} are functions of [LRJ]; plotting $V_{maxAP} \cdot (V_0 - V_{maxAP})^{-1}$, a straight line is obtained that passes through the origin, so $F \cdot (1 - F)^{-1} = 0$; see Problem 2.12). $V_{max} = 25\ \mu mol \cdot min^{-1} \cdot mL^{-1}$; $K_M = 0.7\ mM$; $K_{IN} = 4.5\ mM$; $K'_{IN} = 2.8\ mM$.

2.35 The enzyme α-glucosidase (EC 3.2.1.20) catalyzes the hydrolysis of maltose into glucose. α-glucosidase is competitively inhibited by the product glucose and inhibited at high maltose concentrations in a partial uncompetitive mode.

a) Determine a kinetic rate expression in terms of the dissociation constants for: the secondary enzyme–substrate complex (K_M), the secondary enzyme–product complex (K_P), the tertiary enzyme–substrate–substrate complex (K'), and the maximum reaction rates of product formation from the enzyme–substrate active complex (V_{max}) and the enzyme–substrate–substrate partially active complex (V'_{max}). The molar concentrations of maltose and glucose are [M] and [G], respectively.

Answer:

$$v = \frac{\left(V_{max} + V'_{max} \cdot \frac{[M]}{K'}\right) \cdot [M]}{K_M \left(1 + \frac{[G]}{K_P}\right) + [M]\left(1 + \frac{[M]}{K'}\right)}$$

b) Demonstrate that the maltose concentration that maximizes the initial rate of maltose hydrolysis, $[M]^*$, is:

$$[M]^* = \frac{K_M + K_M \cdot \sqrt{1 + \frac{K'}{K_M} \cdot \frac{V_{max}}{V'_{max}} \left(\frac{V_{max}}{V'_{max}} - 1\right)}}{\frac{V_{max}}{V'_{max}} - 1}$$

2.36 The enzyme methylmalonyl mutase (MMM) (EC 5.4.99.2) catalyzes the conversion of methyl malonic acid (M) into succinic acid (S). Fumaric acid (F) is an inhibitor of the enzyme. The following results were obtained with a purified preparation of this enzyme:

[M] (mM)	v (μmoles S $\cdot$ min^{-1} $\cdot$ mg^{-1})			
	[F] = 0 mM	[F] = 10 mM	[F] = 15 mM	[F] = 20 mM
0.5	12.80	12.15	11.86	11.57
1	23.42	21.36	20.45	19.63
2	40.06	34.37	32.09	30.09
3	52.48	43.12	39.59	36.60
4	62.11	49.42	44.84	41.03
5	69.80	54.17	48.71	44.25
10	92.75	67.04	58.88	52.49

v: initial rate of succinic acid synthesis

Identify the mechanism of inhibition exerted by F on MMM, calculate the values of the kinetic parameters and discuss about the accuracy of the values obtained.

Answer: Uncompetitive inhibition (total). $V_{max} = 137.4$ μmol $\cdot$ min^{-1} $\cdot$ mg^{-1}; $K_M = 4.86$ mM; $K'_I = 17.6$ mM. Values of K_M and K'_I are within the range of the substrate and inhibitor concentrations used, respectively. More data points are advisable. Inhibitor concentrations are few, so the total character of inhibition should be confirmed with data covering a broader range of [F].

2.37 Demonstrate that in the case of mixed-type inhibition, the point of intersection of the straight lines obtained in a double reciprocal plot of initial reaction rate versus substrate concentration at different inhibitor concentrations corresponds to the reciprocal negative value of the dissociation constant of the enzyme–inhibitor–substrate tertiary complex into enzyme–inhibitor secondary complex and substrate.

2.38 Working with a purified L-malic enzyme (L-malate:TPN oxidoreductase decarboxylating, EC(1.1.1.40) from *Escherichia coli*, a K_M of 0.15 mM and a V_{max} of $5\,\mu mol \cdot sec^{-1} \cdot mg_{enzyme}^{-1}$ were obtained. The strain was subjected to directed evolution to obtain a mutant L-malic enzyme whose affinity for L-malate was doubled, but the reactivity was reduced by 70% with respect to the native enzyme. The enzyme catalyzes the reaction:

$$(S)\text{-malate} + NADP^+ \rightarrow pyruvate + NADPH + CO_2$$

In one kinetic experiment, 1 mg of native enzyme and 1 mg of mutant enzyme were added to an L-malate solution and the initial reaction rate obtained was $385\,\mu moles$ pyruvate $\cdot min^{-1}$. What was the (S)-malate concentration in the solution?

Answer: [(S)-malate] $= 10.25$ mM.

2.39 A certain enzyme has a rather unusual kinetic behavior: it is competitively inhibited by one of the products of reaction (P), but once the inhibitor has bound to the enzyme, an allosteric site is created to bind the substrate (S), so that this enzyme-substrate-product tertiary complex is partially active. Demonstrate that the rate equation can be represented by Equation 2.18 and evaluate the apparent kinetic parameters.

Answer:

$$v = \frac{\dfrac{V_{max} + V'_{max}\dfrac{K_M \cdot [P]}{K_P \cdot K'}}{1 + \dfrac{K_M \cdot [P]}{K_P \cdot K'}} \cdot [S]}{K_M \dfrac{1 + \dfrac{[P]}{K_P}}{1 + \dfrac{K_M \cdot [P]}{K_P \cdot K'}} + [S]}$$

$$V_{maxAP} = \frac{V_{max} + V'_{max}\dfrac{K_M \cdot [P]}{K_P \cdot K'}}{1 + \dfrac{K_M \cdot [P]}{K_P \cdot K'}} \qquad K_{AP} = K\frac{1 + \dfrac{[P]}{K_P}}{1 + \dfrac{K_M \cdot [P]}{K_P \cdot K'}}$$

where V_{max} and V'_{max} are the maximum reaction rates of product formation from the enzyme–substrate complex and the enzyme–substrate–product complex, respectively, and K_M, K_P, and K' are the dissociation constants of the enzyme–substrate complex, the enzyme–product complex, and the enzyme–substrate–product complex, respectively.

2.40 Alkaline phosphatase (AP; EC 3.1.3.1)-producing organisms have been isolated from a water ecosystem, most of them identified as *Bacillus* sp. The kinetic behavior of the APs produced by these strains was determined using *p*-nitrophenyl phosphate (PNPP) as an artificial substrate, measuring the *p*-nitrophenol (NP) released. Most APs were severely inhibited by inorganic phosphate, which is one of the products of hydrolysis. One of the isolated APs showed an interesting kinetic behavior: inorganic phosphate could not bind to it unless the substrate was first bound, and once inorganic phosphate was bound the enzyme was partially inhibited. The initial rates of PNPP hydrolysis (v) at various initial inorganic phosphate (P) concentrations were:

[PNPP] (mM)	v (μmoles NP$\cdot$min$^{-1}\cdot$mL^{-1})				
	[P] = 0 mM	[P] = 1 mM	[P] = 2 mM	[P] = 3 mM	[P] = 4 mM
0.1	3.33	3.82	4.15	4.40	4.59
0.25	5.56	5.68	5.75	5.79	5.82
0.5	7.14	6.77	6.59	6.47	6.39
0.75	7.89	7.24	6.92	6.74	6.61
1	8.33	7.50	7.11	6.88	6.72

Derive the rate expression for the hydrolysis of PNPP with such an AP. Describe how the kinetic parameters can be determined and calculate them.

Answer: The rate equation can be represented by Equation 2.18, being:

$$V_{maxAP} = \frac{\left(V_{max} + V'_{max}\cdot\frac{[P]}{K'_P}\right)}{1+\frac{[P]}{K'_P}} \qquad K_{MAP} = \frac{K_M}{1+\frac{[P]}{K'_P}}$$

where V_{max} and V'_{max} are the maximum reaction rates of product formation from the enzyme–substrate complex and the enzyme–substrate–product complex, respectively, and K_M and K'_P are the dissociation constants of the enzyme–substrate complex and the enzyme–substrate–product complex, respectively.

At [P] = 0, a double reciprocal plot of v^{-1} versus [PNPP]$^{-1}$ will yield V_{max}^{-1} and $-K_M^{-1}$ as intercepts in the y- and the x-axis, respectively. Values of K_{MAP} at different [P] will allow the determination of K'_P as the intercept in the x-axis when plotting K_{MAP}^{-1} versus [P]. Using the values of V_{maxAP} obtained at different values of [P] and plotting the values of $V_{maxAP}\cdot(V_{max}-V_{maxAP})^{-1}$ versus [P]$^{-1}$, a straight line is obtained whose y-axis intercept is $[V'_{max}\cdot V_{max}^{-1}]\cdot[1-V'_{max}\cdot V_{max}^{-1}]^{-1}$, so that V'_{max} can be determined.

$V_{max} = 10\,\mu$mol$\cdot$min$^{-1}\cdot$mL^{-1}; $V'_{max} = 6\,\mu$mol$\cdot$min$^{-1}\cdot$mL^{-1}; $K_M = 0.2$ mM; $K'_P = 1.5$ mm.

2.41 The enzyme inosine nucleosidase (EC 3.2.2.2) catalyzes the hydrolysis of inosine (I) to yield hypoxanthine (H) and ribose (R), H being an enzyme inhibitor. With a preparation of inosine nucleosidase, the following initial rates of hydrolysis (v) were obtained:

[I] (mM)	v (μmol $\cdot$ min^{-1} $\cdot$ mg^{-1})		
	[H] = 0 mM	[H] = 0.5 mM	[H] = 2 mM
0.1	0.455	0.161	0.055
0.4	1.43	0.588	0.213
0.75	2.14	1.00	0.385
1.5	3.00	1.67	0.714
2.5	3.28	1.94	0.872
3.0	3.75	2.50	1.25

Identify the mechanism of inhibition of inosine nucleosidase by hypoxanthine. A mistake was made when measuring the concentration of inosine in one case; identify the wrongly measured inosine concentration and indicate the correct value. Determine the values of the kinetic parameters of the enzyme and discuss the validity of the results obtained.

Answer: Competitive inhibition by hypoxanthine. Wrong value: [I] = 2.5 mM; right value: [I] = 1.9 mM. Maximum reaction rate, V_{max} = 5 μmol $\cdot$ min^{-1} $\cdot$ mg^{-1}; Michaelis constant for inosine: K_M = 1 mM; inhibition constant for hypoxanthine, K_H = 0.25 mM. Values of [H] are few, so there is uncertainty about the value of K_H.

2.42 The following results have been obtained in the kinetically controlled synthesis of amoxicillin from p-hydroxyphenylglycine amide (PHFA) and 6-aminopenicillanic acid (APA) catalyzed by penicillin G acylase (EC 3.5.1.11):

[PHFA] (mM)	v_S (μmol $\cdot$ min^{-1} $\cdot$ g^{-1})			
	[APA] = 10 mM	[APA] = 20 mM	[APA] = 30 mM	[APA] = 40 mM
10	20.0	36.4	50.0	61.5
20	28.6	50.0	66.7	80.0
30	33.3	57.1	75.0	88.9
40	36.4	61.5	80.0	94.1

v_S: initial rate of amoxicillin synthesis

Determine the mechanism of synthesis of amoxicillin catalyzed by penicillin G acylase and evaluate the kinetic parameters describing such a mechanism and the corresponding rate equation. Discuss the accuracy of the results obtained.

Answer: The mechanism is ordered sequential and PHFA is the substrate that binds first to the enzyme. In fact, the acyl–enzyme complex must be formed prior to the nucleophilic attack of such a complex by the β-lactam nucleus (APA) in order to synthesize the β-lactam antibiotic (amoxicillin). This is sustained by the common intercept on the y-axis ($V_{maxAP(A)}^{-1}$) obtained when plotting v_S^{-1} versus [PHFA]$^{-1}$.

From expressions in Table 2.4:

$$V_{max} = 200 \; \mu\text{moles} \cdot \text{min}^{-1} \cdot \text{g}^{-1}$$
$$K'_{MB} = 30 \, \text{mM}$$
$$K_{MA} = 19.6 \, \text{mM}$$

The rate equation is:

$$v = \frac{200 \cdot [PHFA] \cdot [APA]}{588 + 30 \cdot [PHFA] + [PHFA] \cdot [APA]}$$

The ranges of concentration are adequate for both substrates, but more data points are advisable.

2.43 The following results have been obtained with a nitrogenase from *Thiobacillus ferrooxidans* at very high ($>10 \, K_M$) and very low ($<0.1 \, K_M$) substrate concentrations, where the reaction of nitrogen reduction to ammonia was monitored by the oxidation of reduced ferredoxin in the presence of N_2 and ATP:

pH	v_{high}	v_{low}
1	5	2.4
2	46	16.8
3	250	41.7
4	455	49.3
5	499	47.1
6	497	31.3
7	454	7.2
8	249	0.8
9	45	0.08
10	4	not measurable

v_{high}: initial rate of reaction at very high substrate concentration
v_{low}: initial rate of reaction at very low substrate concentration

Determine the approximate values of the ionic equilibrium constants (and the respective pK values) corresponding to the enzyme–substrate complex (K_{ES}) and the free enzyme (K_E).

Answer:

$K_{1ES} = 10^{-3} \, \text{M}$	$pK_{1ES} = 3.0$
$K_{1E} = 3.16 \cdot 10^{-3} \, \text{M}$	$pK_{1E} = 2.5$
$K_{2ES} = 10^{-8} \, \text{M}$	$pK_{2ES} = 8.0$
$K_{2E} = 6.31 \cdot 10^{-7} \, \text{M}$	$pK_{2E} = 6.2$

2.44 Glucoamylase (1,4-α-D-glucan glucohydrolase, EC3.2.1.3) is an exo-hydrolase that catalyzes the release of glucose units from the nonreducing end of starch and related oligosaccharides containing α1-4 linkages. A value of $1.25 \cdot 10^{-3}\,\mathrm{g \cdot min^{-1}} \cdot$ unit of activity^{-1} has been reported for the rate constant of glucose formation from soluble starch (k_T) at 20 °C with a glucoamylase from *Aspergillus niger* [22].

If the energy of activation of such a reaction is assumed to be $20\,\mathrm{kcal \cdot mole^{-1}}$, estimate the value of the rate constant in a range from 10 to 40 °C (consider $R = 1.98\,\mathrm{cal \cdot g\text{-}mole^{-1} \cdot K^{-1}}$ and the validity of the Arrhenius equation in this temperature range).

Answer:

T (°C)	$k_T \cdot 10^3$ $(\mathrm{g \cdot min^{-1} \cdot unit^{-1}})$
10	0.37
15	0.69
20	1.25
25	2.23
30	3.90
35	6.70
40	11.31

2.45 The lipase (triacylglycerol acylhydrolase, EC 3.1.1.3) from *Thermomyces lanuginosus* is of current industrial interest for both the hydrolysis of triglycerides and the synthesis of sugar fatty acid esters. The following results have been obtained in the hydrolysis of *p*-nitrophenyl butyrate (pNPB) to butyrate and *p*-nitrophenol (pNP) in an aqueous medium at 40 and 50 °C:

[pNPB] (mM)	v (μmoles pNP $\cdot$ min^{-1} $\cdot$ mg$_{protein}^{-1}$)	
	$T = 40\,°C$	$T = 50\,°C$
5	28.4	37.3
10	48.6	64.0
20	75.6	99.6
40	104.6	137.9
80	129.5	170.7

v: initial rate of pNP formation

Enzyme inactivation in the buffered aqueous medium was followed at 40 and 50 °C. The results obtained were:

t (h)	$A \cdot A_0^{-1} \cdot 100$	
	$T = 40\,°C$	$T = 50\,°C$
0	100	100
20	87.0	66.0
40	75.8	43.5
80	57.5	18.9
160	33.0	3.6

A: enzyme activity
A_0: initial enzyme activity
t: incubation time

Determine the values of the energy of activation of the hydrolysis of pNPB and the energy of activation of lipase inactivation.

Answer: The energy of activation of the hydrolysis of pNPB is $2652 \, \text{cal} \cdot \text{mol}^{-1}$ ($11.1 \, \text{kJ} \cdot \text{mole}^{-1}$). The energy of activation of lipase inactivation is $21\,972 \, \text{cal} \cdot \text{mole}^{-1}$ ($92.0 \, \text{kJ} \cdot \text{mole}^{-1}$).

Values are valid only in the temperature range from 40 to 50 °C and should not be extrapolated beyond that range.

2.46 A certain β-amylase (1,4-(-D-glucan maltohydrolase, EC 3.2.1.2) from malted barley seeds catalyzes the hydrolysis of maltotetraose into maltose. The energy of activation of this reaction in the temperature range from 20 to 60 °C was $25\,000 \, \text{kJ} \cdot \text{mole}^{-1}$. On the other hand, the energy of activation of inactivation of this enzyme in this temperature range was $105\,000 \, \text{kJ} \cdot \text{mole}^{-1}$.

Draw curves showing the variation in the catalytic rate constant (k) of the reaction of maltotetraose hydrolysis and the first-order inactivation rate constant (k_D) of the β-amylase. Assume a baseline value of 1 for both constants at 20 °C.

Draw a curve representing the variation with temperature of the "destabilization factor" (DF) (relative half-life value at a certain temperature with respect to a reference temperature; assume the reference temperature to be 20 °C).

Answer: Curves for k, k_D and DF variation with temperature are shown in Figure 2.5.

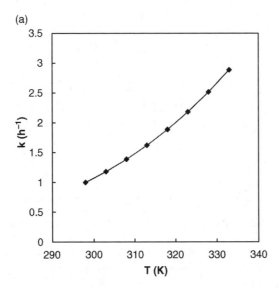

Figure 2.5 *Effect of temperature on the rate constants of hydrolysis of (a) maltotetraose (k), (b) rate of β-amylase inactivation (k_D), and (c) destabilization factor (DF).*

(b)

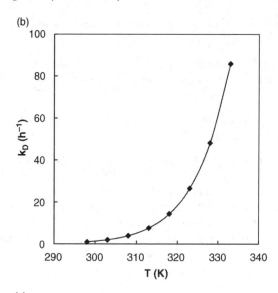

(c)

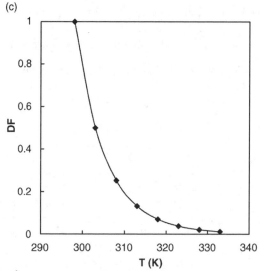

Figure 2.5 *(Continued)*

References

1. Mattiasson, B. and Mosbach, K. (1976) Assay procedures for immobilized enzymes. *Methods in Enzymology*, **44**, 335–353.
2. Bisswanger, H. (2004) *Practical Enzymology*, Wiley VCH, Weinheim.
3. Henri, V. (1902) Théorie générale de l'action de quelques diastases. *Comptes Rendus Hebdomadaires des Séances de l'Académie des Sciences*, **135**, 916–919.
4. Michaelis, L. and Menten, M.L. (1913) Die Kinetik der Invertinwirkung. *Biochemische Zeitschrift*, **49**, 333–369.

5. Briggs, G.E. and Haldane., J.B.S. (1925) A note on the kinetic of enzyme action. *Biochemical Journal*, **19**, 338–339.
6. King, E.L. and Altman, C. (1956) A schematic method of deriving the rate laws for enzyme-catalyzed reactions. *Journal of Physical Chemistry*, **60**, 1375–1378.
7. Myers, D. and Palmer, G. (1985) Microcomputer tools for steady-state enzyme kinetics. *Computer Applications in the Biosciences*, **1**, 105–110.
8. Cleland, W.W. (1963) The kinetic of enzyme-catalysed reactions with two or more substrates or products. *Nomenclature and rates, Biochimica et Biophysica Acta*, **67**, 104–137.
9. Marangoni, A.G. (2003) *Enzyme Kinetics: A Modern Approach*, Wiley Interscience, Hoboken.
10. Santacoloma, P.A., Sin, G., Gernaey, K.V., and Woodley, J.M. (2011) Multienzyme catalyzed processes: next generation biocatalysis. *Organic Process Research and Development*, **15**, 203–212.
11. Cornish-Bowden, A. (2003) *Fundamentals of Enzyme Kinetics*, 3rd edn, Portland Press, London.
12. Ramachandran, P.A., Krishna, R., and Panchal, C.B. (1976) Analysis of immobilised multi-enzyme reaction systems with Michaelis-Menten kinetics. *Journal of Applied Chemistry and Biotechnology*, **26**, 214–224.
13. Ishii, N.Y., Suga, A., Hagiya, A. *et al.* (2007) Dynamic simulation of an in vitro multi-enzyme system. *FEBS Letters*, **581**, 413–420.
14. Goldman, R. and Katchalski, E. (1971) Kinetic behavior of a two-enzyme membrane carrying out a consecutive set of reactions. *Journal of Theoretical Biology*, **32**, 243–257.
15. Krishna, R. and Ramachandran, P.A. (1975) Analysis of diffusional effects in immobilised two-enzyme systems. *Journal of Applied Chemistry and Biotechnology*, **25**, 623–640.
16. Michaelis, L. and Davidsohn, H. (1911) The theory of the isoelectric effect. *Biochemische Zeitschrift*, **30**, 143–150.
17. Illanes, A. (1999) Stability of biocatalysts. *Electronic Journal of Biotechnology*, **2**, 1–9.
18. Berteau, O., McCort, I., Goasdoue, N. *et al.* (2002) Characterization of a new alpha-L-fucosidase isolated from the marine mollusc *Pectan maximus* that catalyzes the hydrolysis of alpha-L-fucose from algal fucoidan (*Ascophyllum nodosum*). *Glycobiology*, **12**, 273–282.
19. Aramori, I., Fukagawa, M., Tsumura, M. *et al.* (1991) Isolation of soil strains producing new cephalosporin acylases. *Journal of Fermentation and Bioengineering*, **72**, 227–231.
20. Whiteley, C. (1999) Enzyme kinetics: partial and complete non-competitive inhibition. *Biochemical Education*, **27**, 15–18.
21. Davis, B.G., Maughan, M.A.T., Green, M.P. *et al.* (2000) Glycomethanethiosulfonates: powerful reagents for protein glycolysation. *Tetrahedron Asymmetry*, **11**, 245–262.
22. Polakovič, M. and Bryjak, J. (2004) Modelling of potato starch saccharification by an *Aspergillus niger* glucomylase. *Biochemical Engineering Journal*, **18**, 57–63.

3

Enzyme Kinetics in a Heterogeneous System

3.1 Introduction

The principles of enzyme kinetics were developed based on homogeneous systems in which both the enzyme protein and the substrates and products of reaction are dissolved in the (aqueous) reaction medium. Under such conditions, the rate of substrate conversion is solely determined by the catalytic activity, since the frequency of collision between the enzyme and substrate molecules is high enough to preclude any mass-transfer limitations. The situation may be quite different when the enzyme is attached to or embedded in a solid matrix, creating a heterogeneous system in which the enzyme is in a different phase to the one in which the substrates and products reside and the course of the reaction is being monitored. In that case, the substrate conversion rate may well be limited by mass-transfer rates, so that the actual catalytic potential of the enzyme is not expressed. Heterogeneous enzyme catalysis acquired technological relevance with the advent of enzyme immobilization in the late 1960s [1]. Its technological benefits with respect to efficiency of catalyst use must then be balanced against the reduction in catalytic potential due to mass-transfer limitations and configurational effects. The field of enzyme immobilization has bloomed since the 1970s and its technological relevance has been highlighted again in recent years as many enzyme-catalyzed organic synthesis reactions are conducted in harsh nonconventional (nonaqueous) reaction media [2]. The subject of enzyme immobilization is reviewed in Section 3.2 as a necessary background to an analysis of heterogeneous enzyme kinetics.

3.2 Immobilization of Enzymes

From a process perspective, enzyme immobilization has traditionally been considered a strategy for allowing continuous reactor operation and delivery of catalyst-free products.

Problem Solving in Enzyme Biocatalysis, First Edition. Andrés Illanes, Lorena Wilson and Carlos Vera.
© 2014 John Wiley & Sons, Ltd. Published 2014 by John Wiley & Sons, Ltd. Companion Website:
http://www.wiley.com/go/illanes-problem-solving

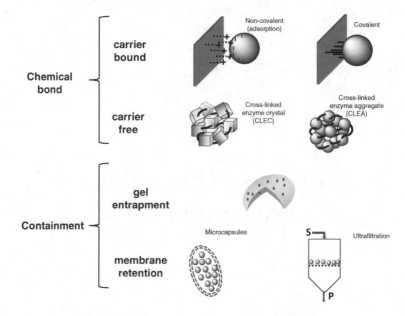

Figure 3.1   *Strategies of enzyme immobilization.*

However, a major issue with immobilization is enzyme stabilization, and enzyme immobilization technology is now strongly oriented in that direction, especially with the current trend in biocatalysis toward the use of enzymes in harsh, nonconventional reaction media.

Various enzyme immobilization strategies are represented in Figure 3.1.

3.2.1   Immobilization on Solid Supports (Carrier-Bound Systems)

The range and diversity of organic and inorganic materials used for enzyme immobilization is impressive. Organic polymers are flexible and, since many are products of chemical synthesis, can be tailored with precision for a given purpose; their mechanical properties are not always adequate for reactor operation, however, and the support is barely recoverable after enzyme exhaustion. Inorganic supports, on the other hand, are less flexible but are endowed with good mechanical properties and can be recovered after enzyme exhaustion, even if covalently linked to the enzyme, by being subjected to combustion to remove the organic matter. There is no ideal support and the best choice must be assessed in each individual case, based mostly on mechanical properties, chemical stability, compatibility with enzyme use, recoverability, protein loading capacity, micro-structure, availability and cost.

3.2.1.1   *Covalent Immobilization to Solid Supports*

In this strategy, the enzyme is linked to the support by a covalent bond involving functional groups in the amino acid residues (i.e. hydroxyl, sulfhydryl, amino, carboxy) of the protein molecule. The chemistry of immobilization may be complex, the enzyme activity lost during immobilization can be significant, the support is usually not recoverable after enzyme exhaustion and the kinetic properties of the enzyme may be changed considerably. However, the methodology is quite flexible, allowing directed immobilization, and the stability of the catalyst is high, making this strategy quite appealing from a process perspective [3]. Multiple-point covalent attachment to functionalized supports, such as glyoxyl agarose, has allowed the stability of a large number of enzymes of industrial interest to be increased significantly [4].

3.2.1.2 *Noncovalent Immobilization to Solid Supports*

Many different kinds of enzyme–support interaction can be used, such as ionic bonds, hydrophobic interactions, and van der Waals forces. In the enzyme immobilization terminology, these interactions are frequently referred to as "adsorption." The immobilization procedure is simple, the enzyme activity lost during immobilization may be nil and the support is recoverable after enzyme exhaustion. However, the catalyst stability is usually poor, the enzyme being prone to desorption during operation, especially when using aqueous reaction media [5].

3.2.2 Immobilization by Containment

3.2.2.1 *Entrapment*

In this case, the enzyme is retained in the inner cavities of a polymeric matrix so that a chemical network is formed which retains the enzyme molecule. Several organic polymers, such as alginate, polyacrylamide, polyurethane, polyvinyl alcohol, and κ-carrageenan, have been used, as have sol–gel systems for both organic and inorganic (silica) matrices [6]. Gel entrapment is a commonly used strategy for cell immobilization; however, with free enzymes, desorption is almost unavoidable and can be quite significant. This problem can be circumvented by entrapping previously aggregated enzymes [7], as presented in Section 3.2.3.

3.2.2.2 *Membrane Retention*

In this case, the enzyme is retained by microencapsulation or containment in a membrane device, allowing the passage of reaction substrates and products. The enzyme may or may not be associated to the membrane structure and more than one enzyme can be retained, allowing cascade reactions to be performed. Containment by ultrafiltration membranes is particularly relevant to processes in which coenzyme-requiring enzymes are involved and both the enzyme and the derivatized coenzyme are retained [8]. The membrane can also act as a molecular sieve, separating the products from the unreacted substrate in the case of depolymerizing hydrolases.

3.2.3 Immobilization in Carrier-Free Systems

In this case, the enzyme protein constitutes its own support so that high specific activities can be obtained. This has the additional advantage of the absence of an inert support, which can sometimes be even more expensive than the enzyme itself [9]. Carrier-free immobilized enzymes are prominent catalysts in processes requiring high productivity and yield, as well as with labile enzymes insufficiently stabilized by immobilization to solid supports. Carrier-free immobilized enzymes are prepared by cross-linking the protein containing the enzyme with bifunctional reagents, mostly glutaraldehye. It is convenient if the protein is already aggregated, by either crystallization or precipitation. Although these catalysts lack a solid support, they can be recovered by conventional separation from the reaction medium, very much as in the case of carrier-bound enzymes.

3.2.3.1 Cross-Linked Enzyme Crystals

Cross-linked enzyme crystals (CLECs) are produced from enzyme crystals from a highly pure enzyme solution. Although CLECs are excellent catalysts, being highly stable and having very high specific activity, their cost is also extremely high, as a consequence of the requirement for a pure protein starting material, meaning their applications are restricted to very high-added-value processes and sophisticated nonindustrial uses. Cutting these costs and tailoring existing CLECs to suit particular requirements for high-added-value products may become an attractive option for some industrial processes of organic synthesis [10].

3.2.3.2 Cross-Linked Enzyme Aggregates

Cross-linked enzyme aggregates (CLEAs) are produced from enzyme protein aggregates from conventional nondenaturing protein precipitation techniques. This strategy represents a breakthrough in enzyme biocatalysis as CLEAs have the positive properties of carrier-free systems (high stability and specific activity) without their drawback of high production costs, since no purified enzyme is required for the starting material [11]. In fact, several conventional techniques used for protein purification are applicable to CLEAs and high activity yields can be obtained using very simple immobilization protocols. It is thus no wonder that CLEAs from a large number of technologically important enzymes have been produced in recent years. These should have an industrial impact in the near future, and special reactor configurations have been developed for their proper recovery and handling [12]. CLEAs are particularly well suited to the co-immobilization of more than one enzyme (combi-CLEAs), allowing the catalysis of cascade reactions [13].

3.2.4 Parameters of Enzyme Immobilization

Enzyme immobilization is a multiple-variable process, and more than one parameter must therefore be taken into consideration when evaluating it. The most important are: enzyme immobilization yield (Y_E), representing the fraction of contacted activity expressed in the catalyst; protein immobilization yield (Y_P), representing the fraction of contacted protein bound to the support; and catalyst specific activity (a_{sp}), representing the enzyme activity expressed per unit mass (or unit volume) of biocatalyst. As a result of enzyme inactivation, steric hindrance, or mass-transfer limitations, the bound enzyme protein expresses only a fraction of the expected activity, while the unbound enzyme protein may be partly inactive. Taking this into consideration:

$$Y_E = \frac{E_I}{E_C} = \frac{E_I}{E_I + E_R + E_L} \tag{3.1}$$

Although Equation 3.1 looks like a material balance, it actually refers to activities, not masses, and thus the term E_L (enzyme activity not expressed) has been included. E_L may be attributed either to the inactivation of the bound and unbound enzyme or to mass-transfer limitations and steric hindrances of the immobilized enzyme. Note that in certain cases Y_E

can be greater than 1, denoting the well-reported phenomenon of enzyme hyperactivation, which is frequently seen in lipases, some of which are activated by interfacial contact with hydrophobic surfaces [14].

The protein immobilization yield, defined by the material balance in Equation 3.2, may help elucidate the meaning of E_L:

$$Y_P = \frac{P_I}{P_C} = \frac{P_C - P_R}{P_C} \tag{3.2}$$

If $Y_P \gg Y_E$, mass-transfer limitations and steric hindrances are likely to be significant; however, immobilization may be positively or negatively selective for the enzyme with respect to the whole protein and this may also influence the relationship between Y_P and Y_E.

The specific activity of the catalyst, defined by Equation 3.3, is also a very important parameter of enzyme immobilization and will have a direct impact on reactor operation (see Chapter 6):

$$a_{sp} = \frac{E_I}{M_{cat}} \tag{3.3}$$

The catalyst can also be characterized in terms of its specific protein load, as defined by Equation 3.4, but this parameter will give no insight into the potency of the catalyst, although it is a useful parameter for evaluating the loading capacity of the support:

$$p_{sp} = \frac{P_C - P_R}{M_{cat}} \tag{3.4}$$

When the immobilization protocol for a certain enzyme is well standardized, determination of the previously mentioned parameters at the end of the immobilization process allows for its proper evaluation. However, if such a protocol is not well standardized, preliminary trials using different supports and immobilization strategies and conditions will be required and the time course of immobilization will provide important information for a proper assessment of the immobilization protocol. Measurement of the activity in solution (activity in the supernatant after sample filtration) and total activity (i.e. the activity of the soluble plus immobilized enzyme: activity in the suspension) should be monitored throughout immobilization, and a control will be required in order to determine the stability of the soluble enzyme under the conditions of immobilization (pH, temperature, presence of additives) so as to elucidate the effect of the immobilization conditions on enzyme stability.

3.2.5 Optimization of Enzyme Immobilization

Optimization of an immobilization protocol implies determining the best values of the protocol's most important variables (i.e. temperature, pH, time of contact, enzyme-protein-to-support-mass ratio, conditions for support activation if required). The best immobilization conditions for immobilization yield and for mass (or volumetric) specific activity of the catalyst are usually different, with rather low specific activity being obtained under the conditions that maximize yield and vice versa. At this point, the operational stability of the catalyst may also come into play, and this is not trivial because immobilization variables may have an effect on the stability of the catalyst produced. For example, more contact time during immobilization may produce a more stable catalyst despite not producing an increase in immobilization yield or specific activity. Therefore, a third parameter may be added in order to build up an objective function that can be optimized. In order to optimize the immobilization protocol, these parameters, which have different impacts on process economics, should be properly weighted in a cost-related objective function. This is by no means a simple task, especially in the early stages of process development when the catalyst is being developed. To complicate things further, the economic impacts of these mentioned parameters run in different directions, as yield is mostly related to the production cost of the catalyst and specific activity and stability are mostly related to the cost of the enzyme reactor operation (see Chapter 6). A weighted objective function that considers these parameters is advisable, but the problem is how to properly weight them. A preliminary appraisal of which parameter is the most significant (e.g. if the base enzyme is very expensive, the immobilization yield is likely to be more significant) may enable optimization by allowing that parameter to be optimized, with the other one used as a restriction parameter by defining its minimum acceptable value. A sensitivity analysis is advisable, with varying weights given to the parameters in the objective function and the variability in the optimum values of the immobilization variables being appraised. With carrier-bound enzymes, optimization should also consider the cost of the support; disregarding this aspect can be quite misleading. In certain cases, the cost of the support is quite important and may even be much higher than the cost of the base enzyme. Under such circumstances, the use of recoverable supports (supports with a productive lifespan higher than that of the enzyme) that are not covalently linked to the enzyme may be the right choice, as illustrated by the paradigmatic case of the production in Japan of L-amino acids from racemic mixtures using L-aminoacylase immobilized in ion exchangers. The use of carrier-free immobilized enzymes is an obvious potential advantage here.

3.3   Mass-Transfer Limitations in Enzyme Catalysis

As a consequence of enzyme aggregation, attachment, or containment within a matrix, both conformational and micro-environmental effects can affect enzyme kinetics. "Conformational effects" refer to the enzyme protein structural changes produced by immobilization, and also to steric hindrances due to the close proximity of the enzyme molecule to the surface of the support; in either case, the kinetic behavior will differ from that of the free enzyme. "Micro-environmental effects" refer to partition and mass-transfer limitations. The partition of substrates and products (and other species) will produce a discontinuity in the corresponding profiles at the matrix–medium interface. Mass-transfer limitations will imply that the reaction rate will not be solely determined by the enzyme catalytic potential, but also by substrate(s) mass-transfer rates from the bulk reaction medium to the enzyme site and product(s) mass-transfer rates from that site back into the reaction medium. Under such conditions, species concentration profiles will be developed in the immediate vicinity of the catalyst surface, and on its inside if the enzyme is hosted there, as shown in Figure 3.2, so that the effective species concentrations at the catalyst site will differ from those measured in the bulk liquid medium.

The intrinsic kinetics of an immobilized enzyme represents its proper behavior, but it will only be revealed in the absence of partition and mass-transfer limitations—a situation

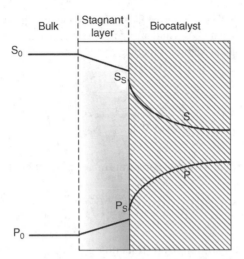

Figure 3.2 *Schematic representation of external and internal diffusional restrictions.*

that is hardly attainable except under conditions promoted purposely in order to avoid such effects. Even so, the kinetic behavior will not necessarily be the same as that of the free enzyme because of the conformational changes previously mentioned.

The inherent kinetics of an immobilized enzyme is that obtained in the absence of mass-transfer limitations and will reflect partition effects if present; otherwise, intrinsic and inherent behaviors will coincide.

The effective kinetics of an immobilized enzyme is the one directly determined from its observed behavior; however, it does not reflect its actual catalytic potential, being also referred as "apparent". This information is valid only under the precise conditions in which the experiments were performed, so it has little meaning for design and scale-up purposes, where intrinsic (or inherent) behavior must be assessed, as presented in the following sections.

3.3.1 Partition Effects

Partition effects may be important if the physicochemical properties of the bulk medium and the enzyme environment within the support are significantly different from each other, as they occur when the support is electrically charged; in such cases, the reactive charged species will be partitioned just as would a charged substrate (or product) or proton, thereby affecting the pH profile of the immobilized enzyme.

Partition can be described by means of the electrostatic partition coefficient, as given by Equations 3.5 and 3.6 for substrates and protons, respectively:

$$K_{P,S} = \frac{s}{s_0} = \exp\left(\frac{-Z \cdot \in \cdot \Psi}{k_B \cdot T}\right) \tag{3.5}$$

$$K_{P,H^+} = \frac{h^+}{h_0^+} = \exp\left(\frac{- \in \cdot \Psi}{k_B \cdot T}\right) \tag{3.6}$$

For a substrate, partition (see Figure 3.2) will be favorable if its charge is opposite to that of the support ($\psi < 0$, so that s > s$_0$) and unfavorable if their charges are alike ($\psi > 0$, so that < s$_0$). Then, for Michaelis–Menten kinetics:

$$v = \frac{V_{max}}{\dfrac{K_M}{s_0 \cdot \exp\left(\dfrac{-Z \cdot \in \cdot \Psi}{k_B \cdot T}\right)} + 1} \tag{3.7}$$

so the partition effect can be visualized in terms of an apparent Michaelis constant (K_{MAP}):

$$K_{MAP} = K_M \cdot \exp\left(\frac{Z \cdot \in \cdot \Psi}{k_B \cdot T}\right) \tag{3.8}$$

In the case of proton partition, from Equation 3.6:

$$pH - pH_0 = 0.43 \frac{\in \cdot \Psi}{k_B \cdot T} \tag{3.9}$$

If the net charge of the support is positive, $\psi > 0$ and pH > pH$_0$; if it is negative, $\psi < 0$ and pH < pH$_0$. The pH profile of enzyme activity is then displaced to the left (lower pH) for cationic supports and to the right (higher pH) for anionic supports. The magnitude of pH displacement is reduced by increasing the ionic strength of the medium, which is a good way to avoid partition effects and thus reveal the intrinsic behavior of the enzyme. This magnitude can in principle be used to estimate the value of ψ, which is otherwise quite difficult to determine.

3.3.2   External Diffusional Restrictions in Impervious Biocatalysts

External diffusional restrictions (EDR) occur in the presence of a stagnant liquid layer surrounding the solid catalyst particle. In such a stagnant layer, in the absence of convection, transport of solutes will occur only by molecular diffusion, so that the substrate (and product) profile will develop across the layer for so long as the substrate (or product) diffusion limits the catalytic potential of the enzyme, as shown in Figure 3.2. At steady state, the mass-transfer rate and reaction rate should be equal in magnitude at the catalyst surface, so:

$$r' = v' = J \tag{3.10}$$

where, for Michaelis–Menten kinetics:

$$v' = \frac{V'_{max} \cdot s_S}{K_M + s_S} \tag{3.11}$$

where V'_{max} and K_M are the intrinsic parameters of the enzyme (assuming no partition effects) and:

$$J = h(s_0 - s_S) = h' \cdot A_S(s_0 - s_S) \tag{3.12}$$

Then, from Equations 3.10 to 3.12:

$$h(s_0 - s_S) = \frac{V'_{max} \cdot s_S}{K_M + s_S} \tag{3.13}$$

This is more conveniently expressed in dimensionless form:

$$\beta_0 - \beta_S = \frac{\alpha \cdot \beta_S}{1 + \beta_S} = \alpha \cdot \upsilon \tag{3.14}$$

The dimensionless Damkoehler number (α) is:

$$\alpha = \frac{V'_{max}}{K_M \cdot h} \tag{3.15}$$

This represents the relative magnitudes of the enzyme catalytic potential (first-order rate constant $V'_{max} \cdot K_M^{-1}$) and substrate mass-transfer rate (h), meaning that high values of α ($\gg 1$) imply that EDR are relevant and small values ($\ll 1$) imply that it is negligible. A system is free of EDR when $h > V'_{max} \cdot K_M^{-1}$.

From Equation 3.14, β_S is obtained as a function of measurable β_0 and α:

$$\beta_S = \frac{-(1 + \alpha - \beta_0) + \sqrt{\left[(1 + \alpha - \beta_0)^2 + 4\beta_0\right]}}{2} \tag{3.16}$$

A graphical representation of Equation 3.14 is presented in Figure 3.3.

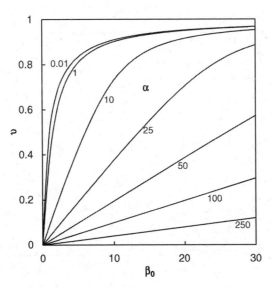

Figure 3.3   *Kinetics of immobilized enzymes under external diffusional restrictions. υ: dimensionless initial reaction rate; β_0: dimensionless bulk substrate concentration; α: Damkoehler number.*

A convenient way of representing the magnitude of EDR is by means of the effectiveness factor, which is defined as the ratio of effective (apparent) reaction rate to intrinsic reaction rate. For simple Michaelis–Menten kinetics:

$$\eta = \frac{\dfrac{V'_{max} \cdot s_S}{K_M + s_S}}{\dfrac{V'_{max} s_0}{K_M + s_0}} = \frac{\beta_S(1 + \beta_0)}{\beta_0(1 + \beta_S)} \tag{3.17}$$

The value of η describes the impact of EDR quite neatly, since it represents the fraction of the catalytic potential of the enzyme that is expressed when a certain condition is met under the influence of EDR.

From Equations 3.16 and 3.17, η can be expressed in terms of measurable entities (β_0 and α):

$$\eta = \frac{(1 + \beta_0)\left[1 + \alpha + \beta_0 - \sqrt{(1 + \alpha - \beta_0)^2 + 4\beta_0}\right]}{2\beta_0\alpha} \tag{3.18}$$

A convenient way of representing Equation 3.18 is shown in Figure 3.4. At low values of α (insignificant EDR), behavior is typically Michaelian, but as α increases the dependence of η on β_0 becomes progressively more linear. However, whatever the magnitude of α, η will approach 1 (no EDR) as β_0 increases. This is limited in practice by the substrate solubility in the reaction medium.

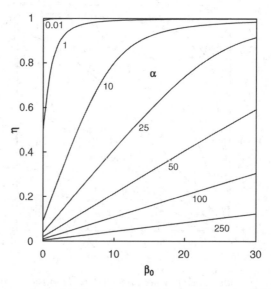

Figure 3.4   *Effectiveness factor (η) of an enzyme catalyst under external diffusional restrictions. β_0: dimensionless bulk substrate concentration; α: Damkoehler number.*

3.3.3   Internal Diffusional Restrictions in Porous Biocatalysts

When the enzyme is immobilized in the inside of a gel or a microporous matrix, high protein loading is attainable, but this is not necessarily reflected by a high specific activity as severe mass-transfer limitations may occur. In that case, the substrate will have to diffuse from the bulk reaction medium through the stagnant liquid layer on to the catalyst surface and thence inside the catalyst internal structure at diffusion rates lower than those in the liquid medium (the reverse pathway will occur for the product). As a consequence of internal diffusional restrictions (IDR), a substrate (and product) profile will be developed as shown in Figure 3.1. In this case, unlike in EDR, enzyme molecules within the support will be exposed to different environmental conditions according to their position within the catalyst matrix, so that a differential analysis is required to properly describe the system and determine the profiles within the catalyst particle.

Such an analysis will vary according to the geometry of the catalyst particle, so a generalized analysis, like that carried out for EDR, is no longer possible. The cases of the catalyst slab (resembling a membrane) and catalyst spherical particle (a quite common catalyst shape) are amenable to a rigorous mathematical analysis, while cases of more complex geometries or particles of undefined geometry can allow approximate solutions to be derived. The slab and the sphere can be considered extreme geometrical forms with an infinite and a minimum radius of curvature, respectively, with other geometries lying somewhere in between.

The analysis is carried out in three steps. In the first, differential equations are developed considering enzyme reaction kinetics and mass-transfer rates, whose solution yields the substrate (and product) profile within the biocatalyst particle; in the second, the distribution of local effectiveness factors inside the catalyst particle is obtained from the substrate profile; in final one, a global effectiveness factor is obtained by properly averaging this distribution; this global effectiveness factor represents the behavior of the catalyst particle as a whole.

From a differential mass balance around a proper catalyst section and the corresponding mass-transfer equation (Fick's first law of diffusion), the following differential equations for the slab and spherical geometries can be derived given Michaelis–Menten kinetics [15]:

Catalyst Geometry	Differential Equation	Boundary Conditions	
Slab	$$D_{eff} \cdot \frac{d^2s}{dx^2} - \frac{V''_{max} \cdot s}{K_M + s} = 0$$	(I) $\begin{array}{c} x = 0 \\ s = s_0 \end{array}$	
		(II) $\begin{array}{c} x = \dfrac{L}{2} \\ \dfrac{ds}{dx} = 0 \end{array}$	(3.19)
Sphere	$$D_{eff} \frac{d^2s}{dr^2} + \frac{2D_{eff}}{r} \frac{ds}{dr} - \frac{V''_{max} \cdot s}{K_M + s} = 0$$	(I) $\begin{array}{c} r = R_P \\ s = s_0 \end{array}$	
		(II) $\begin{array}{c} r = 0 \\ \dfrac{ds}{dr} = 0 \end{array}$	(3.20)

These equations can be more conveniently expressed in dimensionless form:

Catalyst Geometry	Differential Equation	Boundary Conditions
Slab	$$\dfrac{d^2\beta}{dz^2} - \Phi^2 \cdot \dfrac{\beta}{1+\beta} = 0$$	(I) $\begin{aligned} z &= 0 \\ \beta &= \beta_0 \end{aligned}$ $z = 0.5$ (3.21) (II) $\dfrac{d\beta}{dz} = 0$
Sphere	$$\dfrac{d^2\beta}{d\rho^2} + \dfrac{2}{\rho}\dfrac{d\beta}{d\rho} - 9\Phi^2 \dfrac{\beta}{1+\beta} = 0$$	(I) $\begin{aligned} \rho &= 1 \\ \beta &= \beta_0 \end{aligned}$ $\rho = 0$ (3.22) (II) $\dfrac{d\beta}{d\rho} = 0$

Note that boundary conditions (I) assumes that EDR are negligible compared to IDR. In many cases this will be a reasonable assumption, but if it is not then the substrate concentration at the liquid–particle interface will differ from that in the bulk and the boundary conditions (I) will have to be modified, as presented ahead.

The dimensionless Thièle modulus (Φ) represents the relative impact of the reaction on mass-transfer rates in the case of IDR and can be conveniently defined as [16]:

$$\Phi = L_{eq} \sqrt{\frac{V''_{max}}{K_M \cdot D_{eff}}} \tag{3.23}$$

A high value of Φ means that either $V''_{max} \cdot K_M^{-1} \gg D_{eff}$ or else L_{eq} is big enough so that the system is mass-transfer limited; a small value of Φ means that either $D_{eff} \gg V''_{max} \cdot K_M^{-1}$ or else L_{eq} is small enough to make the system free of IDR and limited by the catalytic potential of the enzyme. L_{eq} is the equivalent length of the catalyst particle (i.e. its ratio of volume to surface area) so, for slabs and spheres:

$$\text{Slab} \quad L_{eq} = L \tag{3.24}$$

$$\text{Sphere} \quad L_{eq} = \frac{R_P}{3} \tag{3.25}$$

Equations 3.21 and 3.22 are highly nonlinear, so must be solved numerically, which can be done using available software. Resolution is then provided as a set of values for the dependent variable β at different values of the independent variable z (or ρ) and the parameter Φ:

$$\text{Slab} \quad \beta = f(\beta_0, z, \Phi) \tag{3.26}$$

$$\text{Sphere} \quad \beta = f(\beta_0, \rho, \Phi) \tag{3.27}$$

Equations 3.26 and 3.27 represent the corresponding substrate profiles within the slab and spherical catalyst particles, as shown in Figures A.7 and Figure A.8, respectively.

The next step is to determine the local effectiveness factor profile (η), which by definition is:

$$\eta = \frac{\dfrac{V''_{max} \cdot s}{K_M + s}}{\dfrac{V''_{max} s_0}{K_M + s_0}} = \frac{\beta(1 + \beta_0)}{\beta_0(1 + \beta)} \tag{3.28}$$

By replacing Equations 3.26 and 3.27 into Equation 3.28, the distributions of η inside the slab and spherical catalysts, respectively, are obtained:

$$\text{Slab} \quad \eta = f(\beta_0, z, \Phi) \tag{3.29}$$

$$\text{Sphere} \quad \eta = f(\beta_0, \rho, \Phi) \tag{3.30}$$

The final step is to determine a global effectiveness factor (η_G) that adequately represents such a distribution and the behavior of the catalyst particle as a whole. Since the distribution of η is highly nonlinear with respect to z (or ρ), the mean integral value of the η distribution function over the volume of the biocatalyst particle is adequate to represent this distribution. Then:

$$\text{Slab} \quad \eta_G = \frac{\int_0^1 \eta \cdot A \cdot dz}{\int_0^1 A \cdot dz} = \int_0^1 \eta \cdot dz \tag{3.31}$$

$$\text{Sphere} \quad \eta_G = \frac{\int_0^1 \eta \cdot \rho^2 d\rho}{\int_0^1 \rho^2 d\rho} = 3\int_0^1 \eta \cdot \rho^2 d\rho \tag{3.32}$$

From Equations 3.31 and 3.32, η_G is obtained as a function of the dimensionless bulk substrate concentration and Thièle modulus:

$$\eta_G = f(\beta_0, \Phi) \tag{3.33}$$

This correlation is represented as a surface-of-response plot in Figure 3.5a and b for the slab and spherical catalysts, respectively. η_G is a strong function of Φ; however, the magnitude of Φ at which such dependence occurs strongly depends on the magnitude of β_0.

Michaelis–Menten kinetics reduces to first-order kinetics at $s \ll K_M$ ($\beta \ll 1$), so that the rate expression is simply:

$$v'' = \frac{V''_{max}}{K_M} \cdot s \tag{3.34}$$

and the resulting dimensionless differential equations are linear:

$$\text{Slab} \quad \frac{d^2\beta}{dz^2} - \Phi^2 \cdot \beta = 0 \tag{3.35}$$

$$\text{Sphere} \quad \frac{d^2\beta}{d\rho^2} + \frac{2}{\rho}\frac{d\beta}{d\rho} - 9\Phi^2 \cdot \beta = 0 \tag{3.36}$$

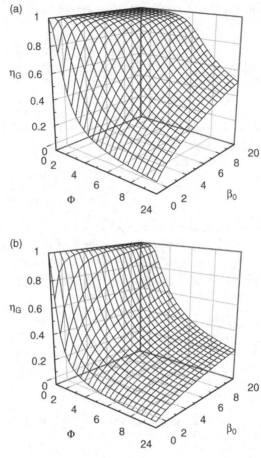

Figure 3.5 *Global effectiveness factor (η_G) of an immobilized enzyme with intrinsic Michaelis–Menten kinetics under internal diffusional restrictions as a function of bulk substrate concentration (β_0) and Thièle modulus (Φ). (a) Catalytic slab. (b) Catalytic sphere.*

They can thus be solved analytically using the same boundary conditions as in Equations 3.21 and 3.22 (see Sections A.3.3 and A.3.4) to obtain:

Slab

$$\beta = \beta_0 \frac{\cosh[\Phi(z - 0.5)]}{\cosh[0.5\Phi]}$$

$$\eta = \frac{\beta}{\beta_0} = \frac{\cosh[\Phi(z - 0.5)]}{\cosh[0.5\Phi]}$$

(3.37)

$$\eta_G = \frac{\tanh\dfrac{\Phi}{2}}{\dfrac{\Phi}{2}}$$

(3.38)

Sphere

$$\beta = \beta_0 \frac{\sinh[3\Phi \cdot \rho]}{\rho \cdot \sinh[3\Phi]}$$

$$\eta = \frac{\beta}{\beta_0} = \frac{\sinh[3\Phi \cdot \rho]}{\rho \cdot \sinh[3\Phi]}$$

(3.39)

$$\eta_G = \frac{1}{\Phi} \cdot \left[\frac{1}{\tanh(3\Phi)} - \frac{1}{3\Phi} \right]$$

(3.40)

First-order kinetics ($s \ll K_M$) is never attained in an enzyme reactor (even at outlet or final conditions, s will be on the order of magnitude of K_M) so the analytical solution is mostly of academic interest. Even so, an analytical solution is frequently reported, but if it is applied then the enzyme catalytic potential will be underestimated (see dotted line in Figures A.7 and Figure A.8), which may lead to a significant error of overdimensioning.

This analysis, presented for simple irreversible Michaelis–Menten-type kinetics, can be extended to more complex kinetics involving reversible Michaelis–Menten kinetics and product and substrate inhibition kinetics. The interested reader is referred to the article published by Jeison *et al.* [17].

For catalyst particles of complex or undefined geometries, the problem may be extremely difficult to solve. An approximate solution can be obtained by introducing the actual equivalent length of the particle and solving the equations derived for the particle of defined geometry (i.e. slab or sphere) that most closely resembles it.

It has been assumed until now that the effect of EDR on enzymes immobilized in the inner space of the catalyst matrix is negligible and that therefore the concentration of substrate at the catalyst surface is no longer equal to that in the bulk reaction medium. This assumption is often but not always valid, so in the case of significant EDR the combined effect of EDR and IDR can be described by replacing the boundary conditions (I) in the corresponding differential Equations 3.21 and 3.22 with an equation of continuity at the liquid–catalyst interface:

Particle geometry	Boundary Condition (I)	
Slab	$h'(s_0 - s_S) = -D_{eff} \cdot \frac{ds}{dx}\big\|_{x=0}$ $Bi \cdot (\beta_0 - \beta_S) = -\frac{d\beta}{dz}\big\|_{z=0}$	(3.41)
Sphere	$h'(s_0 - s_S) = D_{eff} \cdot \frac{ds}{dr}\big\|_{r=R}$ $Bi \cdot (\beta_0 - \beta_S) = \frac{d\beta}{d\rho}\big\|_{\rho=1}$	(3.42)

Bi is the dimensionless Biot number:

$$Bi = \frac{h' \cdot L_{eq}}{D_{eff}}$$

(3.43)

and represents the relative impact of EDR compared to IDR. If $Bi \gg 1$, EDR are negligible, so little error is introduced through consideration of the former boundary conditions (I). If $Bi \ll 1$, EDR are significant, so differential equations must be solved using Equations 3.41 and 3.42 as boundary conditions (I); this is done as before, by using numerical methods. In the case of first-order kinetics, the problem can be solved analytically (asymptotic solution). The following expressions for the global effectiveness factors are then obtained:

$$Slab \quad \eta_G = \frac{\tanh \dfrac{\Phi}{2}}{\dfrac{\Phi}{2}\left(1 + \dfrac{\Phi}{2} \cdot \tanh \dfrac{\dfrac{\Phi}{2}}{Bi}\right)} \tag{3.44}$$

$$Sphere \quad \eta_G = \frac{Bi\left(\dfrac{1}{\tanh(3\Phi)} - \dfrac{1}{3 \cdot \Phi}\right)}{\Phi\left(Bi - 1 + \dfrac{3\Phi}{\tanh(3\Phi)}\right)} \tag{3.45}$$

If $Bi \gg 1$ (negligible EDR), Equations 3.44 and 3.45 are reduced to Equations 3.38 and 3.40, respectively.

3.4 Determination of Intrinsic Kinetic and Mass-Transfer Parameters

When working with immobilized enzymes, mass-transfer limitations are likely to occur no matter what method of immobilization is used. It is recommended that a very simple diagnostic experiment be carried out to assess the presence of mass-transfer limitations. Simply gather sufficient data on initial reaction rates (v) at different bulk substrate concentrations (s_0) and plot them in a double reciprocal plot. If a straight line is obtained throughout the substrate concentration range (choose a range covering the expected conditions of the process for which the enzyme is the intended catalyst), this is indicative that no significant mass-transfer limitations exist; a monotonic curve of progressively increasing slope denotes that EDR are significant; a curve with an inflection point denotes that IDR are also significant.

3.4.1 EDR

Intrinsic kinetic parameters (V'_{max}, K_M) and film mass-transfer coefficient (h or h′) must be determined in order to evaluate the behavior of immobilized enzymes under EDR. They are likely to be significant when immobilizing the enzyme in the surface of an impervious support, but this does not preclude them from also being significant when immobilizing the enzyme in the inside of a solid matrix, as analyzed in Section 3.3.3. Intrinsic kinetic parameters can be determined experimentally from initial reaction rate data, while film mass-transfer coefficients can be determined both in the same way and also by empirical correlations. Experimental evaluation of kinetic and mass-transfer parameters will be presented in this section. For information on the use of empirical correlations to determine

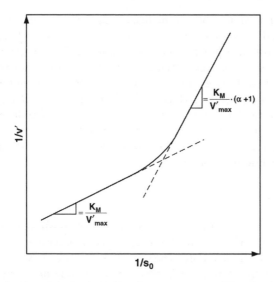

Figure 3.6 *Determination of the intrinsic kinetic parameters and Damkoehler number (α) of an immobilized enzyme subjected to external diffusional restrictions by linearization of the initial rate data in a double reciprocal plot. v': initial reaction rate; s_0: bulk substrate concentration; K_M: Michaelis constant; V'_{max}: maximum reaction rate.*

mass-transfer coefficients, the reader is referred to articles published by Rovito and Kittrell [18] and Buchholz [19].

Experimental determination of intrinsic kinetic and mass-transfer parameters is based on the determination of initial reaction rates (v') within a broad range of bulk substrate concentrations (s_0), as proposed by Buchholz [19]. Equation 3.13 is nonlinear with respect to measurable s_0, but it is easy to demonstrate that at the extreme ranges of bulk substrate concentration ($s_0 \gg K_M$ and $s_0 \ll K_M$), linear correlations will be obtained in a double reciprocal plot (immobilized enzymes) of v'^{-1} versus s_0^{-1}, as shown in Figure 3.6:

Zone of s_0	Equation	Slope	Intercept y-axis	Intercept x-axis
(1) $s_0 \gg K_M$	$\dfrac{1}{v'} = \dfrac{K_M}{V'_{max}} \cdot \dfrac{1}{s_0} + \dfrac{1}{V'_{max}}$ (3.46)	$\dfrac{K_M}{V'_{max}}$	$\dfrac{1}{V'_{max}}$	$-\dfrac{1}{K_M}$
(2) $s_0 \ll K_M$	$\dfrac{1}{v'} = \dfrac{K_M}{V'_{max}} \cdot (\alpha + 1) \cdot \dfrac{1}{s_0}$ (3.47)	$\dfrac{K_M}{V'_{max}}(\alpha + 1)$	0	0

Intrinsic kinetic parameters are then obtained as the intercepts of both axes of the straight line obtained in zone 1 ($s_0 \gg K_M$), while α (and therefore h) is obtained from the slopes of both zones as:

$$\alpha = \frac{Slope\ zone\ 2}{Slope\ zone\ 1} - 1 \qquad (3.48)$$

This method, although straightforward, must be used with caution as important restraints apply. As presented in Section 2.6, the regions of s_0 in which linear correlations are obtained are the most error-prone (very low v' values in the low s_0 range and very small differences in v' values in the high s_0 range), so a careful analysis of statistical significance of the data is required. Besides, the substrate concentration is limited by the substrate solubility in the reaction medium, which may be a significant restraint on data gathering in zone 1. Another constraint of this method is that the results are dependent on the fluid regime, so the conditions must be carefully controlled and specified, especially the agitation rate. A reasonable suggestion is to scale down the conditions of the enzyme reactor used to perform the kinetic experiments.

3.4.2 IDR

Intrinsic kinetic parameters (V''_{max}, K_M) and effective substrate diffusion coefficients within the catalyst matrix (D_{eff}) must be determined in order to evaluate the behavior of immobilized enzymes under IDR. They are likely to occur when immobilizing the enzyme within a solid matrix, be it a gel or a microporous solid particle. Several strategies have been proposed by which to estimate intrinsic kinetic parameters. A very convenient method is based on comminuting the matrix to obtain catalyst particles small enough to approach a condition where IDR are negligible (very low Φ; recall Equation 3.23). If a sound correlation is obtained between the apparent kinetic parameters and the particle size, extrapolation to size zero may give an estimate of the intrinsic kinetic parameters. It can be argued to what extent the enzyme in the comminuted biocatalyst suffers sufficient structural changes to poorly represent its structure in the intact catalyst particle. We have determined that for enzymes immobilized in porous matrices of a high degree of surface congruency, the values of the apparent kinetic parameters for the comminuted catalyst very much approach the values of the kinetic parameters of the free enzyme, so the method seems applicable [20]. Another approach to estimating the values of the intrinsic kinetic parameters is to reduce the protein load in the support so that the mass-specific activity of the biocatalyst (units of activity per unit mass of support) is low enough to move the system away from kinetic control (very low Φ; recall Equation 3.23); it may also be argued to what extent this biocatalyst resembles in its kinetic behavior the one with a high protein load. Experiments to determine intrinsic parameters of immobilized enzymes under IDR should be performed under conditions at which EDR are negligible, so the agitation rate is varied to the point where it has no effect on the reaction rate. It has also been proposed that a procedure similar to the one used for the determination of intrinsic parameters in the case of EDR be used to measure v'' at $s_0 \gg K_M$ (zone 1) and at $s_0 \ll K_M$ (zone 2). V''_{max} and K_M can be determined from the linear zone 1 in a double reciprocal plot, while the value of η_G can be estimated from the y-axis intercept of the extrapolated straight line obtained in zone 2, which corresponds to $(V''_{max} \cdot \eta_G)^{-1}$ [21]. The drawbacks mentioned for EDR are even more pronounced in the case of IDR.

The effective diffusion coefficient (D_{eff}) of the substrate within the biocatalyst particle should be experimentally determined, since the values of diffusion coefficients in water (D_0), available for a large number of substances in chemical or biochemical handbooks, may differ significantly from those inside the catalyst matrix. In the case of porous matrices, a simplified correlation has been proposed:

$$D_{eff} = D_0 \frac{\varepsilon}{\zeta} \tag{3.49}$$

This shows that $D_0 < D_{eff}$, since by definition $\varepsilon < 1$ and $\zeta > 1$. In practice, the value of D_{eff} may lie between 20 and 80% of D_0, so using the latter value will overestimate reaction rates and therefore the system will be undersized.

Values of D_{eff} may be determined experimentally for the biocatalyst particle by working with a chemically or radioactively labeled substrate. A more simple and very interesting procedure is the one proposed by Grünwald [22], in which back diffusion of the solute (be it the substrate or the product) is determined by measuring the amount of solute released in time from the solute-saturated catalyst matrix. For spherical particles of radius R_P, the following simplified expression is deduced:

$$\ln\left(\frac{S_\infty - S_t}{S_\infty}\right) = -\frac{\pi^2 \cdot D_{eff}}{R_P^2} \cdot t + constant \tag{3.50}$$

where the subscripts t and ∞ refer to variable time and infinite time, respectively.

Once the intrinsic kinetic parameters and D_{eff} are evaluated, the Thièle modulus is determined from Equation 3.23.

Solved Problems

3.1 A β-galactosidase is immobilized in a heterofunctional agarose support. The immobilization yield in terms of protein (Y_P) is 95% and in terms of expressed activity (Y_E) is 85%. The immobilization procedure consists in contacting 10 g of support (density $0.7\,g \cdot mL^{-1}$) with 9 mL of soluble β-galactosidase preparation and 93 mL of bicarbonate buffer under stirring for 3 hours. At the end of the process, the expressed activity in the immobilized catalyst is $286\,IU \cdot g^{-1}$, the remaining activity in solution is $2.6\,IU \cdot mL^{-1}$ and the remaining protein in solution is $0.18\,mg \cdot mL^{-1}$.

Determine: the activity lost during immobilization, the protein load in the support (mass of protein immobilized per unit mass of support) and the specific activity of the soluble β-galactosidase preparation in $IU \cdot mg_{protein}^{-1}$.

Resolution: Volume of remaining solution $= 93 + 9 = 102$ mL

$$E_R = 2.6(IU \cdot mL^{-1}) \cdot 102(mL) = 265.2(IU)$$
$$E_I = 286(IU \cdot g^{-1}) \cdot 10(g)^{-1} = 2860(IU)$$

From Equation 3.1 and $Y_E = 0.85$:

$$E_L = 239.5\ (IU)$$
$$E_c = E_I + E_R + E_L = 2860 + 265.2 + 239.5 = 3364.7\ (IU)$$
$$P_R = 0.18(mg \cdot mL^{-1}) \cdot 102(mL) = 18.36(mg)$$

From Equation 3.2 and $Y_P = 0.95$:

$$P_C = 367.2(mg)$$
$$P_I = P_C - P_R = 348.84(mg)$$

Answer:

$$\text{Activity lost} = 239.5(IU)$$

$$\text{Support load} = \frac{348.84}{10} = 34.8(mg \cdot g^{-1})$$

$$\text{Specific activity of enzyme preparation} = \frac{3364.7}{367.2} = 9.16(IU \cdot mg^{-1})$$

3.2 Pancreatic lipase (triacylglycerol acylhydrolase, EC 3.1.1.3) is immobilized in glyoxyl agarose and octadecyl Sepabeads by contacting 50 mL of a liquid lipase preparation with $20\,g \cdot L^{1}$ of protein and a specific activity of $10\,IU \cdot mg^{-1}_{protein}$ with 10 g of each. After immobilization on glyoxyl agarose, a solid catalyst (GAL) with a specific activity of $350\,IU \cdot g^{-1}_{support}$ and a supernatant with $2\,mg_{protein} \cdot mL^{-1}$ and $30\,IU \cdot mL^{-1}$ are obtained. After immobilization on octadecyl Sepabeads, a solid catalyst (SBL) with a specific activity of $1200\,IU \cdot g^{-1}_{support}$ is obtained, while no protein or enzyme activity is detected in the supernatant.

Determine the protein immobilization yield (Y_P) and enzyme activity yield (fraction of the contacted activity expressed in the catalyst after immobilization) (Y_E), discussing the meaning of the results obtained. Assume that the volume occupied by the solids is negligible.

Resolution: Initial conditions:

$$\text{Protein mass} = 50(mL) \cdot 20(mg \cdot mL^{-1}) = 1000(mg)$$

$$\text{Activity} = 10(IU \cdot mg^{-1}_{protein}) \cdot 1000(mg_{protein}) = 10000(IU)$$

Supernatant:

Catalyst	Protein Mass*	Enzyme Activity
GAL	$50(mL) \cdot 2(mg \cdot mL^{-1}) = 100(mg)$	$50(mL) \cdot 30(IU \cdot mL^{-1}) \cdot = 1500(IU)$
SBL	—	—

*Determined by the difference between initial and supernatant

Solid catalyst:

Catalyst	Protein Mass*	Enzyme Activity
GAL	$1000 - 100 = 900(\text{mg})$	$350(\text{IU} \cdot \text{g}^{-1}) \cdot 10(\text{g}) = 3500(\text{IU})$
SBL	$1000(\text{mg})$	$1200(\text{IU} \cdot \text{g}^{-1}) \cdot 10(\text{g}) = 12000(\text{IU})$

*Determined by the difference between initial and supernatant

Answer: From Equations 3.1 to 3.3:

Catalyst	Y_P (%)	Y_E (%)	$a_{sp}(\text{IU} \cdot \text{mg}^{-1}_{\text{protein}})$
GAL	90	35	3.9
SBL	100	120	12.0

In the case of GAL, Y_E is low and specific activity is less than that of the free enzyme, so the support has a negative selectivity for the lipase. Activity loss is significant (65%), which may be a result of inactivation during immobilization or steric hindrance and/or mass-transfer limitations; the former possibility is more likely since glyoxyl agarose offers a congruent surface for the immobilization of the enzyme and steric hindrances and mass-transfer limitations are seldom severe. In the case of SBL, the activity recovered is higher than initial, meaning that hyper-activation has occurred, a phenomenon usually found when lipases are immobilized in hydrophobic supports [23].

3.3 A recombinant esterase from *Pseudomonas fluorescens* is immobilized by adsorption on Celite 545 at pH 6.0 and various temperatures after a contact time of 6 hours. The following results are obtained:

T (°C)	IE (%)	RE (%)	NEE (%)
20	24.8	25.3	48.9
30	30.2	20.1	48.0
35	35.9	15.8	45.4
40	39.2	12.2	43.8
50	44.5	5.1	37.7
60	44.2	—	26.4

IE: immobilized enzyme activity (% of contacted)
RE: residual enzyme activity in liquid phase after immobilization (% of contacted)
NEE: enzyme activity not expressed in the catalyst due to conformational effects (% of contacted)

Assume that, aside from conformational effects, the loss of activity is caused by inactivation during immobilization. Based on this information, estimate the energy of activation of inactivation of the esterase.

Resolution: The activity lost due to inactivation (LE) is:

$$LE = 100 - IE - RE - NEE \tag{3.51}$$

so, from the data provided:

T (°C)	20	30	35	40	50	60
LE (%)	1.0	1.7	2.9	4.8	12.7	29.4

Given that the enzyme inactivates according to first-order kinetics (Equation 2.39):

$$\frac{100 - LE}{100} = \exp(-k_D \cdot 6) \tag{3.52}$$

T (°C)	$k_D \cdot 10^3$ (h^{-1})	$T^{-1} \cdot 10^3$ (K)$^{-1}$	ln k_D
20	1.68	3.41	−6.39
30	2.86	3.30	−5.86
35	4.91	3.25	−5.32
40	8.20	3.20	−4.80
50	22.64	3.10	−3.79
60	58.02	3.00	−2.85

Answer: From these data, given that the Arrhenius equation is valid (Equation 2.40), by linear regression of the last two columns:

$$\frac{E_{ia}}{R} = 9007$$

$$E_{ia} = 17,834(\text{cal} \cdot \text{mol}^{-1})$$

The energy of activation of inactivation is 17.8 (Kcal · mol^{-1}), which is a reasonable value for a protein. The correlation coefficient is 0.992, which is acceptable. The best temperature for immobilization is around 50 °C.

Comment: Along the period of immobilization (6 hours), the immobilized enzyme may adopt different configurations, so the value of E_{ia} reported cannot be associated with a particular catalyst configuration.

3.4 Bovine trypsin (EC 3.4.21.4) is immobilized in the surface of magnetic particles. The enzyme catalyzes the hydrolysis of benzoyl arginine ethyl ester (BAEE) into benzoyl arginine (BA) and ethyl alcohol (A) and the following kinetic data are gathered under conditions at which external diffusional restrictions are negligible:

[BAEE] (mM)	v (mmol $\cdot$ h^{-1} $\cdot$ cm^{-2})	
	[A] = 0 mM	[A] = 30 mM
15	27.78	24.19
30	35.71	32.61
45	39.47	36.89
60	41.67	39.47
90	44.12	42.45

Determine the intrinsic kinetic parameters of the enzyme and analyze the accuracy of the results obtained. If under the conditions of use, the film transport coefficient of BAEE is 10 cm $\cdot$ min^{-1} and the concentration of BAEE in the bulk liquid medium is 120 mM, determine the effectiveness factor of the catalyst (assume that inhibition by A is negligible).

Resolution: By linear regression of the data using Lineweaver–Burk or Eaddie–Hofstee plots (see Section 2.6), the following results are obtained:

- Ethyl alcohol is a competitive inhibitor of trypsin.
- Maximum reaction rate of BAEE hydrolysis: $V_{max} = 50$ mmol $\cdot$ h^{-1} $\cdot$ cm^{-2}.
- Michaelis constant for BAEE: $K_M = 12$ mM.
- Competitive inhibition constant by ethanol $K_I = 91.5$ mM (inhibition is mild).
- Results may have significant error: the K_M value obtained is outside the substrate range covered and K_I is determined from only one measurement (at 30 mM ethyl alcohol). The substrate concentration should be extended to the 1–10 mM range and more ethyl alcohol concentrations (at least three more) are recommended.

Data for the determination of the effectiveness factor of the catalyst:

$$h' = 10 \text{ cm} \cdot \text{min}^{-1}$$
$$V_{max} = 50 \text{ mmoles} \cdot \text{h}^{-1} \cdot \text{cm}^{-2} = 0.833 \text{ mmoles} \cdot \text{min}^{-1} \cdot \text{cm}^{-2}$$
$$K_M = 12 \text{ mM} = 0.012 \text{ mmoles} \cdot \text{cm}^{-3}$$
$$\beta_0 = \frac{s_0}{K_M} = \frac{120}{12} = 10$$

From Equation 3.15:

$$\alpha = \frac{0.833}{10 \cdot 0.012} = 6.94$$

From Equation 3.16:

$$\beta_S = \frac{2.06 + \sqrt{44.24}}{2} = 4.36$$

From Equation 3.17:

$$\eta = \frac{4.36 \cdot 11}{10 \cdot 5.36} = 0.895$$

Answer: The effectiveness factor is rather high (close to 90%), despite a rather large α. This is because the substrate concentration is high. The following table shows how η varies according to β_0:

β_0	β_S	η
1	0.14	0.246
2.5	0.43	0.421
5	1.21	0.657
10	4.36	0.895
15	8.77	0.958
20	13.54	0.978

At low bulk substrate concentrations the system is severely limited by EDR.

3.5 Pectinesterase (pectin methylesterase, EC 3.1.1.11) catalyzes the hydrolysis of pectin (P) into pectate and methanol [24]. The following results are obtained with a pectinesterase from *Penicillium occitani* immobilized on the surface of spherical particles of activated zirconia:

[P]	0.05	0.15	0.25	0.5	1	1.5	2.5	5	25	37.5	50	75	100
v	0.76	2.21	3.63	6.98	12.68	16.97	22.09	26.41	29.38	29.59	29.69	29.79	29.85

[P]: pectin concentration in the bulk reaction medium $(g \cdot L^{-1})$
v: initial reaction rate $(mg_{methanol} \cdot min^{-1} \cdot g_{catalyst}^{-1})$

Determine the intrinsic kinetic parameters and evaluate the effectiveness factor of the catalytic particle as a function of bulk substrate concentration in the liquid medium.

Resolution: Intrinsic kinetic parameters are to be determined at conditions under which EDR are negligible, so data obtained at high bulk substrate concentrations should be used (see Section 3.4.1). The value of [P] over which EDR are negligible is not apparent. If only the highest values of [P] are considered, much error will be introduced, since the differences in reaction rates under such conditions are quite small. If more data points down to lower [P] are considered, EDR will exist and the kinetic parameters determined will be apparent. It is thus necessary to make a judicious choice of the range of [P] that will be considered in determining the intrinsic parameters. From an inspection of the data provided, EDR are significant below $25 \, g \cdot L^{-1}$, but the fewer points considered above $25 \, g \cdot L^{-1}$, the higher the error. The best choice is thus the range of [P] from 25 to $100 \, g \cdot L^{-1}$. The rate data for this range are:

$[P] \, (g \cdot L^{-1})$	$v(mg_{methanol} \cdot min^{-1} \cdot g_{catalyst}^{-1})$	$[P]^{-1} \, (L \cdot g^{-1})$	$v^{-1}(g_{catalyst} \cdot min \cdot mg_{methanol}^{-1})$
25	29.38	0.0400	0.0340
37.5	29.59	0.0267	0.0338
50	29.69	0.0200	0.0337
75	29.79	0.0133	0.0336
100	29.85	0.0100	0.0335

By linear regression of the data in a double reciprocal plot (see Section 3.4.1), the following values for the intrinsic kinetic parameters are obtained:

$$V_{max} = 30.0 (mg_{methanol} \cdot g_{catalyst}^{-1} \cdot min^{-1})$$

$$K_M = 0.48 \, g \cdot L^{-1}$$

$$Slope_{zone1} = 0.016$$

with a correlation coefficient of 0.995, which is acceptable.

Considering that the enzyme is saturated with P in this range, the enzyme-specific activity can be calculated as:

$$activity = \frac{30 \, mg_{methanol} \cdot g_{catalyst}^{-1} min^{-1}}{32 \, mg_{methanol} \cdot mmole^{-1}} = 938 \, IU \cdot g_{catalyst}^{-1}$$

where IU is defined as the amount of enzyme producing 1 μmole of methanol per minute.

In order to assess the impact of EDR and determine the value of α, data in the lower range of [P] should be used. According to the procedure described in Section 3.4.1, the slope of the linear correlation obtained in such a range will allow α to be determined according to Equation 3.47. By inspecting the data provided, a sound linear correlation is obtained at values of [P] up to $1.5 \, g \cdot L^{-1}$. The rate data for this range are as follows:

[P] $(g \cdot L^{-1})$	$v(mg_{methanol} \cdot min^{-1} \cdot g_{catalyst}^{-1})$	$[P]^{-1} (L \cdot g^{-1})$	$v^{-1}(g_{catalyst} \cdot min \cdot mg_{methanol}^{-1})$
0.05	0.76	20	1.316
0.15	2.21	6.667	0.453
0.25	3.63	4	0.276
0.5	6.98	2	0.143
1	12.68	1	0.0789
1.5	16.97	0.667	0.0589

By linear regression of the data in a double reciprocal plot (see Section 3.4.1), one gets:

$$Slope_{zone\,2} = \frac{K_M}{V'_{max}}(\alpha + 1) = 0.065$$

and since:

$$Slope_{zone\,1} = \frac{K_M}{V'_{max}} = 0.016$$

$$\alpha = 3.1$$

Answer: Substituting in Equation 3.18:

β_0	0.1	0.5	1	2.5	5	10	20	50
η	0.215	0.288	0.375	0.599	0.823	0.952	0.988	0.998

At substrate concentrations higher than $20 \, K_M$, the system is essentially free of EDR.

3.6 Carbonic anhydrase (carbonate hydro-lyase, EC 4.2.1.1) is a Zn^{++}-containing lyase that catalyzes the reversible hydration of carbon dioxide: $CO_2 + H_2O \rightarrow HCO_3^- + H^+$. The following kinetic results have been obtained with bovine carbonic anhydrase immobilized in the surface of chitosan-covered magnetite particles at various CO_2 concentrations:

$[CO_2]$ (μM)	20	40	60	80	100
v ($\mu mol \cdot h^{-1} \cdot cm^{-2}$)	5.5	8.9	10.9	12.3	13.3

v: initial rate of CO_2 hydration

Using the Hanes linearization method ($[CO_2]/v$ versus $[CO_2]$), determine the kinetic parameters of the enzyme.

If these parameters can be considered intrinsic, determine the effectiveness factor of the immobilized carbonic anhydrase if the film mass-transfer coefficient of CO_2 is $2 \, cm \cdot min^{-1}$ and the concentration of dissolved CO_2 is $50 \, \mu M$.

Determine the magnitude of the Damkoehler number if the effectiveness factor of the catalytic particle cannot be less than 0.7.

Resolution: The values of the kinetic parameters, determined by linearization according to Hanes (see Section 2.6), are:

$$V_{max} = 19.9 \, \mu moles \cdot h^{-1} \cdot cm^{-2}$$
$$K_M = 49.8 \, mM$$

with a correlation coefficient close to 1. The Damkoehler number is:

$$\alpha = \frac{19.9 \, \mu moles \cdot h^{-1} \cdot cm^{-2}}{0.0498 \, \mu moles \cdot cm^{-3} \cdot 120 \, cm \cdot h^{-1}} = 3.33$$

$$\beta_0 = \frac{50}{49.8} = 1.00$$

Answer: From Equation 3.18:

$$\eta = 0.434$$

From Equation 3.16:

$$\beta_S = \frac{-\alpha + \sqrt{\alpha^2 + 4}}{2}$$

From Equation 3.17:

$$0.7 = \frac{2 \cdot \beta_S}{1 + \beta_S}$$

$$\beta_S = 0.539 = \frac{-\alpha + \sqrt{\alpha^2 + 4}}{2}$$

$$\alpha = 1.32$$

The maximum value of the Damkoehler number that will attain an effectiveness factor of 0.7 is 1.32. To comply with this, either the film mass-transfer coefficient must be increased (i.e. by increasing the agitation rate) or the catalytic potential of the enzyme $V'_{max} \cdot K_M^{-1}$ must be reduced (i.e. by immobilizing less protein, which may be inconvenient from a technological perspective).

3.7 The enzyme β-lactamase catalyzes the hydrolysis of penicillin G to harmless penicilloic acid, which is a well-known mechanism for resistance to penicillins. A β-lactamase from *Bacillus cereus* is immobilized in the surfaces of nonporous polystyrene beads; the enzyme is inhibited at high concentrations of penicillin G. Working with moderate penicillin G concentrations and at high agitation rates, the initial reaction rates of penicillin G hydrolysis are measured and the intrinsic values of V'_{max} and K_M are determined by nonlinear regression to be $500\ \mu\text{moles} \cdot \text{h}^{-1} \cdot \text{g}_{catalyst}^{-1}$ and $10\,\text{mM}$, respectively. The maximum reaction rate attainable is $264\ \mu\text{moles} \cdot \text{h}^{-1} \cdot \text{g}_{catalyst}^{-1}$. Determine the intrinsic inhibition constant of the immobilized β-lactamase for penicillin G.

Under moderate agitation, the system becomes diffusion limited, with a Damkoehler number of 5. Find the mathematical expression for the dimensionless penicillin G concentration at the surface of the catalyst as a function of the bulk liquid penicillin G concentration.

Resolution: The kinetic expression of uncompetitive inhibition at high substrate concentrations is represented by Equation 2.57 and the maximum initial reaction rate corresponds to the value obtained by setting $\dfrac{dv'}{ds} = 0$, so:

$$s^* = \sqrt{K_M \cdot K_S} \qquad v'^* = \frac{V'_{max} \cdot s^*}{K_M + s^* + \frac{s^{*2}}{K_S}}$$

where s^* is the substrate concentration maximizing v' (v'^*). Then:

$$v'^* = \frac{V_{max} \cdot \sqrt{K_M \cdot K_S}}{K_M + \sqrt{K_M \cdot K_S} + K_M} \qquad 264 = \frac{500 \cdot \sqrt{10 \cdot K_S}}{20 + \sqrt{10 \cdot K_S}}$$

$$K_S = 50\,\text{mM}$$

In the case of uncompetitive inhibition by high substrate concentration, Equation 3.13 becomes:

$$h(s_0 - s_S) = \frac{V'_{max} \cdot s_S}{K_M + s_S + \frac{s_S^2}{K_S}} \qquad \beta_0 - \beta_S = \frac{\alpha \cdot \beta_S}{1 + \beta_S + \kappa \cdot \beta_S^2}$$

$$(\beta_0 - \beta_S) \cdot (1 + \beta_S + 0.2 \cdot \beta_S^2) = 5 \cdot \beta_S$$

Answer: The intrinsic inhibition constant for penicillin G is $50\,\text{mM}$. This value can be considered intrinsic to being obtained from rate data at a high agitation rate.

From the preceding equation, the mathematical expression for the dimensionless penicillin G concentration at the surface of the catalyst (β_S) as a function of its bulk liquid concentration (β_0) is:

$$0.2 \cdot \beta_S^3 + \beta_S^2 \cdot (1 - 0.2 \cdot \beta_0) + \beta_S \cdot (6 - \beta_0) = 5$$

This is a third-grade equation, so three values of β_S will be obtained for each value of β_0. The correct value of β_S must be determined from a stability analysis. Hysteresis will occur, so the correct value will depend on the trajectory of β_S; in the case of reactor operation, β_S will always decrease, either in time (in a batch operation) or with respect to the position in the catalyst bed (in the case of a continuous-column reactor).

3.8 The intrinsic kinetic parameters of an immobilized β-galactosidase (EC 3.2.1.23) from *Aspergillus oryzae* in polyacrylamide gel are: $V''_{max} = 250\ \mu$moles of hydrolyzed lactose $\cdot$ min$^{-1} \cdot$ g$^{-1}_{catalyst}$ and $K_M = 58$ mM. The diffusion coefficient of lactose in the gel is $5.1 \cdot 10^{-2}$ cm$^2 \cdot$ min^{-1}. Determine the maximum size of catalyst particle that can be used to obtain a global effectiveness factor greater than 90%. The density of the polyacrylamide gel is 1.1 g $\cdot$ mL^{-1}. Assume first-order kinetics.

Resolution:

$$K_M = 0.058 \text{ mmoles} \cdot \text{cm}^{-3}$$
$$V''_{max} = 0.25 \text{ mmoles} \cdot \text{min}^{-1} \cdot \text{g}^{-1} = 0.275 \text{ mmoles} \cdot \text{min}^{-1} \cdot \text{cm}^{-3}$$
$$D_{eff} = 5.1 \cdot 10^{-2} \text{ cm}^2 \cdot \text{min}^{-1}$$

From Equation 3.23:

$$\Phi = \frac{R_P}{3} \sqrt{\frac{0.275}{0.058 \cdot 5.1 \cdot 10^{-2}}} = 3.214 R_P$$

From Equation 3.40:

$$0.9 < \frac{1}{3.214 \cdot R_P} \cdot \left[\frac{1}{\tanh(9.642 \cdot R_P)} - \frac{1}{9.642 \cdot R_P} \right]$$

Answer: Solving the inequation gives $R_P < 1.38$ mm.

3.9 The enzyme β-glucosidase (EC 3.2.1.21) catalyzes the reaction:

$$\text{Cellobiose} + \text{H}_2\text{O} \rightarrow 2\text{Glucose}$$

A thermostable β-glucosidase from *Pyrococcus furiosus* is immobilized in controlled-pore ceramic particles and the following kinetic data are obtained at 70 °C:

[C]	20	50	100	250	400	500
v	60.9	120.7	179.5	253.6	282.8	294.1

[C]: cellobiose concentration in the bulk liquid (mM)
v: initial rate of cellobiose hydrolysis (μmoles $\cdot$ min$^{-1} \cdot$ g$^{-1}_{catalyst}$)

Using nonlinear regression of such data, determine the apparent kinetic parameters of the immobilized enzyme and discuss the accuracy of the results obtained.

If V''_{max} is not affected by immobilization but the apparent K_M increases by 30% with respect to the intrinsic value, determine the maximum catalyst particle volume that can be used if the global effectiveness factor must be at least 0.8. The diffusion coefficient of lactose in water at that temperature has been determined as $9 \cdot 10^{-2}\,cm^2 \cdot min^{-1}$. The controlled-pore ceramic has a density of $1.8\,g \cdot mL^{-1}$, a void fraction of 0.6, and a pore relative tortuosity of 1.2. Assume in this case first-order kinetics with respect to cellobiose concentration.

Resolution: The apparent kinetic parameters are determined by nonlinear regression of the data, assuming Michaelis–Menten kinetics:

$$K_M = \frac{95}{1.3} = 73.1\,mM = 0.0731\,mmoles \cdot cm^{-3}$$
$$V''_{max} = 0.63\,mmoles \cdot min^{-1} \cdot cm^{-3}$$

From Equation 3.49:

$$D_{eff} = 9 \cdot 10^{-2} \cdot \frac{0.6}{1.2} = 4.5 \cdot 10^{-2}\,cm^2 \cdot min^{-1}$$

From Equation 3.23:

$$\Phi = \frac{R_P}{3}\sqrt{\frac{0.63}{0.0731 \cdot 4.5 \cdot 10^{-2}}} = 4.61\,R_P$$

From Equation 3.40:

$$0.8 < \frac{1}{4.61 \cdot R_P} \cdot \left[\frac{1}{\tanh(13.83 \cdot R_P)} - \frac{1}{13.83 \cdot R_P}\right]$$

Answer:

$$K_{MAPP} = 95\,mM = 0.095\,mmoles \cdot cm^{-3}$$
$$V''_{max} = 0.35\,mmoles \cdot min^{-1} \cdot g^{-1} = 0.63\,mmoles \cdot min^{-1} \cdot cm^{-3}$$

The substrate range is adequate, but more data points are recommended.

Solving the inequation gives $R_P < 1.48\,mm$.

3.10 An enzyme electrode, designed for the determination of penicillin during fermentation, is constructed based on penicillinase (penicillin amido-β-lactamhydrolase, EC 3.5.2.6) from *Bacillus cereus* immobilized in a polypropylene membrane of dimensions $1 \times 0.5 \times 0.1\,cm$ coupled to a glass electrode (the enzyme catalyzes the hydrolysis of penicillin to penicilloic acid). The V''_{max} and K_M obtained with this enzyme are $1\,\mu mol \cdot mL^{-1} \cdot s^{-1}$ and $50\,mM$, respectively. The diffusion coefficient for penicillin in the membrane is $10^{-10}\,m^2 \cdot s^{-1}$. Determine the global effectiveness factor, assuming first-order kinetics for penicillin hydrolysis. Make any other assumptions explicit.

Resolution:

$$K_M = 50\,\text{mM} = 0.05\ \text{mmoles} \cdot \text{cm}^{-3}$$
$$V''_{max} = 1\,\mu\text{mole} \cdot \text{cm}^{-3} \cdot \text{s}^{-1} = 0.001\ \text{mmoles} \cdot \text{cm}^{-3} \cdot \text{s}^{-1}$$
$$D_{eff} = 10^{-6}\ \text{cm}^2 \cdot \text{s}^{-1}$$
$$L_{eq} = \frac{\text{membrane volume}}{\text{membrane surface area}} = \frac{1 \cdot 0.5 \cdot 0.1}{2(1 \cdot 0.5 + 1 \cdot 0.1 + 0.5 \cdot 0.1)} = 0.0385\ \text{cm}$$

Given that the kinetic parameters are intrinsic, from Equation 3.23:

$$\Phi = 0.0385\sqrt{\frac{0.001}{0.05 \cdot 10^{-6}}} = 5.45$$

An approximate global effectiveness factor can be determined if the electrode membrane shape is approached to the catalytic slab, assuming first-order kinetics for the reaction of penicillin hydrolysis. Then, from Equation 3.38:

$$\eta_G = \frac{\tanh 2.725}{2.725} = 0.364$$

Answer: The global effectiveness factor of the electrode membrane can be estimated at 0.364. Severe IDR are present, which is an asset in this case because it allows a more linear electrode response with respect to penicillin concentration.

3.11 Galactose oxidase (EC 1.1.3.9) can catalyze the oxidation by molecular oxygen of glycerol to optically pure S(-)glyceraldehyde. Working with a galactose oxidase preparation from *Dactylium dendroides* immobilized in controlled-pore glass spherical particles, the first-order kinetic rate constant is $0.1\,\text{s}^{-1}$. The estimated diffusion coefficient of glycerol in the pores of the catalyst is $2 \cdot 10^{-5}\ \text{cm}^2 \cdot \text{s}^{-1}$ and the film mass-transfer coefficient of glycerol in the liquid is $3.3 \cdot 10^{-4}\ \text{cm} \cdot \text{s}^{-1}$. Within a range of particle diameters between 0.1 and 1.0 mm, determine the global effectiveness factors of the catalyst particles and compare their values with those obtained under negligible EDR. Make any assumptions explicit.

Resolution: In this case, both EDR and IDR are present. From the data and Equations 3.23 and 3.43:

$$\Phi = \frac{d_P}{6}\sqrt{\frac{0.1}{2 \cdot 10^{-5}}} = 11.785 \cdot d_P$$
$$Bi = \frac{3.3 \cdot 10^{-4} \cdot d_P}{6 \cdot 2 \cdot 10^{-5}} = 2.75 \cdot d_P$$

where d_P is the diameter (in cm) of the catalyst particle.

From Equation 3.45:

$$\eta_G = \frac{2.75\left(\dfrac{1}{\tanh(35.36 \cdot d_P)} - \dfrac{1}{35.36 \cdot d_P}\right)}{11.785\left(2.75 \cdot d_P - 1 + \dfrac{35.36 \cdot d_P}{\tanh(35.36 \cdot d_P)}\right)}$$

In the absence of significant EDR, from Equation 3.40:

$$\eta_{G0} = \frac{1}{11.785 \cdot d_P} \cdot \left[\frac{1}{\tanh(35.36 \cdot d_P)} - \frac{1}{35.36 \cdot d_P}\right]$$

Answer: From the expressions of the global effectiveness factors in the presence (η_G) and absence (η_{G0}) of significant EDR:

d_P (cm)	0.01	0.02	0.03	0.04	0.05	0.06	0.07	0.08	0.09	0.10
η_G	0.414	0.253	0.182	0.141	0.116	0.098	0.085	0.075	0.067	0.060
η_{G0}	0.992	0.968	0.932	0.888	0.839	0.789	0.740	0.693	0.650	0.610
$\eta_{G0} \cdot \eta_G^{-1}$	2.40	3.83	4.78	6.30	7.23	8.05	8.71	9.24	9.70	10.17

The term $\eta_{G0} \cdot \eta_G^{-1}$ represents the relative magnitude of EDR; this magnitude increases with the increase in particle size, meaning that EDR are more significant at higher particle sizes. Of course, the magnitude of EDR will vary according to the fluid regime in the system and can be reduced by increasing the agitation rate.

3.12 The enzyme succinate dehydrogenase (EC 1.3.99.1) catalyzes the reduction of succinate into fumarate and is inhibited competitively by malonate. For the production of fumarate, whole cells of *Escherichia coli* immobilized in the outer layer of spherical polyacrylamide-coated ceramic particles are used. Determine the expression of the effectiveness factor as a function of the Damkoehler number for a system in which the concentrations of succinate and malonate in the bulk liquid are equal to the respective dissociation constants of their enzyme secondary complexes.

Resolution: For an enzyme subjected to product competitive inhibition, Equation 3.13 now becomes:

$$h(s_0 - s_S) = \frac{V'_{max} \cdot s_S}{K_M\left(1 + \dfrac{p}{K_P}\right) + s_S} \tag{3.53}$$

where p is the inhibitory product concentration (malonate in this case) and K_P the corresponding product competitive inhibition constant. In dimensionless terms:

$$\beta_0 - \beta_S = \frac{\alpha \cdot \beta_S}{1 + \gamma_S + \beta_S} \tag{3.54}$$

Assuming there is no product accumulation in the catalyst surface: $\gamma_S = \gamma_0 = 1$ and $\beta_0 = 1$. From Equation 3.54:

$$\beta_S^2 + \beta_S \cdot (\alpha + 1) - 2 = 0 \tag{3.55}$$

The effectiveness factor under product inhibition (η_P) can be conveniently defined as the ratio of effective reaction rate to reaction rate in the absence of diffusional restrictions and inhibition. So:

$$\eta_P = \frac{\beta_S(1 + \beta_0)}{\beta_0(1 + \gamma_S + \beta_S)} = \frac{2 \cdot \beta_S}{2 + \beta_S} \tag{3.56}$$

Answer: From Equations 3.55 and 3.56:

α	0	0.1	0.3	0.5	1	2	5	10	50
β_S	1	0.967	0.906	0.851	0.732	0.562	0.317	0.179	0.039
η_P	0.67	0.65	0.63	0.60	0.53	0.44	0.27	0.16	0.04

The value of η_P at $\alpha = 0$ denotes the impact of product inhibition alone, since no EDR are present. The effect of mass-transfer limitations becomes less pronounced under enzyme inhibition, as a consequence of the lower activity of the catalyst that moves the system away from mass-transfer limitations.

Suppose now that the same cells are immobilized in spherical polyacrylamide gel particles of a diameter equivalent to 18 times the length of the stagnant liquid layer at the outside of the particle under the operating conditions. The diffusion coefficient of succinate in polyacrylamide gel can be estimated as 50% of its value in water. Evaluate the global effectiveness factor as a function of the Thièle modulus. Assume the same bulk substrate concentration as before and that first-order kinetics applies.

Resolution: The film mass-transfer coefficient h' is now expressed as:

$$h' = \frac{D_0}{\delta}$$

In the presence of inhibition, the Thièle modulus (Φ_P) for the spherical particles can be expressed as:

$$\Phi_P = \frac{d_P}{6} \sqrt{\frac{V''_{max}}{K_{MAP} \cdot D_{eff}}}$$

In this case, K_{MAP} refers to the value for product inhibition, so:

$$K_{MAP} = K_M(1 + \gamma)$$

$$\frac{\Phi_P}{\Phi} = \sqrt{\frac{K_M}{K_{MAP}}} = \sqrt{\frac{1}{1 + \gamma}} = 0.707$$

From Equation 3.43:

$$Bi = \frac{D_0 \cdot L_{eq}}{\delta \cdot D_{eff}} = \frac{D_0}{D_{eff}} \cdot \frac{d_P}{6 \cdot \delta} = 2 \cdot \frac{18}{6} = 6$$

Assuming first-order kinetics, from Equation 3.45:

$$\eta_G = \frac{Bi\left(\dfrac{1}{\tanh(3\Phi_P)} - \dfrac{1}{3 \cdot \Phi_P}\right)}{\Phi_P\left(Bi - 1 + \dfrac{3\Phi_P}{\tanh(3\Phi_P)}\right)} = \frac{6}{0.707 \cdot \Phi} \cdot \frac{\dfrac{1}{\tanh(2.12\Phi)} - \dfrac{1}{2.12\Phi}}{5 + \dfrac{2.12\Phi}{\tanh(2.12\Phi)}} \tag{3.57}$$

Answer: From Equation 3.57:

Φ	0.1	0.5	1	2	5	10	15	20	50
η_G	0.994	0.881	0.659	0.351	0.099	0.031	0.015	0.009	0.0015

3.13 An invertase (β-fructofuranosidase, EC 3.2.1.26) is immobilized in cubic particles of polyvinyl alcohol gel. The diffusion coefficient of sucrose inside the gel is $7 \cdot 10^{-9}$ $m^2 \cdot s^{-1}$, while the film volumetric mass-transfer coefficient is $8 \cdot 10^{-3} cm^3 \cdot s^{-1}$. Determine the Biot number as a function of the cube edge.

Resolution:

$$\begin{aligned} D_{eff} &= 7 \cdot 10^{-9}(m^2 \cdot s^{-1}) = 4.2 \cdot 10^{-3}(cm^2 \cdot min^{-1}) \\ h &= 8 \cdot 10^{-3}(cm^3 \cdot s^{-1}) = 0.48(cm^3 \cdot min^{-1}) \\ A_{cube} &= 6 \cdot a^2 (cm^2) \\ L_{eq,cube} &= \frac{a^3}{6 \cdot a} = \frac{a}{6}(cm) \end{aligned}$$

Answer: From Equation 3.43:

$$Bi = \frac{h' \cdot L_{eq}}{D_{eff}} = \frac{h \cdot L_{eq}}{A_{cube} \cdot D_{eff}} = \frac{0.48 \cdot \dfrac{a}{6}}{6 \cdot a^2 \cdot 4.2 \cdot 10^{-3}} = \frac{3.175}{a}$$

where a is in cm. As can be seen, the Biot number (Bi) will be much higher than 1 (negligible EDR) for the usual dimensions of a catalyst particle. To make the Biot number much lower than 1 (i.e. 0.1—significant EDR), the catalyst cubic particle should have an edge greater than 31.75 mm, which is certainly large.

3.14 Determine the global effectiveness factor for a spherical immobilized enzyme particle with Michaelis–Menten intrinsic kinetics as a function of the bulk dimensionless substrate concentration and Thièle modulus.

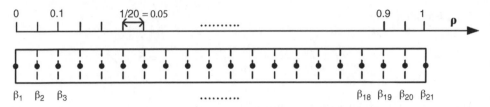

Scheme 3.1 *Finite difference approach for a catalytic sphere.*

Resolution: The first step is to determine the substrate concentration profile inside the catalytic sphere (see Equation 3.27). To do so, Equation 3.22 must be solved using the finite difference method, as described in Section A.4.1. Scheme 3.1 represents the system given 21 nodes.

Using backwards finite differences and applying the boundary condition for node 1 gives:

$$\beta_1 = \beta_2$$

Using centered finite differences for nodes 2–20 gives:

$$\frac{\beta_{i+1} - 2 \cdot \beta_i + \beta_{i-1}}{h^2} + \frac{2}{\rho} \cdot \frac{(\beta_{i+1} - \beta_{i-1})}{2 \cdot h} - 9 \cdot \Phi^2 \cdot \frac{\beta_i}{1 + \beta_i} = 0$$

Applying the boundary condition for node 21 gives:

$$\beta_{21} = \beta_0$$

In this way, finite differences build up a system of nonlinear equations that can be solved by Newton's method (see Section A.4.1). The following table shows the results obtained with $\beta_0 = 10$ and $\Phi = 3$:

Position	1	2	3	4	5	6	7	8	9	10	11
ρ	0.00	0.05	0.10	0.15	0.20	0.25	0.30	0.35	0.40	0.45	0.50
β	0.487	0.487	0.532	0.606	0.712	0.853	1.034	1.259	1.535	1.866	2.257
η_i	0.360	0.360	0.382	0.415	0.458	0.506	0.559	0.613	0.666	0.716	0.762
Position	12	13	14	15	16	17	18	19	20	21	
ρ	0.55	0.60	0.65	0.70	0.75	0.80	0.85	0.90	0.95	1	
β	2.711	3.232	3.822	4.483	5.217	6.024	6.906	7.862	8.893	10	
η_i	0.804	0.840	0.872	0.899	0.923	0.943	0.961	0.976	0.989	1.000	

Once the substrate profile inside the catalyst particle has been determined, the local effectiveness factor for each node i is determined according to Equation 3.28:

$$\eta_i = \frac{\beta_i(1 + \beta_0)}{\beta_0(1 + \beta_i)}$$

The global effectiveness factor is then calculated according to Equation 3.32. Given an odd number of nodes (21), the integration of such an equation is carried out using

Simpson's 1/3 rule:

$$\int_{x_1}^{x_n} f(x)dx = \frac{h}{3} \cdot \left(f(x_1) + 4 \cdot \sum_{i=2,4,6...}^{n-1} f(x_i) + 2 \cdot \sum_{i=3,5,7...}^{n-2} f(x_i) + f(x_n) \right)$$

Therefore the global effectiveness factor is calculated as:

$$\eta_G = 3 \cdot \int_0^1 \eta \cdot \rho^2 \, d\rho \approx 3 \cdot \left(\frac{0.05}{3} \cdot (0.36 \cdot 0^2 + 4 \cdot (0.36 \cdot 0.05^2 + 0.415 \cdot 0.15^2 + \cdots + 0.989 \cdot 0.95^2) \right.$$
$$\left. + 2 \cdot (0.382 \cdot 0.1^2 + 0.458 \cdot 0.2^2 + \cdots + 0.976 \cdot 0.9^2) + 1 \cdot 1^2) \right) = 0.896$$

This means that under the conditions evaluated ($\beta_0 = 10$ and $\Phi = 3$), the catalytic potential of the enzyme is 90% of that in the absence of IDR.

Answer: Repeating this procedure for values of β_0 from 0.3 to 20, results of η_G at different values of Φ are obtained, as shown in Figure 3.7. Note that η_G decreases with the course of the reaction as X increases. Φ is considered constant throughout the reaction, since it only depends on kinetic and mass-transfer parameters. This will no longer be the case when enzyme inactivation is considered (see Section 6.4).

Numerical methods allow problems such as Equation 3.22 to be solved without using an analytical method, simply by computer calculation. However, the limitations of numerical solutions must always be considered. For instance, when using Simpson's 1/3 rule to solve Equation 3.32, η_G will approach the value of h for high values of Φ, since the last node will always have a value of $\eta = 1$ because of the boundary condition at $\rho = 1$. If 21 nodes are used to estimate the η_G of a catalyst following first-order kinetics with respect to substrate concentration and a value of $\Phi = 100$, a value close to 0.05 is obtained according to

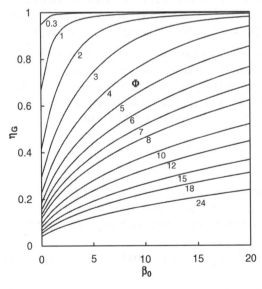

Figure 3.7 *Global effectiveness factor (η_G) of a spherical catalyst particle with an immobilized enzyme with intrinsic Michaelis–Menten kinetics. β_0: dimensionless bulk substrate concentration.*

Simpson's 1/3 rule, which is five times the true value. This difference is caused by the small number of nodes considered. As a general rule, the number of nodes must be greater than $2 \cdot \Phi$ for $\Phi > 10$, which means at least 200 nodes must be considered.

3.15 A β-galactosidase of *Aspergillus oryzae* is immobilized in glyoxyl agarose spherical gel particles. In order to evaluate the immobilization process, catalysts are characterized in terms of their hydrolytic activity on the artificial substrate o-nitrophenyl-galactoside (*o*-NPG), measured at 15 mM *o*-NPG, pH 4.5, and 40 °C. The catalyst has $3000 \, \mathrm{IU} \cdot \mathrm{g}^{-1}$ of support (where 1 IU is the amount of enzyme that hydrolyzes 1 μmole of *o*-NPG under these conditions), a density of $0.96 \, \mathrm{g} \cdot \mathrm{cm}^{-3}$, and an average particle size of 150 μm. The soluble β-galactosidase has a V_{max} of $196\,000 \, \mu\mathrm{mol} \cdot \mathrm{g}^{-1} \cdot \mathrm{min}^{-1}$ and a K_M of 1.5 mM. It is suspected that the results are influenced by the presence of IDR. Estimate the catalyst's global effectiveness factor.

Resolution: First, the value of Φ must be determined, which means the values of the intrinsic kinetic parameters of the enzyme and the effective diffusion coefficient of the substrate must be known. Precise experimental determination of these parameters can be cumbersome, but it is possible to estimate them by making some conservative assumptions in order to reach a first approximation of a solution to the problem. The diffusion coefficient of *o*-NPG in water (D_0), in $\mathrm{cm}^2 \cdot \mathrm{s}^{-1}$, can be estimated according to the Wilke and Chang correlation [25]:

$$D_0 = \frac{5.065 \cdot 10^{-7} \cdot T}{\mu \cdot v_s^{0.6}} \tag{3.58}$$

where T is the temperature in K, μ is the viscosity of the solution in centipoises and v_s is the molar volume of the solute at the normal boiling point, in $\mathrm{cm}^3 \cdot \mathrm{mol}^{-1}$. The viscosity of the solution can be approximated to the viscosity of water, which under such conditions is 0.65 centipoises. The value of v_s is estimated at $293 \, \mathrm{cm}^3 \cdot \mathrm{gmol}^{-1}$ from the available data [25] and the empirical formula of *o*-NPG ($C_{12}H_{15}NO_8$) with one benzene ring.

Since glyoxyl agarose is a porous matrix, D_{eff} can be estimated from Equation 3.49. According to Engasser [26], typical values of ε are in the range from 0.4 to 0.8, while typical values of ζ are in the range from 1 to 2. Assuming $\varepsilon = 0.6$ and $\zeta = 1.5$, an estimate of D_{eff} of $3.2 \cdot 10^{-6} \, \mathrm{cm}^2 \cdot \mathrm{s}^{-1}$ can be obtained from Equation 3.49. V''_{max} can be estimated by approximating it to the value of the measured activity:

$$V''_{max} = 3000 \, \mu\mathrm{mol} \cdot \mathrm{g}^{-1} \cdot \mathrm{min}^{-1} \cdot 0.96 \, \mathrm{g} \cdot \mathrm{cm}^{-3} \cdot \frac{1}{60} \cdot \mathrm{min} \cdot \mathrm{s}^{-1} = 48 \, \mu\mathrm{mol} \cdot \mathrm{cm}^{-3} \cdot \mathrm{s}^{-1}$$

If the intrinsic Michaelis constant of the immobilized enzyme is assumed to be equal to the K_M of the soluble enzyme (this is arguable, but a reasonable assumption for a hydrophilic support with high geometric congruence such as glyoxyl agarose):

$$K_M = 1.5 \, \mathrm{mmol} \cdot \mathrm{L}^{-1} \cdot \frac{1}{1000} \, \mathrm{L} \cdot \mathrm{cm}^{-3} \cdot 1000 \, \mu\mathrm{moles} \cdot \mathrm{mmol}^{-1} = 1.5 \, \mu\mathrm{moles} \cdot \mathrm{cm}^{-3}$$

From Equation 3.23:

$$\Phi = \frac{(0.5 \cdot 0.015)}{3} \sqrt{\frac{48}{1.5 \cdot 3.2 \cdot 10^{-6}}} = 7.9$$

Product inhibition and thermal inactivation are irrelevant during activity measurement, so Michaelis–Menten kinetics can be assumed, allowing η_G to be determined from Figure 3.7 as a function of Φ and β_0. In the protocol for the determination of activity, the initial o-NPG concentration is 15 mM, so the dimensionless bulk substrate concentration β_0 is equal to 10.

Answer: From Figure 3.7, with $\Phi = 7.9$ and $\beta_0 = 10$, $\eta_G = 0.47$.

The reaction is under IDR, so an increase in the concentration of o-NPG in the activity assay is recommended so that the catalytic potential of the enzyme and the value of the immobilization yield are not underestimated.

3.16 Develop a model to evaluate the effect of EDR and IDR on the enantioselectivity of the kinetic resolution of a racemic mixture of ibuprofen esters. The hydrolysis of the esters is catalyzed by *Candida antarctica* lipase immobilized in spherical particles of an organic polymer. The reactions for the slow-reacting R-enanantiomer and the fast reacting S-enantiomer can be considered first-order with respect to the corresponding enantiomer concentration:

$Slow$ (R)-ibuprofen ester $\xrightarrow{\text{H}_2\text{O}}$ (R)-ibuprofen + alcohol

$Fast$ (S)-ibuprofen ester $\xrightarrow{\text{H}_2\text{O}}$ (S)-ibuprofen + alcohol

Resolution: A good parameter by which to evaluate the enantioselective potential of the enzyme is enantioselectivity (E), which measures the ratio of catalytic efficiency of the two enantiomers. In other words, E denotes how favored the production of one of the enantiomers is by the enzymatic action.

For first-order kinetics in the presence of diffusional restrictions, the observed E in the bulk liquid medium is defined as:

$$E_{obs} = \frac{\eta_G^S \cdot \left(\dfrac{V_{max}^{S''}}{K_M^S}\right)}{\eta_G^R \cdot \left(\dfrac{V_{max}^{R''}}{K_M^R}\right)} = \frac{\eta_G^S}{\eta_G^R} \cdot E \tag{3.59}$$

where the R and S supraindices represent the corresponding R and S enantiomeric ibuprofen esters. Equation 3.45 allows the η_G of the catalyst to be calculated for each reaction as a function of its Thièle modulus and Biot number:

$$\eta_G^S = \frac{Bi^S \left(\dfrac{1}{\tanh(3\Phi^S)} - \dfrac{1}{3 \cdot \Phi^S}\right)}{\Phi^S \left(Bi^S - 1 + \dfrac{3 \cdot \Phi^S}{\tanh(\Phi^S)}\right)} \qquad \eta_G^R = \frac{Bi^R \left(\dfrac{1}{\tanh(3\Phi^R)} - \dfrac{1}{3 \cdot \Phi^R}\right)}{\Phi^R \left(Bi^R - 1 + \dfrac{3 \cdot \Phi^R}{\tanh(\Phi^R)}\right)}$$

It is reasonable to assume that the effective diffusion coefficient (D_{eff}) and volumetric film mass-transfer coefficients (h) of the R and S esters are similar, so that $Bi^S = Bi^R$. However, the Thièle modulus will be different for the two esters, since it depends on the values of the kinetic parameters of the enzyme. So:

$$\Phi^S = \frac{R_P}{3} \cdot \sqrt{\frac{V_{max}^{S''}}{K_M^S \cdot D_{eff}}} \qquad \Phi^R = \frac{R_P}{3} \cdot \sqrt{\frac{V_{max}^{R''}}{K_M^R \cdot D_{eff}}} \qquad \Phi^R = \frac{\Phi^S}{\sqrt{E}}$$

The rate expressions for enantiomers S and R are represented by Equations 3.60 and 3.61, respectively:

$$\frac{d[S]}{dt} = -\eta_G^S \cdot \left(\frac{V_{max}^{S''}}{K_M^S}\right) \cdot [S] \tag{3.60}$$

$$\frac{d[R]}{dt} = -\eta_G^R \cdot \left(\frac{V_{max}^{R''}}{K_M^R}\right) \cdot [R] \tag{3.61}$$

The production profiles of both enantiomers are obtained by integration of Equations 3.60 and 3.61 through separation of the variables (see Section A.3.1):

$$\ln([S] \cdot [S_0]^{-1}) = -\eta_G^S \cdot (V_{max}^{S''} \cdot K_M^{S-1}) \cdot t \tag{3.62}$$

$$\ln([R] \cdot [R_0]^{-1} = \eta_G^R \cdot \left(V_{max}^{R''} \cdot K_M^{R-1}\right) \cdot t \tag{3.63}$$

Replacing Equations 3.62 and 3.63 into Equation 3.59, and given an initial racemic mixture ($[S_0] = [R_0]$):

$$E_{obs} = \frac{\eta_G^S}{\eta_G^R} \cdot E = \frac{\ln([S] \cdot [S_0]^{-1})}{\ln([R] \cdot [R_0]^{-1})} \tag{3.64}$$

Defining X as the fractional conversion of substrate and ee_S as the enantiomeric excess of the S-enantiomer gives:

$$X = 1 - \frac{[S] + [R]}{[S_0] + [R_0]} \qquad ee_s = \frac{[S] - [R]}{[S_0] + [R_0]}$$

and replacing this into Equation 3.64 gives:

$$E_{obs} = \frac{\eta_G^S}{\eta_G^R} \cdot E = \frac{\ln((1-X) \cdot (1 - ee_s))}{\ln((1-X) \cdot (1 + ee_s))} \tag{3.65}$$

Equation 3.65 allows as assessment of the effect of mass-transfer limitations on the value of ee_S attained in the resolution of racemic mixtures for first-order reaction kinetics. Enantiomeric excess is a very important parameter in the resolution of racemic mixtures and the synthesis of chiral compounds, since it is directly related to the purity of the product: usually only one enantiomer (eutomer) is the desired product, while the other (distomer) is either nonfunctional or, worse, actually detrimental to its intended use.

Answer: Figure 3.8 is obtained from Equation 3.65 and the expressions for η_G^S and η_G^R. As the figure shows, ee_S at a certain X is reduced by mass-transfer limitations, making E_{obs} significantly lower than E; it can also be seen that IDR have a stronger effect than EDR.

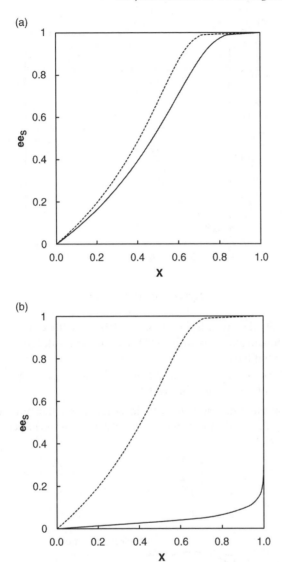

Figure 3.8 *Effect of external and internal diffusional restrictions on the enantiomeric excess (ee$_S$) of the resolution of a racemic mixture of R-S ibuprofen esters. (a) $\Phi = 1$, $Bi = 1$. (b) $\Phi = 10$, $Bi = 1$. (c) $\Phi = 1$, $Bi = 0.1$. $E = 10$. Solid line: ee$_S$ in the presence of diffusional restrictions; dashed line: ee$_S$ in the absence of diffusional restrictions. X: substrate conversion.*

The enzyme load of the catalyst is thus a critical variable in the resolution of racemic mixtures and the synthesis of chiral compounds. A high enzyme load implies an increase in V''_{max} and therefore in Φ (and α), strongly affecting E_{obs} and ee$_S$.

3.17 *Aspergillus niger* glucoamylase immobilized on the surface of glycidyl methacrylate particles is used to hydrolyze thinned starch in a batch recirculating reactor

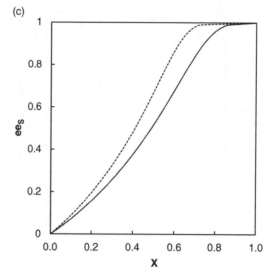

Figure 3.8 (Continued)

(see Figure 3.9). The results (see Figure 3.10) show the effect of the recirculation rate on the initial rate of hydrolysis. Discuss these results from the perspective of mass-transfer limitations.

Answer: Figure 3.10 suggests that the reactor behavior is conditioned by EDR. This assertion is sustained by the following facts:

• High recirculation rates promote an increase in the rate of reaction by reducing the stagnant layer surrounding the catalyst particles, thereby increasing the effectiveness factor.

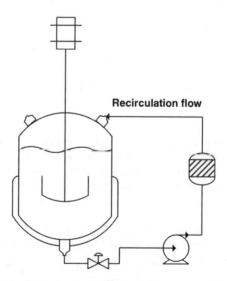

Figure 3.9 *Scheme of operation of a batch recirculating reactor for the hydrolysis of thinned starch with immobilized glucoamylase.*

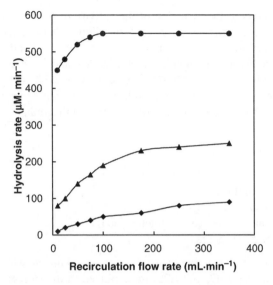

Figure 3.10 *Effect of the recirculation rate on the initial hydrolysis rate at different thinned starch concentrations (g · L^{-1}):* ◆: 1; ▲: 10; •: 100.

- The increase in recirculation rate from 10 to 350 mL · min^{-1} produces an increase in reaction rate of 800, 212, and 22.2% for 1, 10, and 100 g · L^{-1} initial thinned starch concentration, respectively. The lower increase at higher substrate concentration also provides evidence of the presence of EDR, since the incidence of EDR in the reaction rate decreases at higher substrate concentrations, as shown in Figure 3.4.

Supplementary Problems

3.18 Penicillin G acylase (PGA) from recombinant *Escherichia.coli* is immobilized in glyoxyl agarose 6BCL by contacting 4.9 g of activated support (density 0.7 g · mL^{-1}) with 3 mL of a PGA extract previously diluted with 20 mL of 100 mM bicarbonate buffer at pH 10. The enzyme activity and protein concentration obtained during immobilization are presented in the following table:

Time (h)	Activity (IU · mL^{-1})			Protein in Supernatant (mg · mL^{-1})
	Control	Supernatant	Suspension	
0	50	50	50	5
1	50	40	50	3.5
2		32	49	2.5
3	50	25	48	2.1
4		10	47	1.8
5	50	5	46	1.6
6		0	46	1.5
7	50	0	46	1.5

"Control" refers to the evolution of the activity of the free enzyme in solution, "supernatant" refers to the activity of the liquid phase after catalyst removal and "suspension" refers to the activity of the mixture of liquid phase and catalyst. Determine the protein immobilization yield (Y_P) and enzyme activity yield (Y_E), the specific activity of the catalyst (enzyme activity expressed per unit mass), the maximum support load (protein immobilized per unit mass of support), the specific activity of the PGA extract and the time of immobilization.

Answer:

$Y_P = 0.7$
$Y_E = 0.92$
specific activity $= 281.6\,\mathrm{IU \cdot g^{-1}}$
maximum support load $= 21.4\,\mathrm{mg \cdot g^{-1}}$
specific activity of initial enzymatic extract $= 500\,\mathrm{IU \cdot mL^{-1}}$
time of immobilization $= 6$ hours

3.19 The following initial reaction rate (v') data are obtained with a yeast invertase (β-fructofuranosidase, EC 3.2.1.26) immobilized in the surface of chitin flakes at varying initial sucrose concentrations [S]:

[S] (mM)	20	40	60	80	100
v' ($\mathrm{mmol \cdot h^{-1} \cdot cm^{-2}}$)	57	89	109	123	133

Determine the apparent kinetic parameters of the immobilized invertase. If the intrinsic maximum reaction rate is not affected by the immobilization and the apparent K_M is 50% higher than the value of the intrinsic K_M, determine the effectiveness factor of the catalyst if the film mass-transfer coefficient of sucrose under the experimental conditions is $2.5\,\mathrm{cm \cdot min^{-1}}$ and the bulk sucrose concentration is $200\,\mathrm{mM}$.

Answer:

$$V'_{max,app} = 200\,\mathrm{mmoles \cdot h^{-1} \cdot cm^{-2}}$$

$$K_{M,app} = 50\,\mathrm{mM}$$

$$\eta = 0.875$$

The effectiveness factor is rather high, despite a high Damkoehler number ($\alpha = 4$), because the bulk substrate concentration is also high ($\beta_0 = 6$).

3.20 Glucose oxidase (EC 1.1.3.4) immobilized on the surface of titanium oxide particles is characterized according to Michaelis–Menten kinetics. The intrinsic first-order kinetic rate constant is $2\,\mathrm{min^{-1}}$ and the first-order kinetic rate constant at very low glucose concentration is $0.182\,\mathrm{min^{-1}}$.

Evaluate the surface glucose concentration (β_S) and the effectiveness factor (η) of the catalytic particle as a function of the bulk glucose concentration (β_0). What is the percentage error (ε) in determining the glucose concentration at the catalyst surface when considering first-order instead of Michaelis–Menten kinetics?

Answer:

β_0	β_S	η	ε
0.1	0.0092	0.100	1.2
1	0.099	0.180	8.2
2.5	0.285	0.311	20.3
5	0.742	0.511	38.7
10	.70	0.803	66.3
20	10.84	0.961	83.2
30	20.47	0.985	86.7
50	40.24	0.995	88.7
100	90.11	0.999	89.9

The effectiveness factor increases steadily with bulk substrate concentration. The error also increases sharply with bulk substrate concentration as it progressively departs from the condition of validity of first-order kinetics.

3.21 Demonstrate that in the case of an enzyme subjected to uncompetitive inhibition by substrate and external diffusional restrictions, the maximum value of the effectiveness factor is:

$$\eta = (1 + \beta_0^{-1}) \cdot (1 + 2 \cdot \kappa^{0.5})^{-1}$$

where β_0 is the dimensionless substrate concentration in the liquid bulk and κ is the ratio of Michaelis constant to substrate inhibition constant.

3.22 The following results are obtained for the hydrolysis of *p*-nitrophenyl butyrate (*p*NPB) across a wide range of substrate concentrations with a lipase immobilized in octyl agarose:

[*p*NPB]	0.05	0.15	0.25	0.5	1	1.5	2.5
v	0.759	2.208	3.627	6.975	12.681	16.974	22.086
[*p*NPB]	5	25	37.5	50	75	100	
v	26.412	29.376	29.589	29.694	29.790	29.850	

[*p*NPB]: initial millimolar concentration of *p*NPB in the bulk liquid
v: initial rate of *p*-nitrophenol release (μmoles $\cdot$ min^{-1} $\cdot$ g$^{-1}_{catalyst}$)

Determine the intrinsic kinetic parameters of the immobilized enzyme, the value of the Damkoehler number, and the effectiveness factor as a function of the bulk liquid *p*NPB concentration.

Answer:

$$V_{max} = 30 \; \mu\text{moles} \cdot \text{min}^{-1} \cdot \text{g}^{-1}$$
$$K_M = 0.53 \; \text{mM}$$
$$\alpha = 2.7$$

[*p*NPB]	0.05	0.15	0.25	0.5	1	1.5	2.5
η	0.294	0.339	0.383	0.485	0.650	0.766	0.888
[*p*NPB]	5	25	37.5	50	75	100	
η	0.969	0.999	~1	~1	~1	~1	

The system is completely free of external diffusional restrictions at $[p\text{NPB}] > 15\,\text{mM}$.

3.23 The following results are obtained for glucoamylase (endo-1,4-α-D-glucan gluca-nohydrolase, EC 3.2.1.3) from *Aspergillus oryzae* (maltose → 2 glucose) immobilized in the surfaces of magnetic particles of 2 mm diameter in an agitated vessel at 250 rpm:

[M] (mM)	10	25	40	75	100
v ($\mu\text{mol} \cdot \text{min}^{-1} \cdot \text{cm}^{-2}$)	0.583	1.17	1.56	2.10	2.33

[M]: maltose concentration
v: initial rate of maltose hydrolysis

At $[\text{M}] = 10\,\text{mM}$, the following results are obtained:

Agitation Rate (rpm)	50	100	150	250	350
v ($\mu\text{mol} \cdot \text{min}^{-1} \cdot \text{cm}^{-2}$)	0.283	0.425	0.584	0.583	0.583

Using a Hanes plot ([M]/v versus [M]), determine the value of the intrinsic kinetic parameters of the immobilized enzyme and discuss the accuracy of the results obtained. Determine the effectiveness factor of the catalytic particle at bulk liquid [M] of 0.05, 0.1, 0.2, and 0.5 M if the linear film mass-transfer coefficient for maltose under the above conditions is $2.33 \cdot 10^{-4}\,\text{cm}^2 \cdot \text{s}^{-1}$ and discuss the results obtained.

Answer:

$$V_{\max} = 3.5\,\mu\text{moles} \cdot \text{min}^{-1} \cdot \text{cm}^{-2}$$
$$K_M = 50\,\text{mM}$$

[M] (mM)	β_0	β_S	η
50	1	0.19	0.32
100	2	0.45	0.47
200	4	1.24	0.69
500	10	5.74	0.94

From the data given, at 250 rpm the system can be considered free of EDR, so the parameters determined at such an agitation rate can be considered intrinsic. The maltose concentration range is adequate but more rate data is recommended. The Damkoehler number is 5, so EDR are significant at moderate maltose concentrations.

3.24 The enzyme pyruvate decarboxylase (EC 4.1.1.1) catalyzes the decarboxylation of pyruvate into acetaldehyde and CO_2. The following initial reaction rate data are obtained with pyruvate decarboxylase from *Candida utilis* immobilized in the surface of polypropylene particles at various pyruvate concentrations [P]:

[P] (mM)	20	40	60	80	100
v (μmoles $\cdot$ h^{-1} $\cdot$ g$^{-1}_{protein}$)	5.7	8.9	10.9	12.3	13.3

v: initial reaction rate of CO_2 production

The surface protein load density in the catalyst is $20\,mg_{protein} \cdot cm^{-2}$. If the kinetic parameters of the immobilized enzyme obtained from these data are considered intrinsic and the film mass-transfer coefficient for pyruvate is $0.4\,mm \cdot min^{-1}$, determine the effectiveness factor at 25, 100, and 300 mM bulk pyruvate concentration and estimate the bulk pyruvate concentration above which EDR can be considered negligible.

Answer:

β_0	β_S	η
0.5	0.126	0.336
2	0.667	0.600
6	3.422	0.903

EDR are negligible at pyruvate bulk substrate concentrations above 1 M, which is close to the solubility of pyruvate.

3.25 Working with an enzyme immobilized in the surface of an impervious support, the following initial reaction rates in the bulk liquid and at the support–liquid interface are obtained at different agitation rates

β_0	β_S				
	$r_a=10$	$r_a=50$	$r_a=100$	$r_a=250$	$r_a=500$
0.1	0.0092	0.029	0.051	0.078	0.092
1	0.099	0.351	0.618	0.861	0.951
5	0.742	3.109	4.193	4.752	4.917
10	2.702	7.785	9.100	9.728	9.909
20	10.844	17.634	19.050	19.715	19.905

r_a: agitation rate (rpm)

Based on these data, determine the values of the Damkoehler number as a function of agitation rate.

Answer: From a nonlinear regression to Equation 3.14:

r_a (rpm)	10	50	100	250	500
α	10	2.5	1.0	0.3	0.1

Problem Solving in Enzyme Biocatalysis

3.26 A device for measuring blood sugar is constructed based on an enzyme electrode in which glucose oxidase (EC 1.1.3.4) is immobilized in the outer surface of a cellophane membrane. The reaction rate under excess molecular oxygen can be described by Michaelis–Menten kinetics, and under such conditions the following initial rates of glucose oxidation are obtained at varying glucose concentrations in the bulk medium:

$[G_0]$	0.05	0.1	0.3	5	10	20	50	100	150	200
v'	0.0417	0.0833	0.25	4.13	8.09	15.53	32.09	45.45	46.88	47.62

$[G_0]$: bulk liquid glucose concentration (mM)
v': initial reaction rate of glucose oxidation by the catalytic membrane ($\mu mol \cdot min^{-1} \cdot cm^{-2}$)

Evaluate the Damkoehler number of the catalytic membrane and determine the effectiveness factor of the system as a function of glucose concentration in the bulk liquid medium.

Answer: $\alpha = 5$.

$[G]_0$	β_0	β_S	η
0.05	0.005	$8.3 \cdot 10^{-4}$	0.167
0.1	0.01	$1.7 \cdot 10^{-3}$	0.171
0.3	0.03	$5.0 \cdot 10^{-3}$	0.171
1	0.1	$1.7 \cdot 10^{-2}$	0.184
5	0.5	$8.9 \cdot 10^{-2}$	0.245
10	1	0.19	0.319
20	2	0.45	0.466
50	5	1.79	0.770
100	10	5.74	0.937
150	15	10.44	0.973
200	20	15.31	0.986

3.27 An enzyme catalyst is produced by immobilizing a lipase from *Pseudomonas stutzeri* on octadecyl Sepabeads spherical particles (density $= 1.06\,g \cdot cm^{-3}$) with the following size distribution frequency (f) according to their diameter (d_P):

d_P (μm)	50	100	150	200	250	300
f	0.143	0.221	0.357	0.177	0.080	0.022

The intrinsic kinetic parameters of the immobilized enzyme are determined from initial rate data on the hydrolysis of *p*-nitrophenylbutyrate (*p*NPB), and the effective diffusion coefficient of *p*NPB inside the support (D_{eff}) is estimated:

$$V''_{max} = 450\,\mu moles \cdot min^{-1} \cdot g^{-1}$$
$$K_M = 2.53\,mM$$
$$D_{eff} = 4.2 \cdot 10^{-4}\,cm^2 \cdot min^{-1}$$

Evaluate the Thièle modulus of the catalyst, given its particle size distribution.

Answer: $\Phi = 1.63$.

3.28 Glucose isomerase (D-xylose aldose-ketose-isomerase, EC 5.3.1.5) from *Actinoplanes missouriensis* is immobilized in oblate ellipsoidal particles of polyvinyl alcohol gel (LentiKats), whose major and minor axes are 4.0 and 0.4 mm, respectively. The first-order reaction rate constant is $1 \, \text{min}^{-1}$ and the diffusion coefficient of glucose in polyvinyl alcohol gel is $1.42 \cdot 10^{-6} \, \text{cm}^2 \cdot \text{s}^{-1}$. The volume ($V_{PE}$) and surface area ($S_{PE}$) of an oblate ellipsoid are:

$$V_{PE} = \frac{4}{3} \cdot \pi \cdot a \cdot b^2 \qquad S_{PE} = 2 \cdot \pi \cdot b \cdot (a + b)$$

where a and b are the dimensions of the major and minor axis of the oblate ellipsoid, respectively. Calculate the Thièle modulus of the catalyst and the global effectiveness factor of the oblate ellipsoid catalyst, assuming first-order kinetics. By what percentage must the dimension of the LentiKat particle be varied (maintaining the proportion between the axes) in order to obtain an effectiveness factor of 0.9? Make any assumptions explicit.

Answer: Given the equivalent length of the oblate spheroid and assuming that the solution for the spherical particle is an acceptable approximation:

$$\Phi = 2.62$$
$$\eta_G = 0.333$$

Each axe of the Lentikat particle should be reduced by 83.2% to obtain $\eta_G = 0.9$. The volume of the LentiKat particle will have to be reduced by 95.3%.

3.29 Arabinose isomerase (EC 5.3.1.4) catalyzes the conversion of arabinose (A) into ribulose (R). The isomerization reaction is reversible, with an equilibrium constant of 1.25 under the reaction conditions. Cross-linked enzyme aggregates (CLEAs) of the arabinose isomerase from *Geobacillus stearothermophilus* are produced with approximately spherical geometry and $0.524 \, \text{mm}^3$ volume. The initial rates of ribulose isomerization ($v''_{R \rightarrow A}$) obtained at various R concentrations [R] are:

[R] (mM)	1	3	5	10	15	20	30
$v''_{R \rightarrow A}$ ($\text{mmol} \cdot \text{min}^{-1} \cdot \text{cm}^{-3}$)	0.0245	0.0706	0.113	0.207	0.286	0.353	0.462

The enzyme has the same affinity for A as for R.

Determine the rate expression that describes the ratio of isomerization of A into R by CLEAs of arabinose isomerase.

In order to determine the intrinsic kinetic parameters, CLEAs are fragmented into particles 10 times smaller than their initial form. The kinetic parameters obtained with such catalysts are 20% better than those previously determined. Based on this information, determine the value of the Thièle modulus referred to A and R of the nondisrupted CLEAs if the diffusion coefficients of A and R within the catalyst particle are $7 \cdot 10^{-4} \, \text{cm}^2 \cdot \text{min}^{-1}$ and $3 \cdot 10^{-3} \, \text{cm}^2 \cdot \text{min}^{-1}$, respectively. Determine also the global effectiveness factors, assuming in this case that reactions in both directions are first-order with respect to substrate concentration.

***Answer*:**

$$v'' = \frac{0.03125 \cdot [A] - 0.025 \cdot [R]}{0.0208 \cdot [A] + 0.0208 \cdot [R] + 1}$$

where v'' is in $\text{mmol} \cdot \text{min}^{-1} \cdot \text{cm}^{-3}$ and $[A]$ and $[R]$ are in mM.

$$
\begin{aligned}
K_{R,AP} &= K_{A,AP} = 48\,\text{mM} \\
V''_{maxR,AP} &= 1.2\,\mu\text{mole} \cdot \text{min}^{-1} \cdot \text{cm}^{-3} \\
V''_{maxR,AP} &= 1.25 \cdot 1.2 = 1.5\,\text{mmole} \cdot \text{min}^{-1} \cdot \text{cm}^{-3} \\
K_R &= K_{A,} = 40\,\text{mM} \\
V''_{maxR} &= 1.44\,\text{mmole} \cdot \text{min}^{-1} \cdot \text{cm}^{-3} \\
V''_{maxA} &= 1.8\,\text{mmole} \cdot \text{min}^{-1} \cdot \text{cm}^{-3} \\
D_{eff,R} &= 3 \cdot 10^{-3}\,\text{cm}^2 \cdot \text{min}^{-1} \\
D_{eff,A} &= 7 \cdot 10^{-4}\,\text{cm}^2 \cdot \text{min}^{-1} \\
d_p &= 0.1\,\text{cm}
\end{aligned}
$$

From Equation 3.23:

$$\Phi_A = \frac{d_p}{6} \cdot \sqrt{\frac{V''_{max\,A}}{K_A \cdot D_{eff,A}}}$$

$$\Phi_R = \frac{d_p}{6} \cdot \sqrt{\frac{V''_{maxR}}{K_R \cdot D_{eff,R}}}$$

$$
\begin{aligned}
\Phi_A &= 4.23 \\
\Phi_R &= 1.83 \\
\eta_A &= 0.218 \\
\eta_R &= 0.447
\end{aligned}
$$

3.30 A lactate biosensor is constructed based on a 0.5 mm thick polyvinyl sulfonate membrane containing NAD^+-dependent L-lactate dehydrogenase (EC 1.1.1.27) immobilized in its inner structure. Tests with such a membrane yield the following initial reaction rate data:

[L]	2.5	5	10	20	40	80	160	320	640	960	1280
v''	29.4	55.4	99.0	161.3	207.3	263.2	243.9	183.5	117.2	85.6	67.3

where [L] is the bulk L-lactate concentration (mM) and v'' is the initial rate of lactate oxidation into pyruvate, measured by NADH formation ($\mu\text{moles} \cdot \text{min}^{-1} \cdot \text{g}^{-1}_{\text{catalyst}}$).

Identify the reaction mechanism of L-lactate oxidation by L-lactate dehydrogenase and derive the corresponding rate equation using nonlinear regression of the data provided. Determine mathematically the optimum L-lactate concentration and the maximum reaction rate attainable. If the density of the polyvinylsulfonate membrane is $0.6\,\text{g}\cdot\text{cm}^{-3}$, calculate the magnitude of the Thièle modulus and the global effectiveness factor at low L-lactate concentrations. Assume an effective diffusion coefficient of L-lactate in the membrane of $6\cdot10^{-6}\,\text{cm}^2\cdot\text{s}^{-1}$.

Answer:

Uncompetitive inhibition by high L-lactate concentration $v'' = \dfrac{483\cdot[L]}{40+[L]+\frac{[L]^2}{212}}$

$[L]_{\text{optimum}} = 92.08\,\text{mM}$

Maximum attainable $v'' = 258.5\,\mu\text{moles}\cdot\text{min}^{-1}\cdot\text{g}_{\text{catalyst}}^{-1}$

$\Phi = 7.09$

$\eta_G = 0.282$

3.31 β-galactosidase from *Aspergillus oryzae* is immobilized in microporous chitin flakes (apparent density $0.9\,\text{g}\cdot\text{cm}^{-3}$) in the approximate shape of a rectangular parallelepiped of $10\times4\times2\,\text{mm}$. The following initial reaction rates are obtained at various concentrations of the artificial substrate *o*-nitrophenyl β-galactoside:

[*o*NPG]	0.05	0.075	0.15	0.3	0.6	0.8	1.2
$v''\cdot10^3$	2.73	3.91	6.92	11.25	16.36	18.46	21.18

[*o*NPG]: *o*-nitrophenyl β-galactoside mM bulk concentration
v'': initial reaction rate in mmoles of *o*-nitrophenol produced per minute and per gram of catalyst

Assuming Michaelis–Menten kinetics, calculate the apparent kinetic parameters of the catalyst. If diffusional restrictions only affect the enzyme affinity for *o*-NPG, reducing it by 75%, determine the Thièle modulus and the effectiveness factor of the catalyst (assume in this case first-order kinetics). The value of the diffusion coefficient of *o*-NPG in the chitin matrix has been estimated at $5.5\cdot10^{-2}\,\text{cm}^2\cdot\text{min}^{-1}$. By what percentage should the width of the chitin flakes be changed (assuming the other dimensions of the flake to be unchanged) in order to increase the value of the effectiveness factor by 30%?

Answer:

$V''_{\text{maxAP}} = V''_{\text{max}} = 0.03\,\text{mmoles}\cdot\text{min}^{-1}\cdot\text{g}_{\text{catalyst}}^{-1} = 0.027\,\text{mmoles}\cdot\text{min}^{-1}\cdot\text{cm}^{-3}$

$K_{\text{MAP}} = 0.5\,\text{mM}$

$K_M = 0.125\,\text{mM} = 1.25\cdot10^{-4}\,\text{mmol}\cdot\text{cm}^{-3}$

$\Phi = 3.68$

$\eta = 0.515$

The width of the catalyst particle must be reduced by 43.5%.

3.32 The trisaccharide lactosucrose (galactose–glucose–fructose) is hydrolyzed to lactose and fructose by β-fructofuranosidase (EC 3.2.1.26) from *Arthrobacter* sp.

immobilized in spherical particles of Sepabeads of density $0.78\,\text{g}\cdot\text{cm}^{-3}$. The intrinsic maximum reaction rate and Michaelis constant for lactosucrose for the immobilized enzyme are 1.73 mmoles of lactosucrose hydrolyzed per minute per gram of catalyst and 58.3 mM, respectively. The effective diffusion coefficient of lactosucrose inside the gel is $1.35\cdot10^{-5}\,\text{cm}^2\cdot\text{s}^{-1}$.

Determine the variation in the global effectiveness factor of the catalyst (η_G) with respect to the average diameter of the gel particles (d_p) in the range from 0.1 to 1.0 mm. Assume first-order kinetics for lactosucrose concentration.

Answer:

d_p (mm)	0.1	0.2	0.3	0.4	0.5	0.6	0.7	0.8	0.9	1.0
η_G	0.96	0.85	0.73	0.63	0.54	0.48	0.42	0.38	0.34	0.31

3.33 Penicillin G acylase (penicillin amidohydrolase, EC 3.5.1.11) is a key enzyme in the production of semisynthetic β-lactam antibiotics, used mostly for the production of 6-amino penicillanic acid by hydrolysis of penicillin G. Penicillin acylase from recombinant *Escherichia coli* is immobilized in nonsieved glyoxyl agarose particles. In the absence of diffusional restrictions, the following data are gathered:

[PenG]	0.5	1	1.5	2	2.5	5
v''	13.9	21.7	26.8	30.3	32.9	39.7

[PenG]: bulk penicillin G concentration (mM)
v'': initial rate of penicillin G hydrolysis (μmoles phenylacetic acid$\cdot\text{min}^{-1}\cdot\text{mL}_{gel}^{-1}$)

The film mass-transfer coefficient and diffusion coefficient inside the gel for such a biocatalyst are $0.5\,\text{cm}\cdot\text{min}^{-1}$ and $5.33\cdot10^{-5}\,\text{cm}^2\cdot\text{s}^{-1}$, respectively.

The glyoxyl agarose particles have the following size distribution:

d_p (μm)	20	50	100	150	200	250	300	350	400
frequency (%)	7.2	12.9	17.4	27.3	12.9	9.7	6.1	4.4	2.1

Assuming first-order kinetics, determine the global effectiveness factor for the biocatalyst and the percentage error in this factor when disregarding external diffusional restrictions.

Answer:

$$\eta_G = 0.604$$
$$\text{Error} = 58\%$$

3.34 An enzyme is immobilized in the inside of a microporous support in the shape of a cornflake. A proposed approximate solution to determining the global effectiveness factor is to assume a slab configuration; however, it has also been proposed that a spherical configuration be assumed instead. Working at very low substrate concentrations, compare the global effectiveness factors calculated using both geometries at different values of the Thièle modulus.

Answer:

Φ	0.1	0.5	1	2	3	5
η_{Gslab}	0.999	0.980	0.924	0.762	0.603	0.395
$\eta_{Gsphere}$	0.994	0.876	0.672	0.417	0.296	0.187
$\eta_{Gslab}/\eta_{Gsphere}$	1.005	1.119	1.375	1.827	2.037	2.112

It can be demonstrated that at $\Phi > 10$, η_{Gslab} tends to $2 \cdot \Phi^{-1}$ and $\eta_{Gsphere}$ tends to Φ^{-1}, so their ratio tends to 2.

3.35 An α-galactosidase (α-D-galactoside galactohydrolase, EC 3.2.1.22) from *Gibberella fujikuroi* is immobilized in porous glass spherical (GIP) particles and in chitin slabs (GIC) in order to evaluate its potential in the removal of raffinose from soymilk. The intrinsic kinetic parameters of the two immobilized enzymes (V_{max} and K_M), the particle densities (ρ) and the effective raffinose diffusion coefficients (D_{eff}) are determined:

	GIP	GIC
V''_{max} ($\mu mol \cdot min^{-1} \cdot g^{-1}$)	200	180
K_M (mM)	5.0	4.5
ρ ($g \cdot cm^{-3}$)	1.2	0.8
D_{eff} ($cm^2 \cdot min^{-1}$)	$8 \cdot 10^{-4}$	$2 \cdot 10^{-3}$

Assuming first-order kinetics for the hydrolysis of raffinose, evaluate the global effectiveness factors of the two catalysts with respect to particle diameter (d_p) and width (w), respectively.

Answer:

d_p or w (mm)	0.01	0.05	0.1	0.25	0.5	1	2
$\eta_{G,GIP}$	0.999	0.976	0.913	0.664	0.410	0.225	0.117
$\eta_{G,GIC}$	0.999	0.968	0.885	0.581	0.315	0.158	0.079

3.36 The synthesis of the β-lactam antibiotic cefaclor from 7-aminodesacetoxymethyl 3-chlorocephalosporanic acid (7ACCA) and D-phenylglycine methyl ester (PGME) is reported with penicillin G acylase (EC 3.5.1.11) from *Bacillus megaterium*. Using this enzyme, the following initial rates of cefaclor synthesis (v'') are obtained at varying concentrations of 7ACCA and PGME:

[PGME] (mM)	$v''(\mu\text{moles} \cdot \text{min}^{-1} \cdot \text{g}_{\text{catalyst}}^{-1})$				
	[7ACCA] (mM)				
	2	4	6	8	10
2	5.56	10.81	15.79	20.51	25.00
4	8.51	16.33	23.53	30.19	36.36
6	10.35	19.67	28.13	35.82	42.86
8	11.59	21.92	31.17	39.51	47.06
10	12.50	23.53	33.33	42.11	50.00

Determine the reaction mechanism and evaluate all kinetic parameters, analyzing the accuracy of results.

If the same enzyme is now immobilized on spherical particles of polyacrylamide gel, determine the maximum allowable catalyst particle size that will obtain a global effectiveness factor not less than 0.8. Assume in this case that the reaction of synthesis is first-order with respect to 7ACCA concentration. The effective diffusion coefficient of 7ACCA inside the polyacrylamide gel is $2.5 \cdot 10^{-3} \text{ cm}^2 \cdot \text{min}^{-1}$ and the density of the gel is $1.1 \text{ g} \cdot \text{cm}^{-3}$. Make any assumptions explicit.

Answer: From the information in Table 2.4 and the rate data given, the mechanism of synthesis of cefaclor with penicillin G acylase is ordered sequential, with the acyl donor (PGME) being the first substrate to bind to the enzyme to form an acyl–enzyme complex, which is then nucleophilically attacked by the β-lactam nucleus (7ACCA) to synthesize cefaclor.

$$V_{\text{max}} = 200 \ \mu\text{moles} \cdot \text{min}^{-1} \cdot \text{g}_{\text{catalyst}}^{-1}$$
$$K_{\text{M,PGME}} = 5 \text{ mM}$$
$$K'_{\text{M,7ACCA}} = 20 \text{ mM}$$

The rate equation is:

$$v = \frac{200 \cdot [PGME] \cdot [7ACCA]}{100 + 20 \cdot [PGME] + [PGME] \cdot [7ACCA]}$$

where the nomenclature is the same as in Section 2.4, where A stands for PGME and B for 7ACCA

The maximum allowable particles size for $\eta_G = 0.8$ is a sphere 0.6 mm in diameter.

References

1. Chibata, I. and Tosa, T. (1986) Biocatalysis: immobilized cells and enzymes. *Journal of Molecular Catalysis*, **37**, 1–24.
2. Koeller, K.M. and Wong, C.H. (2001) Enzymes for chemical synthesis. *Nature*, **409**, 232–240.

3. Guisán, J.M. (2006) *Immobilization of Enzymes and Cells*, Humana Press, New Jersey.
4. Mateo, C., Palomo, J.M., Fuentes, M. *et al.* (2006) Glyoxyl agarose: a fully inert and hydrophilic support for immobilization and high stabilization of proteins. *Enzyme and Microbial Technology*, **39** (2), 274–280.
5. Brady, D. and Jordaan, J. (2009) Advances in enzyme immobilization. *Biotechnology Letters*, **31**, 1639–1650.
6. Gupta, R. and Chaudhury, N.K. (2007) Entrapment of biomolecules in sol-gel matrix for applications in biosensors: problems and future prospects. *Biosensors and Bioelectronics*, **22** (11), 2387–2399.
7. Wilson, L., Illanes, A., Pessela, B. *et al.* (2004) Encapsulation of crosslinked penicillin G acylase aggregates in Lentikats: evaluation of a novel biocatalyst in organic media. *Biotechnology and Bioengineering*, **86** (5), 558–562.
8. Wichmann, R., Wandrey, C., Bückmann, A.F., and Kula, M.R. (2000) Continuous enzymatic transformation in an enzyme membrane reactor with simultaneous NAD(H) regeneration. *Biotechnology and Bioengineering*, **67** (6), 791–804.
9. Cao, L., Van Rantwijk, F., and Sheldon, R.A. (2000) Cross-linked enzyme aggregates: A simple and effective method for the immobilization of penicillin acylase. *Organic Letters*, **2** (10), 1361–1364.
10. Vaghjiani, J.D., Lee, T.S., Lye, G.J., and Turner, M.K. (2000) Production and characterisation of cross-linked enzyme crystals (CLECs®) for application as process scale biocatalysts. *Biocatalysis and Biotransformation*, **18** (2), 151–175.
11. Sheldon, R.A. (2011) Characteristic features and biotechnological applications of cross-linked enzyme aggregates (CLEAs). *Applied Microbiology and Biotechnology*, **92** (3), 467–477.
12. Sheldon, R.A. (2011) Cross-linked enzyme aggregates as industrial biocatalysts. *Organic Process Research and Development*, **15** (1), 213–223.
13. Dalal, S., Kapoor, M., and Gupta, M.N. (2007) Preparation and characterization of combi-CLEAs catalyzing multiple non-cascade reactions. *Journal of Molecular Catalysis B: Enzymatic*, **44** (3–4), 128–132.
14. Palomo, J.M. and Guisan, J.M. (2012) Different strategies for hyperactivation of lipase biocatalysts. *Methods in Molecular Biology*, **861**, 329–341.
15. Illanes, A. (2008) *Enzyme Biocatalysis: Principles and Applications*, Springer, London.
16. Engasser, J. and Horvath, C. (1973) Effect of internal diffusion in heterogeneous enzyme systems: evaluation of true kinetic parameters and substrate diffusivity. *Journal of Theoretical Biology*, **42**, 137–155.
17. Jeison, D., Ruiz, G., Acevedo, F., and Illanes, A. (2003) Simulation of the effect of intrinsic reaction kinetics and particle size on the behavior of immobilized enzymes under internal diffusional restrictions and steady state operation. *Process Biochemistry*, **39**, 393–399.
18. Rovito, B.J. and Kittrell, J.R. (1973) Film and pore diffusion studies with immobilized glucose oxidase. *Biotechnology and Bioengineering*, **15**, 143–161.
19. Buchholz, K. (1982) Reaction engineering parameters for immobilized biocatalyst, in *Advances in Biochemical Engineering*, vol. **24** (ed. A. Fiechter), Springer-Verlag, Berlin.
20. Valencia, P., Flores, S., Wilson, L., and Illanes, A. (2011) Effect of internal diffusional restrictions on the hydrolysis of penicillin G: reactor performance and specific productivity of 6-APA with immobilized penicillin acylase. *Applied Biochemistry and Biotechnology*, **165**, 426–441.
21. Engasser, J.M. and Horvath, C. (1976) Diffusion and kinetics with immobilized enzymes, in *Immobilized Enzyme Principles* (eds L. Wingard, E. Katchalsky, and L. Goldstein), Academic Press, New York.
22. Grünwald, P. (1989) Determination of effective diffusion coefficient: an important parameter for the efficiency of immobilized biocatalysts. *Biochemical Education*, **17**, 99–102.

23. Wilson, L., Palomo, J.M., Fernández-Lorente, G. *et al.* (2006) Improvement of the functional properties of a thermostable lipase from *Alcaligenes* sp. via strong adsorption on hydrophobic supports. *Enzyme and Microbial Technology*, **38**, 975–980.
24. Jayani, R.S., Saxena, S., and Gupta, R. (2005) Microbial pectinolytic enzymes: a review. *Process Biochemistry*, **40**, 2931–2944.
25. Wilke, C.R. and Chang, P. (1955) Correlation of diffusion coefficients in dilute solutions. *AIChE Journal*, **1**, 264–270.
26. Engasser, J. (1978) A fast evaluation of diffusion effects on bound enzyme activity. *Biochimica et Biophysica Acta*, **526**, 301–310.

4

Enzyme Reactor Design and Operation under Ideal Conditions

4.1 Modes of Operation and Reactor Configurations

Enzymes have a wide spectrum of technological application and are fundamental tools in molecular biology and biotechnological research [1]. Most enzyme technological applications refer to their use as process catalysts [2], representing more than 70% of their global commercial value. Conventional use of enzymes is mostly as catalysts for hydrolysis reactions conducted in aqueous media; this still represents their major application. In recent decades, however, their application has been extended to organic synthesis reactions, mostly performed in nonaqueous media and with immobilized enzymes [3]. Conventional homogeneous catalysis processes in aqueous media, conducted batchwise in stirred-tank reactors with no enzyme recovery, have been superseded for a variety of reactor configurations and modes of operation for heterogeneous systems, whether by the use of a solid catalyst (immobilized enzyme) or a heterogeneous reaction medium (organic solvent, ionic liquid, semisolid medium) [4,5].

The poor stability of enzymes under process conditions is a major obstacle to their widespread use as industrial catalysts. Therefore, the design and production of robust enzyme catalysts is a major issue in enzyme biocatalysis. Immobilized enzymes are powerful process catalysts, stable and recoverable, making for an increased efficiency of use and thus reducing their impact on total operating costs [6]. Immobilized enzymes can be used either in sequential batch operation (with enzyme recovery after each batch) or continuously, going from stirred-tank to column- or tubular-type reactors. On the other hand, the use of homogeneous catalysis with enzymes dissolved in the reaction medium is mostly restricted to batch operation in stirred-tank reactors, with a few exceptions such as starch liquefaction, which is usually performed in continuously operated tubular reactors

Problem Solving in Enzyme Biocatalysis, First Edition. Andrés Illanes, Lorena Wilson and Carlos Vera.
© 2014 John Wiley & Sons, Ltd. Published 2014 by John Wiley & Sons, Ltd. Companion Website:
http://www.wiley.com/go/illanes-problem-solving

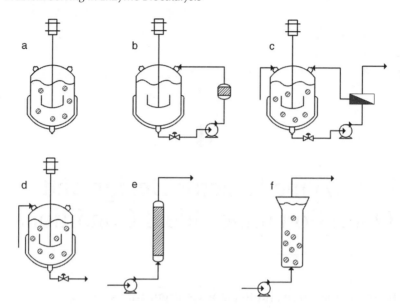

Figure 4.1 *Configurations of reactors with immobilized enzymes: (a) batch-stirred tank reactor; (b) recirculation batch reactor; (c) continuous stirred-tank membrane reactor; (d) continuous stirred-tank reactor; (e) continuous packed-bed reactor; (f) continuous fluidized-bed reactor.*

with soluble α-amylase. Membrane reactors can also be employed to allow the use of soluble enzymes in a continuous mode of operation.

Most common reactor configurations are illustrated in Figure 4.1.

4.2   Definition of Ideal Conditions

Operation under ideal conditions represents a preliminary approach to reactor performance evaluation and design. Ideal conditions imply:

- Ideal flow regime within the reactor; thats is, complete mixing (Peclet number $= 0$) in the case of a stirred-tank reactor and plug-flow regime (Peclet number $= \infty$) in the case of tubular- or column-type reactors.
- Controlled and constant operational conditions, such as temperature and pH.
- No activity losses during reactor operation.
- Absence of mass-transfer limitations and partition in the case of heterogeneous catalysis.

Departure from ideal behavior is unavoidable, however. It can result from one or more of the following causes:

- Incomplete mixing or backmixing, which will occur to some extent in stirred-tank reactors and tubular (column) reactors, respectively.
- Nonisothermal operation, which results from heat-transfer limitations and pH gradients as a consequence of incomplete mixing or H^+ generation/consumption during reaction.

- Enzyme inactivation or leakage during catalyst use.
- Partition or diffusional restrictions, in the case of heterogeneous catalysis (see Chapter 3).

Reactor performance and design will be evaluated under ideal conditions in this chapter. Most of the common causes of departure from ideal behavior will be analyzed in Chapters 5 and 6.

4.3 Strategy for Reactor Design and Performance Evaluation

A modular approach to the development of an enzyme reactor design or performance evaluation model can be envisaged in which there are two fundamental components: a sound kinetic expression (see Chapter 2) and a material balance according to the type of reactor and mode of operation. With these two components, a basic equation describing operation under ideal conditions can be developed. Nonideal conditions can then be incorporated in the form of modules relating to mass-transfer limitations, enzyme inactivation and nonideal flow regime. Mass-transfer limitations and enzyme inactivation will be analyzed in Chapter 5 and Chapter 6, respectively. For nonideal flow regime, the interested reader is referred to the article published by Lübbert and Jørgenssen [7] and the classical text on heterogeneous chemical catalysis by Levenspiel [8].

The equation describing enzyme reactor performance can be used either for design purposes (determining reactor size for a given production task) or to evaluate reactor outputs for determined inputs. Whatever the case, that equation will reduce by one the number of degrees of freedom in the system: these are enzyme load, substrate concentration, and operation time (or flow rate, in the case of continuous operation), which are the input variables; the output response is the substrate-into-product conversion. The system is thus not completely determined and two degrees of freedom will remain, which is very important because it allows optimization (usually by setting up a cost-related objective function).

4.4 Mathematical Models for Enzyme Kinetics, Modes of Operation, and Reactor Configurations under Ideal Conditions

Substrate and product concentrations will vary during the course of reaction. Since they are usually related by the stoichiometry of the reaction, it is convenient to define a variable that includes both, which is the substrate-into-product conversion (X), defined as:

$$X = \frac{s_i - s}{s_i} = \frac{p}{n \cdot s_i} \tag{4.1}$$

For an equimolar reaction (quite usual in enzyme-catalyzed reactions), $n = 1$, so from Equation 4.1:

$$X = \frac{s_i - s}{s_i} = \frac{p}{s_i}$$
$$s = s_i(1 - X) \tag{4.2}$$
$$p = s_i \cdot X$$

4.4.1   Batch Enzyme Reactor

This type of reactor is usually a batch stirred-tank reactor (BSTR). When the enzyme is dissolved in the reaction medium, a one-batch operation is generally carried out, after which the enzyme is inactivated and removed. In the case of an immobilized enzyme, the biocatalyst is stable enough to be recovered by conventional retention after product removal and then used in subsequent batches.

4.4.1.1   Single-Substrate Reactions

Equation 4.3 represents the material balance on the reactor:

$$v(e, X) = -\frac{ds}{dt} = s_i \frac{dX}{dt} \tag{4.3}$$

$$\int_0^X \frac{dX}{v(e, X)} = \int_0^t \frac{dt}{s_i} \tag{4.4}$$

Under ideal conditions, Equation 4.4 can be integrated directly because the enzyme activity is constant and independent of reaction time. Integrating it gives:

$$\frac{k \cdot e}{K_M} t = \frac{m_{cat} \cdot a_{sp}}{K_M} t = f(s_i, X) \tag{4.5}$$

Equation 4.5 represents the model of BSTR operation for single-substrate reactions. $X = f(e, s_i, t)$ will be obtained according to the kinetic expression (see Section 2.3). Given Equation 2.18 and the values of the kinetic parameters in Table 2.3, solving Equation 4.4 with X defined by Equation 4.2 leads to the development of model equations for various kinetics, as presented in Table 4.1. The term in the left-hand side of Equation 4.5 is dimensionless and is sometimes referred to as the "dimensionless operation time"; it involves enzyme activity and reaction time, the most important variables for reactor performance.

For batch operation, τ represents operation time; for continuous operation, it represents fluid residence time.

As pointed out before, there are still two degrees of freedom, so from the three operational variables (m_{cat}, t, and s_i), two can be established separately. Equation 4.5 can be used to determine the enzyme load required (m_{cat}) to obtain the desired conversion (X) from a determined initial substrate concentration (s_i) after a given time, or the time required to attain such conversion at a given enzyme load. As seen from Equation 4.5, for a given X and s_i, $m_{cat} \cdot$ t will be constant (this is obviously not true if enzyme inactivation is considered), meaning that the same performance can be obtained at a high enzyme load for a short reaction time or at a low enzyme load for a prolonged reaction time, which is an interesting proposal for optimization since the impact of the cost of the enzyme and the cost of reactor operation on total operation costs will be different and specific for each particular case. The most common situation is to define operation time according to industrial convenience (i.e. one batch per shift) and determine the enzyme load required to attain the desirable X at that time.

Table 4.1 Models of enzyme reactor performance for single-substrate reactions.

$$\frac{k \cdot e}{K_M}\tau = \frac{k \cdot E}{K_M \cdot F} =$$

Kinetic Model	CPBR or BSTR	CSTR
M-M kinetics	$\dfrac{s_i}{K_M} \cdot X - \ln(1 - X)$	$\dfrac{s_i}{K_M} \cdot X + \dfrac{X}{1 - X}$
CI by P	$s_i \cdot X \cdot \left[\dfrac{1}{K_M} - \dfrac{1}{K_{IC}} \right] - \left[1 + \dfrac{s_i}{K_{IC}} \right] \ln(1 - X)$	$\dfrac{s_i}{K_M} \cdot X + \dfrac{X}{1 - X} + \dfrac{s_i}{K_{NC}} \cdot \dfrac{X^2}{1 - X}$
Total NCI by P	$s_i X \left[\dfrac{1}{K_M} - \dfrac{1}{K_{INC}} \right] - \left[1 + \dfrac{s_i}{K_{INC}} \right] \ln(1 - X) + \dfrac{s_i^2 X^2}{2 K_M K_{INC}}$	$\dfrac{s_i X}{K_M} + \dfrac{X}{1 - X} + \dfrac{s_i X^2}{K_{INC}(1 - X)} + \dfrac{s_i^2 X^2}{K_M K_{INC}}$
Total UCI by S	$\dfrac{s_i \cdot X}{K_M} - \ln(1 - X) + \dfrac{s_i^2 X}{K_M K_S}(1 - 0.5 \cdot X)$	$\dfrac{s_i \cdot X}{K_M} + \dfrac{X}{1 - X} + \dfrac{s_i^2 X}{K_M K_S}(1 - X)$
CI by P_1 and NCI by P_2	$s_i X \left[\dfrac{1}{K_M} - \dfrac{s_i}{K_{IC}} - \dfrac{1}{K_{INC}} \right] -$ $\left[1 + \dfrac{s_i}{K_{IC}} + \dfrac{s_i}{K_{INC}} \right] \ln(1 - X) + \dfrac{s_i^2 \cdot X^2}{2 \cdot K_M \cdot K_{INC}}$	$\dfrac{X}{1 - X} + \dfrac{s_i \cdot X^2}{1 - X} \cdot \left(\dfrac{1}{K_{IC}} + \dfrac{1}{K_{INC}} \right) +$ $\dfrac{s_i \cdot X}{K_M} \cdot \left(1 + \dfrac{s_i \cdot X}{K_{INC}} \right)$

M-M: Michaelis–Menten kinetics; CI: competitive inhibition; NCI: noncompetitive inhibition; UCI: uncompetitive inhibition; CPBR: continuous packed-bed reactor; BSTR: batch stirred-tank reactor; CSTR: continuous stirred-tank reactor S: substrate; P, P1, P2: products of reaction.

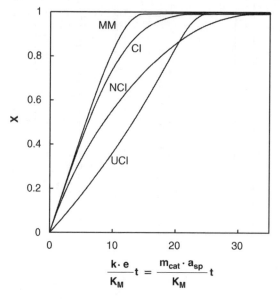

Figure 4.2   *Batch stirred-tank reactor (BSTR) performance under different kinetics at $s_i \cdot K_M^{-1} = 10$ and $K_M \cdot K_I^{-1} = 0.2$. MM: Michaelis–Menten kinetics; CI: competitive inhibition by product, NCI: total noncompetitive inhibition by product; UCI: total uncompetitive inhibition by substrate; X: substrate conversion; k: catalytic rate constant; e: enzyme concentration; t: reaction time; n K_M: Michaelis constant; m_{cat}: catalyst concentration; a_{sp}: specific activity.*

Equation 4.5 may also be used as a design equation, since for a given mass of catalyst the effective reaction volume can be determined simply as:

$$V_{eff} = \frac{M_{cat}}{m_{cat}} \tag{4.6}$$

and the volume of the reactor can be determined by considering the head space and space for internals (usually 20–30% of reactor volume in the case of a stirred-tank reactor). Operation curves for BSTR are presented in Figure 4.2 for the kinetic mechanisms in Table 4.1.

4.4.1.2   Multiple-substrate reactions

Reactions of synthesis involve more than one substrate (usually two). So for the enzyme-catalyzed reaction:

$$A + B \rightarrow Product(s)$$

the material balances over the limiting substrate A and substrate B are represented by Equation 4.7 and Equation 4.8, respectively:

$$v(e, X) = -\frac{da}{dt} = a_i \frac{dX}{dt} \qquad \int_0^X \frac{dX}{v(e, X)} = \int_0^t \frac{dt}{a_i} \tag{4.7}$$

$$v(e, X) = -\frac{db}{dt} = b_i \frac{dX}{dt} \qquad \int_0^X \frac{dX}{v(e, X)} = \int_0^t \frac{dt}{b_i} \tag{4.8}$$

In the case of A limiting (A refers to the first substrate to bind to the enzyme in an ordered sequential (OS) mechanism):

$$a = a_i(1 - X)$$
$$b = b_i - a_i \cdot X \tag{4.9}$$

In the case of B limiting:

$$b = b_i(1 - X)$$
$$a = a_i - b_i \cdot X \tag{4.10}$$

From Equation 2.19 and the values of the kinetic parameters in Table 2.4, and introducing Equation 4.9 or 4.10, the following rate equations are obtained for OS, random sequential (RS), and ping-pong kinetic mechanisms:

OS (A limiting)	$$v = \frac{V_{max} \cdot a_i(1 - X) \cdot (b_i - a_i \cdot X)}{a_i(1 - X)(b_i - a_i \cdot X + K'_{MB}) + K_{MA} \cdot K'_{MB}}$$	(4.11)
OS (B limiting)	$$v = \frac{V_{max} \cdot (a_i - b_i \cdot X) \cdot b_i(1 - X)}{(a_i - b_i \cdot X)(b_i(1 - X) + K'_{MB}) + K_{MA} \cdot K'_{MB}}$$	(4.12)
RS (A limiting)*	$$v = \frac{V_{max} \cdot a_i(1 - X) \cdot (b_i - a_i \cdot X)}{a_i(1 - X)(b_i - a_i X + K'_{MB}) + K'_{MA}(b_i - a_i X) + K_{MA} \cdot K'_{MB}}$$	(4.13)
Ping-pong (A limiting)	$$v = \frac{V_{max} \cdot a_i(1 - X) \cdot (b_i - a_i \cdot X)}{\alpha \cdot K_{MA}(b_i - a_i X) + a_i(1 - X)(K''_{MB} + (b_i - a_i X))}$$	(4.14)
Ping-pong (B limiting)	$$v = \frac{V_{max} \cdot b_i(1 - X) \cdot (a_i - b_i \cdot X)}{\alpha \cdot K_{MA} \cdot b_i(1 - X) + (a_i - b_i X)(K''_{MB} + b_i(1 - X))}$$	(4.15)

*In this case, designation of A and B is arbitrary, so only one rate equation suffices

Replacing these rate equations into Equations 4.7 or 4.8, and incorporating Equation 4.9 or 4.10 according to the case, the following equations for batch reactor performance are obtained:

OS (A limiting) $$\frac{K_{MA}K'_{MB}}{a_i(b_i - a_i)} \ln\left|\frac{b_i - a_i X}{b_i(1 - X)}\right| - \frac{K'_{MB}}{a_i} \ln\left|\frac{b_i - a_i X}{b_i}\right| + X = \frac{k \cdot e}{a_i} t = \frac{m_{cat} \cdot a_{sp}}{a_i} t$$

$$\tag{4.16}$$

OS (B limiting) $\quad \dfrac{K_{MA}K'_{MB}}{b_i(a_i - b_i)} \ln\left|\dfrac{a_i - b_iX}{a_i(1 - X)}\right| - \dfrac{K'_{MB}}{b_i}\ln(1 - X) + X = \dfrac{k \cdot e}{b_i}t = \dfrac{m_{cat} \cdot a_{sp}}{b_i}t$

$$(4.17)$$

RS (A limiting) $\quad \dfrac{K_{MA}K'_{MB}}{a_i(b_i - a_i)} \ln\left|\dfrac{b_i - a_iX}{b_i(1 - X)}\right| - \dfrac{K'_{MB}}{a_i}\ln\left|\dfrac{b_i - a_iX}{b_i}\right| - \dfrac{K'_{MA}}{a_i}\ln(1 - X) + X$

$$= \dfrac{k \cdot e}{a_i}t = \dfrac{m_{cat} \cdot a_{sp}}{a_i}t \qquad (4.18)$$

Ping-Pong (A limiting) $\quad X - \dfrac{\alpha \cdot K_{MA}}{a_i} \cdot \ln|1 - X| - \dfrac{K''_{MB}}{a_i} \cdot \ln\left|\dfrac{b_i - a_i \cdot X}{b_i}\right| = \dfrac{k \cdot e}{a_i}t = \dfrac{m_{cat} \cdot a_{sp}}{a_i}t$

$$(4.19)$$

Ping-Pong (B limiting) $\quad X - \dfrac{K''_{MB}}{b_i} \cdot \ln|1 - X| - \alpha \cdot \dfrac{K_{MA}}{b_i} \cdot \ln\left|\dfrac{a_i - b_i \cdot X}{a_i}\right| = \dfrac{k \cdot e}{b_i}t = \dfrac{m_{cat} \cdot a_{sp}}{b_i}t$

$$(4.20)$$

Figure 4.3 presents a graphical representation of reactor performance for OS (A and B limiting) and RS kinetics.

4.4.2 Continuous Enzyme Reactors

Enzyme immobilization to insoluble supports allows the development of continuous processes. In principle, these are more productive than batch processes because of the lower incidence of unproductive time. Enzymes should be highly stable in order to develop continuous processes, and this can be achieved by immobilization. In this chapter, continuous stirred-tank reactors (CSTRs) and continuous packed-bed reactors (CPBRs) will be analyzed under ideal flow regimes: perfect mixing in the case of CSTR and plug flow in the case of CPBR.

4.4.2.1 Continuous Packed-Bed Reactors

The equation for CPBR performance is obtained from a material balance on a differential element of the catalyst bed:

$$\int\left(\dfrac{dX}{v(e, X)}\right) = \dfrac{\tau}{s_i} \qquad (4.21)$$

where:

$$\tau = \dfrac{V_{eff} \cdot \varepsilon}{F} \qquad (4.22)$$

(a)

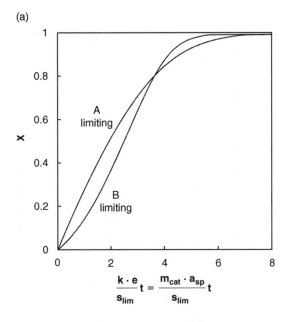

(b)

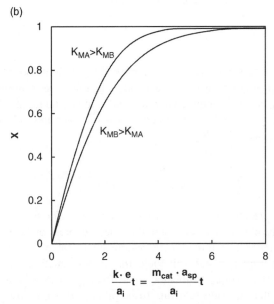

Figure 4.3 *Batch stirred-tank reactor performance for: (a) OS mechanism. Limiting substrate concentration ($s_{lim} = a_i$ or b_i) = 1 mM, substrate in excess concentration (b_i or a_i) = 5 mM, with K_{MA} = 0.5 mM; K'_{MB} = 8 mM. (b) RS mechanism. a_i = 1 mM, b_i = 5 mM, with K_{MA} = 1 mM, K_{MB} = 2 mM, K'_{MA} = 0.5 mM, and K'_{MB} = 1 mM ($K_{MB} > K_{MA}$); and with K_{MA} = 2 mM, K_{MB} = 1 mM, K'_{MA} = 1 mM, and K'_{MB} = 0.5 mM ($K_{MA} > K_{MB}$). (c) Ping-pong mechanism. Limiting substrate concentration ($s_{lim} = a_i$ or b_i) = 1 mM, substrate in excess concentration (b_i or a_i) = 5 mM, with K_{MA} = 0.5 mM, K''_{MB} = 8, and α = 0.5 (see nomenclature in the text).*

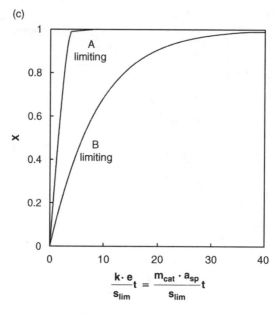

$$\frac{k \cdot e}{s_{lim}} t = \frac{m_{cat} \cdot a_{sp}}{s_{lim}} t$$

Figure 4.3 *(Continued)*

The void fraction (ε) is the fraction of the packed-bed volume occupied by the reaction medium; usual values are between 40 and 60%. In this case, e represents the enzyme concentration referred to the liquid volume. Under ideal conditions, Equation 4.21 can be integrated directly, since the enzyme activity is constant and independent of reaction time. Integrating Equation 4.21 gives:

$$\frac{k \cdot e}{K_M} \cdot \tau = \frac{k \cdot E}{K_M \cdot F} = \frac{M_{cat} \cdot a_{sp}}{K_M \cdot F} = f(s_i, X) \tag{4.23}$$

Equation 4.23 represents the model of CPBR operation for single-substrate reactions. $X = f(M_{cat}, F)$ will be obtained according to the kinetic expression (see Section 2.3). Given Equation 2.18 and the values of the kinetic parameters in Table 2.3, solving Equation 4.23 with X defined by Equation 4.2 will yield the model equations for different kinetics presented in Table 4.1. Notice in this case that for a given X and s_i, $M_{cat} \cdot F^{-1}$ will be constant, meaning that the substrate flux to be processed is directly proportional to the enzyme load in the reactor, which, as for a batch reactor, will allow optimization of operation in terms of cost.

Equation 4.23 may also be used as a design equation, since reactor size is determined by the biocatalyst bed volume, which directly depends on biocatalyst mass:

$$V_{bed} = \frac{M_{cat}}{\rho_{app}} \tag{4.24}$$

As can be seen in Table 4.1, equations for BSTR and CPBR are formally equal, though their physical meanings are different, τ corresponding to the operation time in the former and to the fluid residence time in the latter. However, there are striking similarities: the substrate profile that develops in time in the BSTR is analogous to the one that develops throughout the catalyst bed in the CPBR.

The performances of CPBR and CSTR (see Section 4.4.2.2) for the kinetics of a single-substrate reaction are shown in Figure 4.4. For Michaelis–Menten kinetics (Figure 4.4a), CPBR performance is better than CSTR (for a given X, a lower M_{cat} is required or a higher F can be processed, or else a higher X is obtained for a given $M_{cat} \cdot F^{-1}$ ratio). Differences are even higher in the case of product inhibition (Figure 4.4b), CSTR working at the most unfavorable conditions for the enzyme (higher p, lower s). In the case of inhibition at high s (Figure 4.4c), the best configuration will depend on the operating conditions.

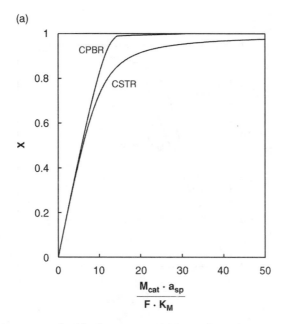

(a)

Figure 4.4 *Continuous packed-bed reactor (CPBR) and continuous stirred-tank reactor (CSTR) performance under different kinetics at $s_i \cdot K_M^{-1} = 10$ and $K_M \cdot K_I^{-1} = 0.2$. (a) Michaelis–Menten kinetics; (b) competitive (CI) and total noncompetitive (NCI) inhibition by product; (c) total uncompetitive inhibition (UCI) by substrate. X: substrate conversion; k: catalytic rate constant; e: enzyme concentration; t: reaction time K_M: Michaelis constant; M_{cat}: catalyst mass; a_{sp}: specific activity; F: feed flow rate; s_i: initial substrate concentration; K_I: inhibition constant.*

(b)

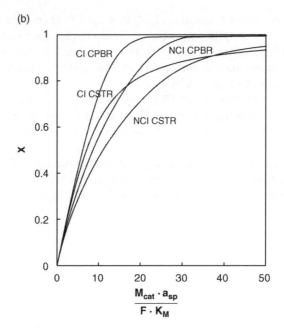

(c)

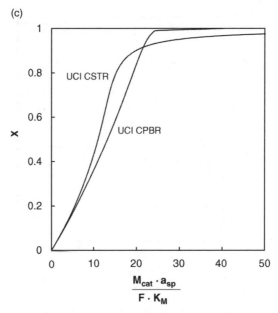

Figure 4.4 *(Continued)*

For a reaction of synthesis with two substrates:

$$A + B \rightarrow Product(s)$$

a material balance on the limiting substrate (A or B) leads to:

$$\int \frac{dX}{v(e, X)} = \frac{\tau}{a_i} \tag{4.25}$$

$$\int \frac{dX}{v(e, X)} = \frac{\tau}{b_i} \tag{4.26}$$

Equations for OS, RS, and ping-pong reaction mechanisms are analogous to Equations 4.16 to 4.20:

OS (A limiting)
$$\frac{K_{MA}K'_{MB}}{a_i(b_i - a_i)} \ln\left|\frac{b_i - a_i X}{b_i(1 - X)}\right| - \frac{K'_{MB}}{a_i} \ln\left|\frac{b_i - a_i X}{b_i}\right| + X = \frac{k \cdot e}{a_i} \tau = \frac{M_{cat} \cdot a_{sp}}{a_i \cdot F} \tag{4.27}$$

OS (B limiting)
$$\frac{K_{MA}K'_{MB}}{b_i(a_i - b_i)} \ln\left|\frac{a_i - b_i X}{a_i(1 - X)}\right| - \frac{K'_{MB}}{b_i} \ln(1 - X) + X = \frac{k \cdot e}{b_i} \tau = \frac{M_{cat} \cdot a_{sp}}{b_i \cdot F} t \tag{4.28}$$

RS (A limiting)
$$\frac{K_{MA}K'_{MB}}{a_i(b_i - a_i)} \ln\left|\frac{b_i - a_i X}{b_i(1 - X)}\right| - \frac{K'_{MB}}{a_i} \ln\left|\frac{b_i - a_i X}{b_i}\right| - \frac{K'_{MA}}{a_i} \ln(1 - X) + X$$
$$= \frac{k \cdot e}{a_i} \tau = \frac{M_{cat} \cdot a_{sp}}{a_i \cdot F} \tag{4.29}$$

Ping-Pong (A limiting)
$$X - \frac{\alpha \cdot K_{MA}}{a_i} \cdot \ln|1 - X| - \frac{K''_{MB}}{a_i} \cdot \ln\left|\frac{b_i - a_i \cdot X}{b_i}\right| = \frac{k \cdot e}{a_i} t = \frac{m_{cat} \cdot a_{sp}}{a_i} t \tag{4.30}$$

Ping-Pong (B limiting)
$$X - \frac{K''_{MB}}{b_i} \cdot \ln|1 - X| - \alpha \cdot \frac{K_{MA}}{b_i} \cdot \ln\left|\frac{a_i - b_i \cdot X}{a_i}\right| = \frac{k \cdot e}{b_i} \tau = \frac{M_{cat} \cdot a_{sp}}{b_i \cdot F} \tag{4.31}$$

4.4.2.2   Continuous Stirred-Tank Reactors

In ideal CSTR operation, every fluid element within the reactor will have the same composition, corresponding to the reactor outlet. So the equation for CSTR performance is obtained from a material balance over the whole reactor, kinetic expression being evaluated at outlet conditions:

$$\frac{X}{v(e, X)} = \frac{\tau}{s_i} \tag{4.32}$$

In this case, the catalyst volume is usually less than 10% of the reaction volume, so ε is over 90%.

Equation 4.32 can be represented as:

$$\frac{k \cdot e}{K_M} \cdot \tau = \frac{k \cdot E}{K_M \cdot F} = \frac{M_{cat} \cdot a_{sp}}{K_M \cdot F} = f(s_i, X) \qquad (4.33)$$

Equation 4.33 represents the model of CSTR operation. Model equations of CSTR performance for different single-substrate reaction kinetics are presented in Table 4.1. Equation 4.33 may also be used as a design equation, since reactor size will be determined from the catalyst concentration in the reactor:

$$V_{eff} = \frac{M_{cat}}{m_{cat}} \qquad (4.34)$$

Catalyst concentration will be limited by the hydrodynamic constraints and maintenance of catalyst integrity; in practice, concentrations higher than $100\,\mathrm{g \cdot L^{-1}}$ are difficult to handle.

For a reaction of synthesis with two substrates:

$$A + B \rightarrow \text{Product(s)}$$

a material balance on the limiting substrate (A or B) leads to:

$$\frac{X}{v(e, X)} = \frac{\tau}{a_i} \qquad (4.35)$$

$$\frac{X}{v(e, X)} = \frac{\tau}{b_i} \qquad (4.36)$$

Replacing the rate expressions for the different mechanisms (Equations 4.11 to 4.15) in Equations 4.35 or 4.36 gives:

$$\text{OS (A limiting)}\quad X\left[\frac{K_{MA}K'_{MB}}{a_i(1-X)(b_i - a_iX)} + \frac{K'_{MB}}{(b_i - a_iX)} + 1\right] = \frac{k \cdot e}{a_i}\tau = \frac{M_{cat} \cdot a_{sp}}{a_i \cdot F} \qquad (4.37)$$

$$\text{OS (B limiting)}\quad X\left[\frac{K_{MA}K'_{MB}}{b_i(1-X)(a_i - b_iX)} + \frac{K'_{MB}}{b_i \cdot (1-X)} + 1\right] = \frac{k \cdot e}{b_i}\tau = \frac{M_{cat} \cdot a_{sp}}{b_i \cdot F} \qquad (4.38)$$

$$\text{RS (A limiting)} \quad X\left[\frac{K_{MA}K'_{MB}}{a_i(1-X)(b_i-a_iX)}+\frac{K'_{MB}}{(b_i-a_iX)}+\frac{K'_{MA}}{a_i(1-X)}+1\right]=\frac{k\cdot e}{a_i}\tau=\frac{M_{cat}\cdot a_{sp}}{a_i\cdot F}$$

$$(4.39)$$

$$\text{Ping-Pong (A limiting)} \quad X\left[\frac{\alpha\cdot K_{MA}}{a_i\cdot(1-X)}+\frac{K''_{MB}}{\cdot(b_i-a_i\cdot X)}+1\right]=\frac{k\cdot e}{a_i}\tau=\frac{M_{cat}\cdot a_{sp}}{a_i\cdot F} \quad (4.40)$$

$$\text{Ping-Pong (B limiting)} \quad X\left[\frac{\alpha\cdot K_{MA}}{(a_i-b_i\cdot X)}+\frac{K''_{MB}}{b_i\cdot(1-X)}+1\right]=\frac{k\cdot e}{a_i}\tau=\frac{M_{cat}\cdot a_{sp}}{b_i\cdot F} \quad (4.41)$$

CPBR and CSTR performance are given in Figure 4.5 for OS (Figure 4.5a and b), RS (Figure 4.5c), and ping-pong (Figure 4.5d) mechanisms. As in the case of single-substrate reaction kinetics, CPBR performance is better than CSTR performance for sequential mechanisms in the case of a two-substrate reaction.

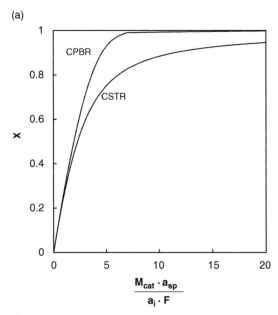

(a)

X

$$\frac{M_{cat}\cdot a_{sp}}{a_i\cdot F}$$

CPBR

CSTR

Figure 4.5 *Continuous packed-bed reactor (CPBR) and continuous stirred-tank reactor (CSTR) performance for: (a) OS mechanism at $a_i = 1\,mM$ and $b_i = 5\,mM$ (substrate A limiting), with $K_{MA} = 0.5\,mM$ and $K'_{MB} = 8$. (b) OS mechanism at $a_i = 5\,mM$, $b_i = 1\,mM$ (substrate B limiting), with $K_{MA} = 0.5\,mM$ and $K'_{MB} = 8$. (c) RS mechanism at $a_i = 1\,mM$, $b_i = 5\,mM$, with $K_{MA} = 2\,mM$, $K_{MB} = 1\,mM$, $K'_{MA} = 1\,mM$, and $K'_{MB} = 0.5\,mM$. (d) Ping-pong mechanism at limiting substrate concentration ($s_{lim} = a_i$ or $b_i) = 1\,mM$, substrate in excess concentration (b_i or $a_i) = 5\,mM$ with $K_{MA} = 0.5\,mM$, $K''_{MB} = 8$, and $\alpha = 0.5$ (see nomenclature in the text).*

(b)

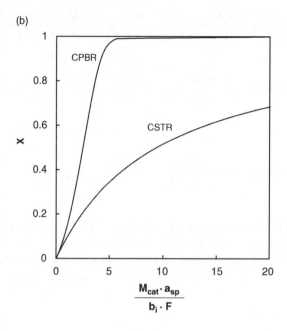

(c)

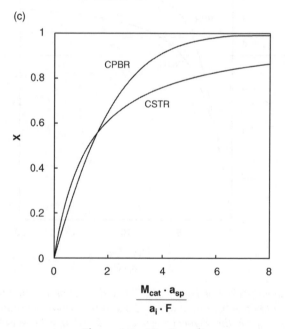

Figure 4.5 *(Continued)*

(d)

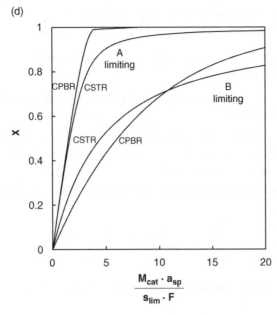

Figure 4.5 (Continued)

Solved Problems

4.1 An immobilized enzyme catalyzes a highly exothermic reaction in which the products of reaction are both inhibitors. What reactor configuration and mode of operation would be most suitable?

Resolution: In principle, immobilized enzymes are well suited to the development of continuous processes. However, in this case, CPBR is not adequate because heat transfer may be a serious problem, so it will be hard to maintain the temperature in the catalyst bed at the desired value. On the other hand, from a kinetic perspective, CSTR is quite inconvenient (see Section 4.4.1). An expanded- or fluidized-bed reactor may be an option because heat transfer is enhanced and from a kinetic perspective, although it is not as favorable as CPBR due to certain degree of backmixing, it is definitely a better option than CSTR. However, in the case of sequential BSTR operating with catalyst recovery, the kinetics will be favorable (just as in the case of CPBR) and high heat transfer rates will allow the temperature to be maintained at the desired value with little or no variation inside the stirred tank.

Answer: The best configuration is a BSTR operating sequentially with catalyst recovery after each batch.

4.2 The trisaccharide raffinose (α-D-galactosylsucrose) is a contaminant in sugar beet juice that retards sucrose crystallization. Raffinose (R) is hydrolyzed by soluble α-galactosidase (α-D-galactoside galactohydrolase, EC 3.2.1.22) from *Aspergillus niger* into sucrose and galactose. Its kinetic parameters are:

$$V_{max} = 2500 \ \mu mol_{hydrolyzed \ raffinose} \cdot g_{enzyme \ protein}^{-1} \cdot min^{-1}$$
$$K_M = 70 \ mM$$

A batch reactor is loaded with $10 \ g \cdot L^{-1}$ of enzyme preparation with 14% protein and $300 \ g \cdot L^{-1}$ of raffinose.

Determine the curve of operation of the batch reactor (raffinose removal versus operation time) and calculate the concentration of enzyme required as a function of the degree of removal of raffinose.

Resolution: According to Equation 4.5, using simple Michaelis–Menten kinetics (see Table 4.1):

$$\frac{[R]_i}{K_M} \cdot X - \ln(1 - X) = \frac{m_{cat} \cdot a_{sp}}{K_M} t \tag{4.42}$$

where X is the substrate conversion, representing raffinose removal in this case.

$$[R]_i = 300 \ g \cdot L^{-1} = 595 \ mM$$

V_{max} can be considered to correspond to the enzyme activity, so:

$$a_{sp} = 2.5 \ mmol_{hydrolyzed \ raffinose} \cdot g_{enzyme \ protein}^{-1} \cdot min^{-1} \cdot 0.14 \ g_{enzyme \ protein} \cdot g_{enzyme \ preparation}^{-1}$$
$$a_{sp} = 0.35 \ mmol_{hydrolyzed \ raffinose} \cdot g_{enzyme \ preparation}^{-1} \cdot min^{-1}$$

From Equation 4.42:

$$8.5X - \ln(1 - X) = 0.05 \cdot t$$

Answer: Solving for X:

t (min)	10	20	50	75	100	150	200	250	300
X	0.052	0.105	0.259	0.384	0.505	0.729	0.903	0.984	0.998

Raffinose removal is virtually complete after 5 hours of reaction; after 2.5 hours, only 73% of raffinose has been removed.

4.3 In order to increase raffinose removal (Problem 4.2), calculate the concentration of catalyst that should be added to the reactor as a function of batch reaction time if 90% removal is considered acceptable.

Resolution: From Equation 4.42:

$$8.5 \cdot 0.9 - \ln 0.1 = \frac{m_{cat} \cdot 0.35}{70} \cdot t_{X=0.9} \qquad m_{cat} = \frac{1990.5}{t_{X=0.9}}$$

Answer:

$t_{x=0.9}$ (mins)	200	100	50	25	20	15
m_{cat} (g·L^{-1})	10.0	19.9	39.8	79.6	99.5	132.7

For hydrodynamic considerations, the reaction time needed to obtain 90% raffinose removal is not less than 20 minutes (about $100\,g\cdot L^{-1}$ of enzyme), since considerable foaming might occur with a high protein concentration in the reactor.

4.4 Based on the kinetics of raffinose hydrolysis, design a reactor (determine the required volume) that will remove 95% of the raffinose after 5 hours of reaction with 100 g of enzyme preparation. Make a sensitivity analysis with respect to initial raffinose concentration down to $1\,g\cdot L^{-1}$ (estimated concentration in sugar beet juice).

Resolution: The design equation (see Equations 4.5 and 4.6) is:

$$\frac{[R]_i}{K_M}\cdot X - \ln(1-X) = \frac{\dfrac{M_{cat}}{V_{eff}}\cdot a_{sp}}{K_M}\cdot t$$

The data provided make this:

$$\frac{[R]_i}{70}\cdot 0.95 - \ln 0.05 = \frac{\dfrac{100}{V_{eff}}\cdot 0.35}{70}\cdot 300$$

$$V_{eff} = \frac{150}{0.013570\cdot [R]_i + 2.996}\cdot 300$$

Answer: Solving this equation gives:

$[R]_i$ (g·L^{-1})	1	2	5	10	50	100	200	300
$[R]_i$ (mM)	1.98	3.97	9.92	19.8	99.2	198	397	595
V_{eff} (L)	49.6	49.2	47.9	46.0	34.6	26.4	17.9	13.6

For the same degree of conversion (raffinose removal), the effective volume of the reactor (reaction volume) will decrease with increasing raffinose concentration. Note that in this case the amount of enzyme added is the same regardless of initial substrate concentration.

4.5 An enzyme is activated according to a mechanism in which the presence of the activator is required for substrate binding to the active site. Product delivery requires formation of a tertiary enzyme–activator–substrate complex and processing of the substrate at the active site. If the activator is the product of reaction, determine the kinetic rate expression and the corresponding kinetic parameters if the following results are obtained at varying substrate and product (activator) concentrations:

s (mM)	v (mmol · min^{-1} · g$_{enzyme\ protein}^{-1}$)			
	p = 2.5 mM	p = 5 mM	p = 10 mM	p = 20 mM
1	3.51	4.55	5.34	5.84
2	5.36	6.49	7.25	7.70
5	7.83	8.71	9.24	9.52
10	9.25	9.84	10.16	10.34
20	10.17	10.52	10.70	10.80
30	10.52	10.77	10.89	10.96

Draw the BSTR operation curve (substrate conversion versus dimensionless operation time, t_{dless}) given an initial substrate concentration of 7.5 mM if the initial substrate conversion is 1%.

Resolution: Given the kinetic mechanism and applying Michaelis–Menten's rapid equilibrium hypothesis (see Chapter 2), the kinetic rate equation (see Equation 2.18) is:

$$v = \frac{V_{max\ AP} \cdot s}{K_{MAP} + s} = \frac{V_{max} \cdot s}{K_M \left(1 + \frac{K_P}{p}\right) + s} \tag{4.43}$$

where V_{max} and K_M have the usual meaning and K_P is the activation constant by product (dissociation constant of the enzyme–product complex). From the data in the table, the double reciprocal (Lineweaver–Burk) linearization method gives:

	p = 2.5 mM	p = 5 mM	p = 10 mM	p = 20 mM
V_{maxAP} (mmol · min^{-1} · g^{-1})	11.30	11.28	11.30	11.30
K_{MAP} (mM)	2.22	1.48	1.11	0.934

$$K_{MAP} = K_M \left(1 + \frac{K_P}{p}\right) = K_M + \frac{K_M K_P}{p}$$

Plotting K_{MAP} versus p^{-1}, a straight line is obtained whose y- and x-axis intercepts are K_M and K_P^{-1}, respectively.

Answer:

$V_{max} = 11.3$ mmol · min^{-1} · g$_{enzyme\ protein}^{-1}$
$K_M = 0.75$ mM
$K_P = 4.94$ mM

$$v = \frac{11.3 \cdot s}{0.75 \left(\frac{4.94}{p} + 1\right) + s}$$

Since at the beginning of the reaction, the product is present in the reaction medium (initial conversion is 0.01), Equation 4.4 is now:

$$\int_{X_0}^{X} \frac{dX}{v(e,X)} = \int_{0}^{t} \frac{dt}{s_i}$$

Given the rate expression for v, with $V_{max} = k \cdot e$, $s = s_i \cdot (1 - X)$ and $p = s_i \cdot X$:

$$\int_{X_0}^{X} \frac{0.75 \left(\frac{4.94}{s_i \cdot X} + 1 \right) + s_i(1 - X)}{(1 - X)} \, dX = k \cdot e \cdot t$$

Integrating and dividing both members by K_M gives:

$$-0.659 \ln \frac{\frac{1-X}{X}}{\frac{1-X_0}{X}} - \ln \frac{1 - X}{1 - X_0} + 10(X - X_0) = \frac{k \cdot e \cdot t}{K_M}$$

With $X_0 = 0.01$:

$$-0.659 \ln \frac{1 - X}{X} - \ln(1 - X) + 2.918 = \frac{k \cdot e \cdot t}{K_M} = t_{dless}$$

Solving this equation gives:

X	0.1	0.2	0.3	0.4	0.5	0.6	0.7	0.8	0.9	0.95	0.99
t_{dless}	2.58	4.23	5.72	7.16	8.61	10.10	11.68	13.44	15.67	17.35	20.45

4.6 Penicillin G is hydrolyzed with immobilized penicillin G acylase (PGA) from *Bacillus megaterium* in order to produce the β-lactam nucleus 6-aminopenicillanic acid (6APA). CPBR and CSTR configurations have been proposed for the treatment of $100 \text{ g} \cdot \text{L}^{-1}$ of penicillin G potassium salt ($C_{16}H_{17}KN_2O_4S$) solution. The product 6APA is a noncompetitive inhibitor of PGA. The Michaelis constant for penicillin G is 0.06 M and the 6APA inhibition constant is 0.25 M. Both reactors are loaded with 2 kg of a PGA catalyst at $500 \text{ IU} \cdot \text{g}^{-1}$ (1 IU of PGA hydrolyzes 1 μmol of penicillin G per minute under these reaction conditions) and operated with a flow rate of $100 \text{ L} \cdot \text{h}^{-1}$.

(a) Calculate the steady-state conversion at the outlet of both reactors and discuss the result. (b) How much catalyst should be used in the poorer-performance reactor in order to match the production of the better one?

Resolution:

Data:

$[PenGK]_i = 100\,g \cdot L^{-1} = 0.27\,M$
$K_M = 0.06\,M = 60\,mM$
$K_{INC} = 0.25\,M = 250\,mM$
$a_{sp} = 0.5\,mmol \cdot min^{-1} \cdot g^{-1} = 30\,mmol \cdot h^{-1} g^{-1}$
$M_{cat} = 2000\,g$
$F = 100\,L \cdot h^{-1}$

According to Equations 4.23 and 4.33, given noncompetitive inhibition by the product (see Table 4.1):

CPBR	$3.42X - 2.08\ln(1 - X) + 2.43X^2 = 10$
CSTR	$4.5X + \frac{X}{1-X} + 1.08\frac{X^2}{1-X} + 4.86X^2 = 10$

Answer (a): Solving each of these equations:

$$X_{CPBR} = 0.906$$
$$X_{CSTR} = 0.711$$

As expected for enzyme inhibition kinetics, the performance of CPBR is better than that of CSTR. In this case, for similar operation conditions and a similar process task, the conversion in CSTR is 21.5% lower than that in CPBR.

Answer (b): Replacing $X = 0.906$ in the CSTR performance equation gives:

$$25.748 = \frac{M_{cat} \cdot 30}{60 \cdot 100}$$

$$M_{cat} = 5427\,g$$

In order to match the conversion of CPBR, the amount of enzyme catalyst used in CSTR must be increased by 171.3%. This highlights that CSTR is not an option for enzymatic reactions subject to product inhibition. Beyond the kinetic considerations, the reactor size will also be significantly bigger, since for a given mass of biocatalyst a much higher volume of reaction will be needed (assume that in CSTR the maximum allowable solid catalyst concentration in the reactor is around $100\,g \cdot L^{-1}$, so that about 90% of the reaction volume is liquid, while in CPBR the volume of liquid will not exceed 60% of the catalyst bed volume.

4.7 For the same system presented in Problem 4.6, assume now that the concentration of penicillin G in the feed stream is going to be varied in the range from 25 to $250\,g \cdot L^{-1}$ of penicillin G potassium salt. How will this affect the steady-state conversion in both reactors?

Resolution: According to Equations 4.23 and 4.33, given noncompetitive inhibition by the product (see Table 4.1):

CPBR:

$$12.67[PenGK]X - \left(1 + \frac{[PenGK]}{0.25}\right)\ln(1 - X) + 33.33[PenGK]^2X^2 = 10$$

CSTR:

$$16.67[PenGK]X + \frac{X}{1-X} + \frac{4[PenGK]X^2}{1-X} + 66.67[PenGK]^2X^2 = 10$$

Answer: Solving each of these equations gives:

[PenGK] $(g \cdot L^{-1})$	[PenGK] (M)	X_{CPBR}	X_{CSTR}	PD_{CSTR} *
25	0.067	0.999	0.877	12.2
50	0.135	0.993	0.834	16.0
100	0.270	0.906	0.711	21.5
150	0.403	0.729	0.574	21.3
200	0.538	0.583	0.463	20.6
250	0.672	0.481	0.384	20.2

*PD_{CSTR}: percentage reduction of X in CSTR with respect to CPBR

4.8 Invertase (β-fructofuranosidase, EC 3.2.1.26) catalyzes the inversion of sucrose, according to:

$$\underset{sucrose}{C_{12}H_{22}O_{11}} + H_2O \rightarrow \underset{glucose}{C_6H_{12}O_6} + \underset{fructose}{C_6H_{12}O_6}$$

A CPBR operates with *Saccharomyces cerevisiae* invertase immobilized in the surface of chitin flakes. The catalyst has a specific activity of $800\,IU \cdot g^{-1}$ (1 IU is defined as the amount of enzyme hydrolyzing 1 μmol of sucrose per minute at reaction conditions). The enzyme is inhibited by glucose. The Michaelis constant is 30 mM and the inhibition constant is 150 mM.

Determine the mass of catalyst required to process $450\,L \cdot h^{-1}$ of sugarcane molasses containing $171\,g \cdot L^{-1}$ of sucrose (S) if 90% steady-state sucrose conversion into glucose plus fructose is required in order to prepare invert sugar for confectionery. Solve assuming that glucose is a competitive inhibitor of invertase and then assuming that it is a total noncompetitive inhibitor of invertase. Discuss the results.

Resolution:

Data:

$a_{sp} = 0.8\,mmol\,min^{-1} \cdot g^{-1} = 48\,mmol \cdot h^{-1} \cdot g^{-1}$
$K_M = 30\,mM$
K_{IC} (or K_{INC}) $= 150\,mM$
$a_{sp} = 0.5\,mmol \cdot min^{-1} \cdot g^{-1} = 30\,mmol \cdot h^{-1} \cdot g^{-1}$
$[S]_i = 171\,g \cdot L^{-1} = 500\,mM$
$F = 450\,L \cdot h^{-1}$

According to Equation 4.23 and Table 4.1:

Assumed Mechanism	Reactor Performance Equation
Competitive	$21.98 = \dfrac{M_{cat} \cdot 48}{30 \cdot 450}$
Total Noncompetitive	$44.48 = \dfrac{M_{cat} \cdot 48}{30 \cdot 450}$

Answer: Solving these equations gives:

$$M_{cat,CPBR} = 6.18\,\text{kg}$$
$$M_{cat,CSTR} = 12.51\,\text{kg}$$

The mass of catalyst required in a CSTR is more than double that in a CPBR, which is a consequence of the very unfavorable conditions under which the enzyme works in CSTR (lowest substrate concentration and highest inhibitor concentration); in addition, the volume of CSTR for the same amount of catalyst is no less than five times that in CPBR, so in this case the volume of CSTR will be about ten times higher than that of CPBR.

4.9 Glucoamylase catalyzes the hydrolysis of maltose according to:

$$\underset{\text{maltose}(S)}{C_{12}H_{22}O_{11}} + H_2O \rightarrow \underset{\text{glucose}(P)}{2C_6H_{12}O_6}$$

The following rate data have been obtained with glucoamylase immobilized in the surface of activated chitin:

s	2.92	14.62	29.2	43.9	58.5	73.1
v_0	37.78	137.21	204.20	244.23	270.52	289.22
v	13.81	60.74	105.37	139.86	166.96	188.95

s: maltose initial concentration (mM)
v_0: initial reaction rate of maltose hydrolysis at 0 initial glucose concentration (μmoles $\cdot$ min^{-1} $\cdot$ g$_{cat}^{-1}$)
v: initial reaction rate of maltose hydrolysis at 166.7 mM initial glucose concentration (μmoles $\cdot$ min^{-1} $\cdot$ g$_{cat}^{-1}$)

The reaction is carried out in CPBR and CSTR in order to process a syrup with $180\,\text{g} \cdot \text{L}^{-1}$ of maltose. Conversion is expected to be 85%. Determine the specific productivity (grams of glucose produced per hour and per gram of catalyst.

Resolution: Given the kinetic mechanism and applying the rapid equilibrium hypothesis of Michaelis–Menten (see Chapter 2), the kinetic rate equation (see Equation 2.18) is:

$$v = \frac{V_{maxAP} \cdot s}{K_{MAP} + s} = \frac{V_{max} \cdot s}{K_M\left(1 + \frac{K_P}{p}\right) + s} \tag{4.44}$$

where K_M is the Michaelis constant for maltose and K_P is the glucose inhibition constant.

From the data provided, by nonlinear regression of the rate data (see Chapter 2):
$V_{max} = 400\mu\text{mol} \cdot \text{min}^{-1} \cdot \text{g}_{cat}^{-1}$
$K_M = 28\,\text{mM}$
$K_P = 87\,\text{mM}$

The K_P value is prone to error, since it was calculated with only one series of data at initial glucose concentration, which in addition was quite a bit lower than the value of K_P determined. It is strongly recommended that more data be generated at initial glucose concentrations of 100 mM or higher.

The volumetric productivity of a process (π) is the amount of product produced per unit time and unit reaction volume. The specific productivity (π_{sp}) is the amount of product produced per unit time and unit mass of catalyst. Given this:

$$\pi = \frac{s_i \cdot X}{\tau} \tag{4.45}$$

$$\pi_{sp} = \pi \cdot \frac{V_{eff}}{M_{cat}} \tag{4.46}$$

From Equation 4.23 (or Equation 4.33):

$$\tau = \frac{f(s_i, X) \cdot K_M \cdot V_{eff}}{M_{cat} \cdot a_{sp}}$$

From Equations 4.45 and 4.46:

$$\pi = \frac{s_i \cdot X \cdot M_{cat} \cdot a_{sp}}{f(s_i, X) \cdot K_M \cdot V_{eff}} \tag{4.47}$$

$$\pi_{sp} = \frac{s_i \cdot X \cdot a_{sp}}{f(s_i, X) \cdot K_M}. \tag{4.48}$$

The values of $f(s_i,X)$ for competitive inhibition are given in Table 4.1 for CPBR and CSTR. With $s_i = 180 \, g \cdot L^{-1}$ (526 mM), $X = 0.85$, and $a_{sp} = 0.4 \, mmol \cdot min^{-1} \cdot g^{-1}$:

CPBR	$f(s_i, X) = 526 \cdot 0.85 \left(\dfrac{1}{28} - \dfrac{1}{87} \right) - \left(1 + \dfrac{526}{87} \right) \cdot \ln 0.15 = 24.2$
CSTR	$f(s_i, X) = 526 \cdot \dfrac{0.85}{28} + \dfrac{0.85}{0.15} + \dfrac{526}{87} \cdot \dfrac{0.85^2}{0.15} = 50.8$

From Equation 4.45:

CPBR	$\pi_{sp} = \dfrac{526 \cdot 0.85 \cdot 0.4}{24.2 \cdot 28} = 0.264$ mmoles maltose hydrolyzed $\cdot \, min^{-1} \cdot g_{cat}^{-1}$
CSTR	$\pi_{sp} = \dfrac{526 \cdot 0.85 \cdot 0.4}{50.8 \cdot 28} = 0.126$ mmoles maltose hydrolyzed $\cdot \, min^{-1} \cdot g_{cat}^{-1}$
CPBR	$\pi_{sp} = 0.264 \cdot 2 \cdot \dfrac{180 \cdot 60}{1000} = 5.7$ g gluose $\cdot \, h^{-1} \cdot g_{cat}^{-1}$
CSTR	$\pi_{sp} = 0.126 \cdot 2 \cdot \dfrac{180 \cdot 60}{1000} = 2.7$ g gluose $\cdot \, h^{-1} \cdot g_{cat}^{-1}$

Answer: The specific productivity in CPBR is 5.7 g glucose per hour per gram of catalyst, while that in CSTR is 2.7 g glucose per hour per gram of catalyst. The specific productivity is CSTR is thus 52.6% lower than that in CPBR, which is to be expected for a reaction inhibited by product.

4.10 β-galactosidases are usually competitively inhibited by the product galactose. However, some authors have claimed that inhibition by galactose is noncompetitive. Assuming the latter holds, determine the optimum concentration of lactose in the feed stream to a CPBR and to a CSTR with immobilized β-galactosidase.

Resolution: It is reasonable to assume that optimum substrate concentration will maximize reactor productivity. Given this, Equation 4.47 can be rewritten as:

$$\pi = C \cdot \frac{s_i}{f(s_i)} \tag{4.49}$$

where:

$$C = \frac{X \cdot M_{cat} \cdot a_{sp}}{K_M \cdot V_{eff}}$$

The condition that be met in order to optimize substrate concentration is:

$$\frac{d\pi}{ds_i} = 0 \tag{4.50}$$

From Equations 4.49 and 4.50:

$$s_i \cdot \frac{df(s_i)}{ds_i} = f(s_i) \tag{4.51}$$

From the corresponding equations in Table 4.1:

$$\text{CPBR} \quad f(s_i) = s_i X \cdot \left[\frac{1}{K_M} - \frac{1}{K_{INC}}\right] - \left[1 + \frac{s_i}{K_{INC}}\right] \cdot \ln(1-X) + \frac{s_i^2 X^2}{2K_M K_{INC}} \tag{4.52}$$

$$\text{CSTR} \quad f(s_i) = \frac{s_i \cdot X}{K_M} + \frac{X}{1-X} + \frac{s_i X^2}{K_{INC}(1-X)} + \frac{s_i^2 X^2}{K_M K_{INC}} \tag{4.53}$$

Answer: Deriving Equations 4.52 and 4.53 and replacing them in Equation 4.51, the optimum concentrations of substrate in the feed stream (s_i^*) for CPBR and CSTR are:

$$\text{CPBR} \quad s_i^* = \frac{\sqrt{-2 \cdot K_M \cdot K_{INC} \cdot \ln(1-X)}}{X}. \tag{4.54}$$

$$\text{CSTR} \quad s_i^* = \frac{\sqrt{K_M \cdot K_{INC} \cdot \frac{X}{1-X}}}{X}. \tag{4.55}$$

To illustrate this, evaluate for a β-galactosidase with $K_M = 50\,mM$ and $K_{INC} = 75\,mM$ how the optimum concentration in the reactor inlet will vary with the steady-state conversion at the reactor outlet in both configurations.

Answer: From Equations 4.54 and 4.55:

X	0.1	0.25	0.4	0.5	0.6	0.7	0.8	0.9	0.95
s_i^* CPBR (mM)	198.7	131.4	109.4	102.0	97.7	96.0	97.1	103.2	111.6
s_i^* CSTR (mM)	204.1	141.4	125.0	122.5	125.0	133.6	153.1	204.1	281.0

In this case, the optimum lactose feed concentration is lower than its solubility, but there may be instances in which these mathematical optima are higher than substrate solubility, in which case it will not be applicable and a concentration close to saturation will be recommended.

4.11 Determine the inversion point (IP) in the operation curves (conversion versus dimensionless time) of CPBR and CSTR for an enzyme subjected to inhibition by high substrate concentration. Assume that the (uncompetitive) inhibition constant of the substrate is twice the value of the Michaelis constant. Give the results for varying dimensionless substrate concentrations in the feed stream to the reactor. The IP is defined as the value of dimensionless time in which both reactor configurations have the same performance. Demonstrate that above the IP, CPBR performance is superior to that of CSTR, but that the opposite holds below the IP.

Resolution: Equaling the performance equations of both reactor configurations (see Table 4.1) gives:

$$\frac{s_i X}{K_M} - \ln(1-X) + \frac{s_i^2 X}{K_M K_S}\left(1 - \frac{X}{2}\right) = \frac{s_i X}{K_M} + \frac{X}{1-X} + \frac{s_i^2 X}{K_M K_S}(1-X) \qquad (4.56)$$

With $K_S/K_M = 2$, Equation 4.56 becomes:

$$\frac{X}{1-X} + \ln(1-X) = \frac{\left(\dfrac{s_i}{K_M} \cdot X\right)^2}{4} \qquad (4.57)$$

$$\frac{s_i}{K_M} = 2\sqrt{\frac{1}{X \cdot (1-X)} + \frac{\ln(1-X)}{X^2}}$$

Answer: From Equation 4.57:

$s_i \cdot K_M^{-1}$	1.5	2	3	4	5	10	20
X_{IP}	0.085	0.413	0.694	0.812	0.872	0.964	0.990
IP	0.31	2.01	5.31	8.78	12.56	38	122

4.12 Invert sugar is produced from sucrose using two continuous reactors in series, with 5 kg each of invertase immobilized in activated chitosan particles. The inetic parameters of the enzyme are $V_{max} = 200$ μmoles $\cdot$ min^{-1} $\cdot$ g$_{cat}^{-1}$ and $K_M = 68.5$ mM. A syrup with 200 g $\cdot$ L^{-1} of sucrose is fed at a flow rate of 180 L $\cdot$ h^{-1}.

(a) If one of the reactors is a CPBR and the other is a CSTR, determine the right sequence of reactors. (b) In what percentage should the catalyst mass in the second reactor increase to make the performance of the unfavorable sequence equal to the one of the favorable sequence?

Resolution (a): Data:

$s_i = 200$ g $\cdot$ L$^{-1} = 584.8$ mM
$V_{max} = 200$ μmol min$^{-1} \cdot$ g$^{-1} = 0.2$ mmol $\cdot$ min$^{-1} \cdot$ g^{-1}
$K_M = 68.5$ mM
$M_{cat} = 5000$ g
$F = 180$ L $\cdot$ h$^{-1} = 3$ L $\cdot$ min^{-1}

In this case, the substrate concentration at the outlet of the first reactor will correspond to the substrate concentration at the inlet of the second. So, for the case where in the second reactor $X_0 \neq 0$, the corresponding equations in Table 4.1 become:

$$\text{CPBR} \quad \frac{s_i}{K_M}(X - X_0) - \ln\frac{1-X}{1-X_0} = \frac{M_{cat} \cdot a_{sp}}{K_M \cdot F} \qquad (4.58)$$

$$\text{CSTR} \quad \frac{s_i}{K_M}(X - X_0) + \frac{X - X_0}{1 - X} = \frac{M_{cat} \cdot a_{sp}}{K_M \cdot F} \qquad (4.59)$$

Given a CPBR followed by a CSTR reactor:

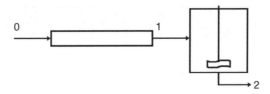

CPBR:

$$\frac{s_i}{K_M}X_1 - \ln(1 - X_1) = \frac{M_{cat,CPBR} \cdot a_{sp}}{K_M \cdot F}$$
$$8.537X_1 - \ln(1 - X_1) = 4.866$$
$$X_1 = 0.491$$

CSTR:

$$\frac{s_i}{K_M}(X_2 - X_1) + \frac{X_2 - X_1}{1 - X_2} = \frac{M_{cat,CSTR} \cdot a_{sp}}{K_M \cdot F}$$
$$8.537(X_2 - 0.491) + \frac{X_2 - 0.491}{1 - X_2} = 4.866$$
$$X_2 = 0.829$$

Given instead a CSTR followed by a CPB reactor:

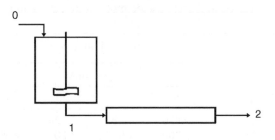

CSTR:

$$\frac{s_i}{K_M}X + \frac{X}{1-X} = \frac{M_{cat,CSTR} \cdot a_{sp}}{K_M \cdot F}$$

$$8.537X_1 + \frac{X_1}{1-X_1-} = 4.866$$

$$X_1 = 0.467$$

CPBR:

$$\frac{s_i}{K_M}(X_2 - X_1) - \ln\frac{1-X_2}{1-X_1} = \frac{M_{cat,CPBR} \cdot a_{sp}}{K_M \cdot F}$$

$$8.537(X_2 - 0.467) - \ln\frac{1-X_2}{1-0.467} = 4.866$$

$$X_2 = 0.871$$

Resolution (b):

$$\frac{s_i}{K_M}(X_2 - X_1) + \frac{X_2 - X_1}{1-X_2} = \frac{M_{cat,CSTR} \cdot a_{sp}}{K_M \cdot F}$$

$$8.537(0.871 - 0.491) + \frac{0.871 - 0.491}{1 - 0.871} = \frac{M_{cat,CSTR} \cdot 0.2a_{sp}}{68.5 \cdot 3}$$

$$M_{cat,CSTR} = 6360\,\text{g}$$

Answer (a): CSTR followed by CPBR is the best configuration, because a higher global conversion is obtained (0.871 versus 0.829).

Answer (b): The mass of catalyst in the second reactor must be increased by 27.2% (6360 versus 5000 g) in order for the CPBR–CSTR configuration to match the performance of the CSTR–CPBR configuration.

4.13 The noncaloric sweetener Aspartame (L-aspartyl-L-phenylalanine methyl ester) is produced by a kinetically controlled synthesis reaction catalyzed by the metalloprotease thermolysin (EC 3.4.24.27) according to:

$$\text{Z-Aspartic acid} + \text{L-phenylalanine methyl ester} \rightarrow \text{Z-Aspartame}$$
$$\quad\;\text{ZA} \qquad\qquad\qquad\qquad \text{PAME} \qquad\qquad\qquad\qquad \text{ZAM}$$

The Z-protecting group (benzyloxy-carbonyl) is then removed chemically in order to yield the product. The following initial reaction rate (v) data of Z-Aspartame synthesis have been obtained using a thermolysin from recombinant *Escherichia coli*:

| | v (μmol $\cdot$ min^{-1} $\cdot$ g^{-1}) | | | |
| | [ZA] (mM) | | | |
[PAME] (mM)	5	10	20	40
2.5	0.550	1.087	2.128	4.082
5	0.980	1.923	3.704	6.900
10	1.613	3.125	5.882	10.526
20	2.381	4.545	8.333	14.286

(a) Determine the probable reaction mechanism and evaluate the corresponding kinetic parameters. (b) Draw the operation curve of a BSTR with an enzyme load of $50\,g \cdot L^{-1}$ at 100 mM ZA and 50 mM PAME initial concentration. Compare it with that obtained at 50 mM ZA and 100 mM PAME.

Resolution (a): Using the double reciprocal of the rate data provided, from Equation 2.19 (assuming PAME to be substrate A and ZA to be substrate B):

| | [B] mM | | | |
	5	10	20	40
$V_{maxAP(B)}$*	4.532	8.333	14.280	22.232
$\Delta_{MAP(B)} = K_{MAP(B)} \cdot V_{maxAP(B)}^{-1}$	4.000	2.000	1.000	0.500
$K_{MAP(B)}$**	18.106	16.667	14.275	11.116

* μmol $\cdot$ min^{-1} $\cdot$ g^{-1}
** mM

| | [A] mM | | | |
	2.5	5	10	20
$V_{maxAP(A)}$*	48.8	50.4	50.0	50.0
$\Delta_{MAP(A)} = K_{MAP(A)} \cdot V_{maxAP(A)}^{-1}$	8.99	5.00	3.00	2.00
$K_{MAP(A)}$**	438.5	252.4	150.0	100.0

* μmol $\cdot$ min^{-1} $\cdot$ g^{-1}
** mM

The mechanism is OS, PAME being the first substrate to bind to the enzyme, since $V_{maxAP(A)}$ is independent of [A] and equal to V_{max}. So (see Table 2.4):

$$V_{maxAP(B)} = \frac{V_{max} \cdot b}{b + K'_{MB}}$$

(4.60)

Linearizing Equation 4.60, from the data in the first table:

$K'_{MB} = 50\,\text{mM}$

$V_{max} = 50\,\mu\text{mol} \cdot \text{min}^{-1} \cdot \text{g}^{-1}$

and:

$$\Delta_{MAP(B)} = \frac{K_{MAP(B)}}{V_{maxAP(B)}} = \frac{K_{MA} \cdot K'_{MB}}{V_{max} \cdot b} \tag{4.61}$$

Linearizing Equation 4.61, from the data in the first table:

$$\frac{K_{MA} \cdot K'_{MB}}{V_{max}} = 20$$

$$K_{MA} = 20\,\text{mM}$$

Since, from Table 2.4:

$$K_{MAP(A)} = K'_{MB} + \frac{K_{MA} \cdot K'_{MB}}{a} \tag{4.62}$$

K_{MA} can also be obtained by linearizing Equation 4.62. From the data in the second table:

$$K_{MA} = 18\,\text{mM}$$

There is a 10% difference, which is a consequence of cumulative error during the calculations.

Answer (a): The mechanism is OS, with PAME being the first substrate to bind to the enzyme, forming an acyl complex with thermolysin, which is then attacked nucleophilically by ZA to yield the product ZAM.

The kinetic parameters are: $V_{max} = 50\,\mu\text{mol} \cdot \text{min}^{-1} \cdot \text{g}^{-1}$; $K'_{MB} = 50\,\text{mM}$; $K_{MA} \approx 20\,\text{mM}$.

The rate equation for ZAM synthesis with thermolysin is then:

$$v = \frac{V_{max} \cdot a \cdot b}{a \cdot b + K'_{MB} \cdot a + K_{MA} \cdot K'_{MB}} = \frac{50 \cdot a \cdot b}{a \cdot b + 50 \cdot a + 1000}$$

Resolution (b): The batch reactor performance when substrate A is limiting is represented by Equation 4.16. Data: $K_{MA} = 20\,\text{mM}$; $K'_{MB} = 50\,\text{mM}$; $a_{sp} = 0.05\,\text{mmol} \cdot \text{min}^{-1} \cdot \text{g}^{-1}$; $a_i = 50\,\text{mM}$; $b_i = 100\,\text{mM}$; $m_{cat} = 50\,\text{g} \cdot \text{L}^{-1}$.

Then:

$$\frac{20 \cdot 50}{50 \cdot 50} \ln \frac{100 - 50 \cdot X}{100(1 - X)} - \frac{50}{50} \ln \frac{100 - 50 \cdot X}{100} + X = \frac{50 \cdot 0.05}{50} \cdot t$$

$$t = 8 \cdot \ln \frac{2 - X}{2 - 2 \cdot X} - 20 \cdot \ln \frac{2 - X}{2} + 20 \cdot X \tag{4.63}$$

The batch reactor performance when substrate B is limiting is represented by Equation 4.17. Now $a_i = 100\,mM$ and $b_i = 50\,mM$, so:

$$t = 8 \cdot \ln\frac{2 - X}{2 - 2 \cdot X} - 20 \cdot \ln(1 - X) + 20 \cdot X \tag{4.64}$$

Answer (b): From Equations 4.63 and 4.64:

X	0.1	0.2	0.3	0.4	0.5	0.6	0.7	0.8	0.9	0.95	0.99
$t_{A\ limiting}$ *	3.5	7.1	10.8	14.8	19.0	23.6	28.8	35.0	43.6	50.7	64.8
$t_{B\ limiting}$ *	4.5	9.4	14.7	20.5	27.1	34.8	44.3	57.0	77.7	97.7	143.3

*t in minutes

As can be appreciated (see Figure 4.6), with an OS mechanism the best option is to limit the reaction with the first substrate to bind to the enzyme. In the current case it is best to limit the reaction with the acyl donor (PAME) and to use an excess of acceptor (ZA). This is the case in strictly kinetic terms, but in a specific situation it might not be so from an economic perspective. If the cost of ZA is significantly higher than that of PAME then using an excess ZA may be inadequate. In the synthesis of β-lactam antibiotics, for example, the acyl donor binds first and then the nucleophile, but the latter is much more expensive than the former so the kinetic advantage will have to be confronted with an economic evaluation.

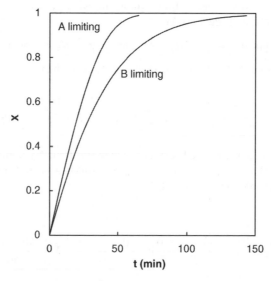

Figure 4.6   *BSTR performance under phenylalanine methyl ester (A) and Z-aspartic acid (B) as limiting substrates. X: limiting substrate conversion; t: reaction time.*

4.14 The enzymatic esterification of ethanol and isovaleric acid in organic solvent media is catalyzed by a commercial lipase from *Candida antarctica* with a protein content of 45% (w/w). The reaction is:

$$\underset{\substack{\text{isovaleric acid} \\ (A)}}{C_5H_{10}O_2} + \underset{\substack{\text{ethanol} \\ (B)}}{C_2H_6O} \rightarrow \underset{\substack{\text{3-methyl butanoic acid ethyl ester} \\ (P)}}{C_7H_{14}O_2} + H_2O$$

The mechanism of reaction has been determined to be ping-pong bi–bi, the acid first binding to the enzyme and the alcohol then binding to the acyl–enzyme complex formed, once water has been released. The experimentally determined kinetic parameters of the reaction are:

$V_{max} = 2.67\ \mu$moles of ester produced per minute per milligram of lipase protein
$V_{max} = 1.2\ \text{mmol} \cdot \text{min}^{-1} \cdot \text{g}^{-1}$ of commercial lipase
$K_{MA} = 62.5\ \text{mM}$
$K''_{MB} = 33.1\ \text{mM}$
$\alpha = 0.27$

(a) Calculate the amount of lipase per liter of reaction volume as a function of limiting substrate conversion required to produce 100 g of product (3-methyl butanoic acid ethyl ester; $C_7H_{14}O_2$) per batch (6 hours' reaction time) if the limiting substrate is the acid and the alcohol is in 50% molar excess. (b) Calculate the same if ethanol is the limiting substrate and isovaleric acid is in 50% molar excess.

Resolution (a): 100 g of P $= 769$ moles; p $= 769$ mM.

Let X be the conversion of A into P at the end of the batch. Then:

$$p = a_i \cdot X$$
$$b_i = 1.5 \cdot a_i$$

From Equation 4.19:

$$X - \frac{\alpha \cdot K_{MA}}{a_i} \cdot \ln(1 - X) - \frac{K''_{MB}}{a_i} \cdot \ln(b_i - a_i \cdot X) = \frac{k \cdot e}{a_i} t = \frac{m_{cat} \cdot a_{sp}}{a_i} t$$

$$X - \frac{0.27 \cdot 62.5 \cdot X}{769} \cdot \ln(1 - X) - \frac{33.1}{769} \cdot X \cdot \ln\left(\frac{1153.5}{X} - 769\right) = \frac{m_{cat} \cdot 1.2 \cdot X}{769} \cdot 360$$

$$(4.65)$$

Resolution (b): Let X be the conversion of B into P at the end of the batch. Then:

$$p = b_i \cdot X$$
$$a_i = 1.5 \cdot b_i$$

From Equation 4.20:

$$X - \frac{K''_{MB}}{b_i} \cdot \ln(1 - X) - \alpha \cdot \frac{K_{MA}}{b_i} \cdot \ln(a_i - b_i \cdot X) = \frac{k \cdot e}{b_i} t = \frac{m_{cat} \cdot a_{sp}}{b_i} t$$

$$X - \frac{33.1}{769} \cdot X \cdot \ln(1 - X) - 0.27 \cdot \frac{62.5}{769} \cdot X \cdot \ln\left(\frac{1153.5}{X} - 769\right) = \frac{m_{cat} \cdot 1.2 \cdot X}{769} \cdot 360$$

$$(4.66)$$

Answer: Solving Equations 4.65 and 4.66 for (a) isovaleric acid limiting and (b) ethanol limiting gives:

X		0.25	0.5	0.6	0.7	0.8	0.9	0.95	0.99
m_{cat} (g·L^{-1})	A limiting	1.16	1.25	1.28	1.31	1.34	1.39	1.43	1.50
	B limiting	1.48	1.55	1.58	1.61	1.65	1.71	1.77	1.90

In this case, it is convenient to limit the reaction with the substrate that reacts first (isovaleric acid) since under comparable conditions the mass of catalyst required for the same performance will be lower.

Supplementary Problems

4.15 Select the most appropriate type of reactor with which to perform the hydrolysis of glutaryl–ACA into 7ACA (7-amino3-(acetyloxy-methyl)-7-amino-8-oxo-5-thia-1-azabicyclo (4.2.0) oct-2-ene-2-carboxylic acid; or simply 7-amino cephalosporanic acid) and glutamic acid by polyacrylamide immobilized glutaryl 7-ACA acylase (EC 3.5.1.93) from recombinant *Escherichia coli*. The reaction is carried out at 25 °C and pH 8.0. Both reaction products are inhibitors of the enzyme, which rapidly inactivates at neutral or acid pH.

Answer: BSTR operating in sequential mode with catalyst recovery. Alternatively, a recirculation batch reactor can be used, where the substrate is pumped to a catalyst module from a reservoir in which the pH is adjusted after each passage. This latter configuration was proposed for the efficient production of 6-aminopenicillanic acid with penicillin G acylase [9].

4.16 A BSTR of 600 L working volume is used for the removal of lactose from spent cheese whey (40 g·L^{-1}) with fungal β-galactosidase at 50 °C and pH 4.5. The enzyme is strongly competitively inhibited by the product, galactose; its kinetic parameters under such conditions are $K_M = 90$ mM and $K_I = 9$ mM. The operation is designed to remove 80% of the lactose after 5 hours of reaction.

(a) Determine the amount of enzyme required in katals (1 katal is the amount of enzyme that hydrolyzes 1 mole of lactose per second under these conditions). (b) A mutant β-galactosidase has been developed through directed evolution, in which K_M is reduced to one-third and K_I is increased three times. What amount of this mutant enzyme will be required to perform the same task?

Answer: (a) $3.952 \cdot 10^{-2}$ katals; (b) $8.237 \cdot 10^{-3}$ katals.

4.17 The following results were obtained with immobilized α-galactosidase (alpha-D-galactoside galactohydrolase, EC 3.2.1.22) from the hyperthermophile *Thermotoga neapolitana* in the hydrolysis of the disaccharide melibiose (D-galactose-α(1 → 6)-D-glucose) into D-glucose and D-galactose at 90 °C:

[M]	0.1	0.3	0.5	1	2	5	10	100	250	500	750	1000	1200
v	2.97	8.74	14.27	27.17	49.34	93.75	125.00	58.824	27.174	14.270	9.673	7.315	6.121

[M]: melibiose mM initial concentration
v: initial reaction rate of melibiose hydrolysis ($\mu mol \cdot min^{-1} \cdot g^{-1}$)

(a) Identify the mechanism of reaction and determine the kinetic parameters of the rate equation. (b) The reaction is carried out in CPBR and CSTR. Determine the geometric locus of the IP (defined as the value of dimensionless time in which both reactor configurations have the same performance; above the IP, CPBR performance is superior to that of CSTR, below it, CSTR performance is superior). How does the IP vary with melibiose concentration in the feed stream in the range from 5 to $50 \, g \cdot L^{-1}$ when both reactors operate under the same conditions of enzyme load and flow rate?

Answer: (a) Total uncompetitive inhibition by high melibiose concentration:
$V_{max} = 300 \, \mu mol \cdot min^{-1} \cdot g^{-1}$
$K_M = 10 \, mM$
$K_S = 25 \, mM$

(b)

[M] ($g \cdot L^{-1}$)	10	15	20	25	30	35	40	45	50
X_{IP}	0.617	0.805	0.882	0.920	0.943	0.957	0.967	0.973	0.978
IP	4.22	8.87	14.04	19.87	26.48	33.85	42.05	51.02	60.84

X_{IP}: conversion at the IP.

4.18 Nitrile hydratase catalyzes the synthesis of acrylamide according to:

$$CH_2{=}CH{-}CN + H_2O \rightarrow CH_2CH{=}CONH_2$$
$$\text{acrylonitrile} \qquad\qquad\qquad \text{acrylamide}$$

The enzyme has been immobilized in very small spherical particles of glyoxyl agarose (density $0.7 \, g \cdot cm^{-3}$) and its specific activity is $350 \, IU \cdot mL_{catalyst}^{-1}$. Acrylamide is a non-competitive inhibitor of the enzyme; the Michaelis constant for acrylonitrile is $50 \, mM$ and the inhibition constant of acrylamide is $90 \, mM$ (consider these parameters as the intrinsic parameters). CPBR and CSTR will be evaluated in the synthesis of acrylamide, processing $10.6 \, g \cdot L^{-1}$ of acrylamide solution at a flow rate of $48 \, L \cdot h^{-1}$, with a catalyst mass of 1 kg each.

(a) Determine the steady-state substrate conversion and residual acrylonitrile concentration in the product stream for both reactors. (b) Draw the curves representing steady-state conversion as a function of catalyst mass in the reactor. (c) How much enzyme should be added to the lower-performance reactor in order to match the performance of the other reactor, maintaining the flow rate at $48 \, L \cdot h^{-1}$? (d) If the enzyme load is maintained at 1 kg, how much less flow rate will have to be processed? Assume that no diffusional restrictions are present at the conditions of reactor operation.

Answer:

(a) $X_{CPBR} = 0.897$

   $X_{CSTR} = 0.687$

(b)

M_{cat} (g)	100	250	500	750	1000	1250	1500	2000
X_{CPBR}	0.200	0.405	0.646	0.797	0.897	0.954	0.981	0.997
X_{CSTR}	0.173	0.329	0.496	0.608	0.687	0.744	0.786	0.842

(c) At $F = 48\,L \cdot h^{-1}$:

X	0.200	0.405	0.646	0.797	0.897	0.954	0.987	0.997
$M_{catCPBR}$ (g)	100	250	500	750	1000	1250	1500	2000
$M_{catCSTR}$ (g)	121.3	349.7	840.5	1577	2945	6129	14 132	86 511
% increase	21.3	39.9	68.1	110.3	194.5	390	842	4226

(d) At $M_{cat} = 1$ kg:

X	0.2	0.5	0.6	0.7	0.8	0.9	0.95	0.99
F_{CPBR} ($L \cdot min^{-1}$)	7.99	2.36	1.78	1.37	1.06	0.79	0.65	0.48
F_{CPBR} ($L \cdot min^{-1}$)	6.59	1.58	1.10	0.76	0.50	0.26	0.14	0.03
% decrease	17.5	33.1	38.2	44.5	52.8	67.1	78.5	93.8

The difference in reactor performance increases significantly with X. At very high values of X (above 90%), either the amount of catalyst mass required in CSTR will be impossible to handle or the feed flow rate will be so low that it is devoid of practical meaning.

4.19 An immobilized enzyme is used in CPBR and CSTR configurations, catalyzing a reaction in which the product is a competitive inhibitor and the inhibition constant is 100% higher than the Michaelis constant. Both reactors are fed at a dimensionless substrate concentration of 10.

Calculate the volume ratio of both reactors if the CSTR works with a catalyst load of $100\,g \cdot kg^{-1}_{solution}$ and the apparent density of the catalyst bed in the CPBR is $0.4\,kg \cdot L^{-1}$. Evaluate in the conversion range from 0.1 to 0.9. Then calculate the same if the substrate concentration in the feed stream is reduced by 90%.

Answer:

X	0.1	0.2	0.3	0.4	0.5	0.6	0.7	0.8	0.9
$\left(\dfrac{V_{CSTR}}{V_{CPBR}}\right)_{s_i/K_M=10}$	4.12	4.28	4.47	4.74	5.11	5.65	6.53	8.20	12.78
$\left(\dfrac{V_{CSTR}}{V_{CPBR}}\right)_{s_i/K_M=1}$	4.17	4.37	4.63	4.97	5.43	6.09	7.14	9.10	14.29

4.20 A CSTR with immobilized xylose isomerase (D-xylose ketoisomerase, EC 5.3.1.5) must be designed to treat $24\,\mathrm{m^3 \cdot h^{-1}}$ of a wood hydrolysate containing $18\,\mathrm{g \cdot L^{-1}}$ of xylose in order to convert it into xylulose, which is a fermentable sugar to many ethanol-producing yeasts. The expected conversion is 50%. The immobilized enzyme has a specific activity of $600\,\mathrm{IU \cdot g^{-1}}$, the Michaelis constant for xylose is 15 mM, and the xylulose competitive inhibition constant is 90 mM.

(a) Design a column reactor under ideal conditions (plug-flow regime, absence of mass-transfer limitations; assume that the reversible reaction is not significant under the conditions of reactor operation). Assume a catalyst bed height-to-diameter ratio of 5, a head space of 10% of the catalyst bed height, and an apparent density of the catalyst bed of $0.2\,\mathrm{g \cdot cm^{-3}}$.

(b) A mutant xylose isomerase was developed by directed evolution, which after immobilization produced a catalyst with 25% higher specific activity, 20% higher xylose affinity and 20% lower xylulose affinity than the parent enzyme. This enzyme made it possible to conduct the reaction at up to 85% conversion. What will the dimensions of the new reactor be (given that the remaining conditions are not changed)?

Answer: (a) Reactor diameter is 37.8 cm; bed height is 1.89 m; reactor height is 2.08 m. (b) New reactor diameter is 43.3 cm; bed height is 2.17 m; reactor height is 2.39 m.

4.21 A plant has been designed to produce a total output of $1000\,\mathrm{L \cdot h^{-1}}$ of invert sugar beet syrup from sugar beet syrup with $600\,\mathrm{g \cdot L^{-1}}$ of sucrose, using immobilized invertase from *Saccharomyces cerevisiae*. The plant has been designed to produce equal volumes of two different qualities of syrup: one with 90% inversion and one with 50% inversion. Kinetic studies have revealed that the enzyme used (invertase) is inhibited at high sucrose concentrations, given kinetic parameters of $K_M = 50\,\mathrm{mM}$ and $K_S = 200\,\mathrm{mM}$, with a kinetic rate constant of 1 mmol of sucrose hydrolyzed per hour per IU of invertase activity (1 IU is defined as the amount of invertase that hydrolyes $1\,\mathrm{\mu mol}$ of sucrose per minute under the reaction conditions).

The plant manager proposes a sequential operation scheme in which sucrose is hydrolyzed in a CPBR to 50% conversion and then half of it is further hydrolyzed to 90% conversion in a CSTR (conversions are defined with respect to the sucrose syrup entering the plant). You argue that it is better to put the CSTR first and then the CPBR.

Justify in quantitative terms your proposal and indicate the percentage of enzyme saved by this method.

Answer:

Reactor Sequence	Enzyme Consumption (Thousand IU)		
	First Reactor	Second Reactor	Total
CPBR – CSTR	6650	756	7406
CSTR – CPBR	4753	1309	6062

You are right: the data in the table show that the enzyme consumption is reduced by 18.2% when using the sequence CSTR–CPBR.

4.22 Penicillin G acylase catalyzes the kinetically controlled synthesis of the antibiotic amoxicillin from *p*-hydroxyphenylglycine methyl ester (HPGME) and 6-amino penicillanic acid (6APA). The following results have been obtained with a commercial preparation of immobilized penicillin G acylase:

| [HPGME] (mM) | v_s (μmol $\cdot$ min$^{-1} \cdot$ g^{-1}) | | | |
	[6APA] (mM)			
	10	20	30	40
10	20.0	36.4	50.0	61.5
20	28.6	50.0	66.7	80.0
30	33.3	57.1	75.0	88.9
40	36.4	61.5	80.0	94.1

v_s: initial reaction rate of amoxicillin synthesis

(a) Evaluate the rate equation of amoxicillin synthesis. (b) Determine the mass concentration of catalyst required to produce a 96% conversion after 2 hours of reaction of a mixture containing 180 mM of 6APA and a 200% molar excess of HPGME. (c) If the supplier of the enzyme allows a ±20% variation in specific activity among enzyme production batches, calculate the conversion variation in that range if the same amount of catalyst is used in each batch of amoxicillin synthesis.

Answer:

(a) $v_s = \dfrac{500 \cdot [HPGME] \cdot [6APA]}{6000 + 30 \cdot [HPGME] + [HPGME] \cdot [6APA]}$

(b) 7.88 $g_{cat} \cdot L_{reaction}^{-1}$.

(c) +20% : X ≈ 1;   − 20% : X = 0.872.

4.23 The enzyme α-amino acid ester hydrolase of *Pseudomonas melanogenum* immobilized in polyvinyl alcohol gel particles catalyzes the synthesis of cefadroxil (Cx) by the kinetically controlled reaction of *p*-hydroxyphenylglycine amide (HFGA) and the β-lactam nucleus 7-amino-3-deacetoxy cephalosporanic acid (7ADCA). The kinetics of the reaction has been modeled as RS and the following parameters have been determined: $K_A = 1.25$ mM, $K'_A = 2.35$ mM, and $K'_B = 0.51$ mM, where A and B are HFGA and 7ADCA, respectively.

Determine the conversion of the limiting substrate into Cx as a function of flow rate in CPBR and CSTR with inlet concentrations of 9.35 and 12.18 mM of HFGA and 7ADCA, respectively. The reactors are loaded with 100 mL of catalyst, whose specific activity is 530 IU $\cdot$ mL^{-1}.

Answer:

X	0.1	0.25	0.5	0.6	0.7	0.8	0.9	0.95	0.99
F_{CPBR} (L $\cdot$ min^{-1})	43.11	16.89	8.03	6.51	5.37	4.46	3.61	3.12	2.46
F_{CSTR} (L $\cdot$ min^{-1})	40.73	15.37	6.47	4.81	3.47	2.29	1.15	0.58	0.12

CPBR is a better configuration for sequential mechanisms. This is reflected in the higher flow rate that can be processed. Differences between CPBR and CSTR performance increase with the increase in conversion. Performance of CSTR at high conversions is always low because of the very low substrate concentrations to which the enzyme is exposed.

4.24 Pyranose-2 oxidase (EC 1.1.3.10) catalyzes the oxidation of D-glucose (G) by molecular oxygen (O_2) to yield the corresponding keto-aldose (2-ketoD-glucose) and hydrogen peroxide. The kinetic mechanism is ping-pong bi–bi, with glucose binding first to the enzyme-FAD^+ to produce oxidized to 2-keto-D-glucose, and enzyme-FADH then reducing molecular oxygen in the presence of water to yield hydrogen peroxide, regenerating the enzyme-FAD^+. The following kinetic parameters are determined for a pyranose-2 oxidase from *Phanerochaete chrysosporium* immobilized in a macroprous silica matrix:

$$K_{MG} = 1.43 \, mM$$
$$K_{MO2} = 0.083 \, mM$$
$$\alpha = 0.58$$

The specific activity of the biocatalyst is $150 \, IU \cdot g^{-1}$ (1 IU being the amount of enzyme oxidizing 1 µmole of glucose per minute under the reaction conditions). A laboratory CSTR loaded with 5 g of catalyst is operated at a flow rate of $100 \, mL \cdot h^{-1}$. Substrate inlet concentrations are $9 \, mg \cdot L^1$ for molecular oxygen and $2 \, g \cdot L^{-1}$ for glucose. Due to the flow regime conditions, only 75% of the activity is expressed.

(a) Calculate the steady-state conversion.

(b) A mutant enzyme has been developed by directed evolution such that after five rounds its affinity for glucose and oxygen increase by 20% each, although no improvement is observed in the reactivity of the enzyme. The expected further improvements per round are 5% for each enzyme in the next five rounds. What steady-state conversions are to be expected with these improved enzymes?

Answer: (a) 0.732.

(b)

Round #	6	7	8	9	10
X	0.769	0.776	0.784	0.791	0.798

References

1. vanBeilen, J.B. and Li, Z. (2002) Enzyme technology: an overview. *Current Opinion in Biotechnology*, **13**, 338–344.
2. Kirk, O., Borchert, T.V., and Fuglsang, C. C. (2002) Industrial enzyme applications. *Current Opinion in Biotechnology*, **13**, 345–351.

3. Nestl, B.M., Nebel, B.A., and Hauer, B. (2011) Recent progress in industrial biocatalysis. *Current Opinion in Chemical Biology*, **15**, 187–193.
4. Mateo, C., Palomo, J.M., Fernandez-Lorente, G. *et al.* (2007) Improvement of enzyme activity, stability and selectivity via immobilization techniques. *Enzyme and Microbial Technology*, **40**, 1451–1463.
5. Hari Krishna, S. (2002) Developments and trends in enzyme catalysis in nonconventional media. *Biotechnology Advances*, **20**, 239–267.
6. Illanes, A. (1999) Stability of biocatalysts. *Electronic Journal of Biotechnology*, **2** (1), 7–15.
7. Lübbert, A. and Jørgenssen, S. (2001) Bioreactor performance: a more scientific approach for practice. *Journal of Biotechnology*, **85**, 187–212.
8. Levenspiel, O. (1999) *Chemical Reaction Engineering*, 3rd edn, Wiley, New York.
9. Vandamme, E. J. (1988) Immobilized biocatalyst and antibiotic production. biochemical, genetic, and biotechnical aspects, in *Bioreactor Immobilized Enzymes and Cells: Fundamentals and Applications* (ed. M Moo-Young), Elsevier, New York, USA, pp. 261–286.

5

Enzyme Reactor Design and Operation under Mass-Transfer Limitations

Chapter 4 discussed the design and operation of an enzyme reactor under ideal conditions. In practice, ideal conditions are rarely met, normally due to either catalyst inactivation or mass transfer limitations in the case of heterogeneous catalysis, or both. This chapter deals with the impact of mass-transfer limitations on reactor performance when using immobilized enzymes, while catalyst inactivation will be discussed in Chapter 6.

A convenient way to consider the impact of mass-transfer limitations on enzyme reactor performance is to introduce the effectiveness factor, as defined in Chapter 3 (see Equation 3.17). The effectiveness factor reflects the restriction imposed by mass-transfer rates on enzyme kinetics: a value of one means that the whole catalytic potential of the enzyme will be expressed; departure from one reflects the magnitude of the mass-transfer limitation. As shown in Chapter 3, the effectiveness factor is determined by the bulk substrate concentration and by a module that accounts for the relative magnitudes of mass-transfer rates and enzyme kinetics. Mass-transfer rates are expressed by a mass-transfer coefficient (a film-diffusion coefficient in the case of mass-transfer limitations external to the catalyst matrix or a diffusion coefficient if within the catalyst matrix), while enzyme kinetics is expressed in terms of the intrinsic kinetic parameters of the enzyme catalyst. In the case of external mass-transfer limitations (external diffusional restrictions, EDR; see Section 3.3.2), the module is the Damkoehler number, as defined by Equation 3.15; in the case of internal mass-transfer limitations (internal diffusional restrictions, IDR; see Section 3.3.3) it is the Thièle modulus, as defined by Equation 3.23.

Equations 5.1 and 5.2 express the effectiveness factors in the case of EDR and IDR, respectively:

$$\eta = f(\beta_0, \alpha) \tag{5.1}$$

$$\eta_G = f(\beta_0, \Phi) \tag{5.2}$$

Problem Solving in Enzyme Biocatalysis, First Edition. Andrés Illanes, Lorena Wilson and Carlos Vera.
© 2014 John Wiley & Sons, Ltd. Published 2014 by John Wiley & Sons, Ltd. Companion Website:
http://www.wiley.com/go/illanes-problem-solving

From the definition of substrate conversion (see Equation 4.2):

$$\beta_0 = \beta_{0i} \cdot (1 - X) \tag{5.3}$$

$$\gamma_0 = \gamma_{0i} \cdot X \tag{5.4}$$

where the subindex 0 refers to the bulk reaction medium and the subindex i to the initial (or inlet) condition. Equations 5.1 and 5.2 can then be expressed as:

$$\eta = f(\beta_{0i}, X, \alpha) \tag{5.5}$$

$$\eta_G = f(\beta_{0i}, X, \Phi) \tag{5.6}$$

These functions, which may either be algebraic equations in the case of immobilized enzymes subjected to EDR or numerical solutions in the case of immobilized enzymes subjected to IDR (except in the case of first-order kinetics, where an algebraic expression is obtained), can then be incorporated into the equations describing the performance of enzymatic reactors operating batchwise or continuously.

5.1 Sequential Batch and Continuously Operated Reactors with Immobilized Enzymes

Stabilization can be significantly enhanced when immobilized enzymes are used as process catalysts, and the heterogeneous condition allows a simple separation of the catalyst from the reaction medium. This makes it possible to develop continuous processes in which the enzyme is retained within the reactor boundaries while the substrate is continuously fed to the reactor and the product is removed free of catalyst. If operating batchwise, the immobilized enzyme is stable enough to be recovered after each batch and reused until the point of catalyst replacement (see Chapters 6 and 7). A product stream free of catalyst is a significant advantage of using immobilized enzymes, a point that is usually underappreciated when working at the laboratory scale. A catalyst-containing product might be inadmissible from a quality standpoint and would require the enzyme catalyst to be inactivated, with a serious risk of product damage, and removed, which could be a cumbersome operation at production scale.

The major constraint of using immobilized enzymes as process catalysts is the inevitable presence of mass-transfer limitations, which are inherent to heterogeneous catalysis (see Chapter 4). The magnitude of such limitations will depend on the structure of the enzyme support, the nature of the substrate (and product) and the fluid regime within the reactor. When enzymes are immobilized on the surfaces of impervious supports, mass-transfer limitations will only occur across the stagnant layer surrounding the catalyst particle; this can in principle be avoided, or at least overcome to a significant degree, by maintaining a proper fluid regime in the reactor, but in practice it may imply high agitation rates and shear stress, which can significantly erode the catalyst particles. When enzymes are immobilized in the internal structure of a gel or porous matrix, mass-transfer limitations will not be directly affected by the external fluid regime and will generally be more significant and difficult to overcome: an open pore structure will reduce mass-transfer limitations, but will also reduce the bound protein density and will be less protective of the enzyme than would a smaller pore structure. Internal mass-transfer limitations can be reduced by reducing the particle size, but this again may result in particles too small to be properly handled within the reactor. Immobilization of enzymes in nanoparticles (nanobiocatalysts) is certainly attractive from the standpoint of mass-transfer limitations and may find important applications in biomedicine and in the construction of biosensors; however, proper handling within the reactor is still a problem to be solved.

Sequential operation with immobilized enzymes is usually conducted in batch stirred-tank reactors (BSTRs) containing a conventional device for solid–liquid separation that allows the catalyst to be contained in the reactor while the product stream is being harvested, or else enables conventional solid catalyst recovery by filtration or centrifuga-tion of the product-containing stream. Containment within the reactor is the preferred option, since it avoids catalyst manipulation and losses. The usual configuration is a stainless-steel screen of adequate mesh on the bottom of a basket-type reactor. A recirculation batch reactor is another option, in which the immobilized enzyme is packed to form a shallow catalyst bed, through which the reaction medium is recirculated until the desired conversion is attained; this configuration can be used when tight control of operational variables such as pH or temperature is required.

Immobilized enzymes are particularly well suited to continuous performance, since the reactor can operate for a prolonged period of time without requiring catalyst replacement or any other kind of intervention, meaning the productive time is a significant fraction of the total operation time. Continuously operated reactors are usually either stirred tank (CSTR) or packed bed (CPBR), the latter being more suitable for most of the kinetic mechanisms of enzymatic catalysis (see Chapter 4), although CSTR does allow better control of operational variables and may decrease the impact of mass-transfer limitations.

5.2 Mathematical Models for Enzyme Kinetics, Modes of Operation, and Reactor Configurations under Mass-Transfer Limitations

Equations describing reactor performance under ideal conditions (absence of mass-transfer limitations in the case of immobilized enzymes) were developed in Chapter 4. For an immobilized enzyme whose effectiveness factor (η or η_G) takes account of mass-transfer limitations, Equations 4.4, 4.21, and 4.32, describing BSTRs, CPBRs, and CSTRs, respectively, can now be rewritten as:

$$\text{BSTR} \quad \int \frac{dX}{v_{effective}} = \int \frac{dX}{v(e,X) \cdot \eta(X)} = \frac{t}{s_{0i}} \tag{5.7}$$

$$\text{CPBR} \quad \int \frac{dX}{v_{effective}} = \int \frac{dX}{v(e,X) \cdot \eta(X)} = \frac{\tau}{s_{0i}} \tag{5.8}$$

$$\text{CSTR} \quad \frac{X}{v_{effective}} = \frac{X}{v(e,X) \cdot \eta(X)} = \frac{\tau}{s_{0i}} \tag{5.9}$$

These equations can describe reactor performance under both EDR and IDR; in the latter case, η is replaced by η_G. If there is more than one substrate reaction, X refers to the limiting substrate (A or B) and s_i is the corresponding molar initial or inlet concentration (a_i or b_i). In the case of EDR, $\eta(X)$ is described by an algebraic equation (see Equation 3.18) and Equations 5.7 to 5.9 can be solved analytically, while in the case of IDR, $\eta(X)$ is the numerical solution of the corresponding differential equation (Equation 3.33). So:

$$\eta_G = f(\beta_{0i}, X, \Phi)$$

As an illustrative example, the case of an immobilized enzyme subjected to external diffusional restrictions with intrinsic Michaelis–Menten kinetics will be analyzed.

From Equations 3.18 and 5.3:

$$\eta(X) = \frac{[1 + \beta_{0i}(1 - X)] \lfloor 1 + \alpha + \beta_{0i}(1 - X) - \sqrt{f(X)} \rfloor}{2\alpha\beta_{0i}(1 - X)} \tag{5.10}$$

$$f(X) = \beta_{0i}^2 \cdot X^2 + 2\beta_{0i}X(\alpha - 1 - \beta_{0i}) + \beta_{0i}(\beta_{0i} - 2\alpha + 2) + \alpha(\alpha + 2) + 1$$

The expression for η depends on the Damkoehler number (α) and the bulk initial dimensionless substrate concentration (β_{0i}). Determination of α can be carried out either experimentally or by using empirical correlations (see Section 3.4.1). As long as the enzyme remains fully active during reactor operation, α will be constant, since it will depend only on the kinetic parameters of the enzyme and the substrate film mass-transfer coefficient (see Equation 3.15). Note that V_{max} depends on the active enzyme concentration, so when considering enzyme inactivation during reactor operation, α will no longer be constant (see Chapter 6). Evaluation of β_{0i} implies knowledge of the intrinsic Michaelis constant (K_M), which must be determined experimentally (see Section 3.4.1).

From Equation 2.7, representing in this case the intrinsic rate expression for Michaelis–Menten kinetics:

$$\upsilon(X) = \frac{v(X)}{k \cdot e} = \frac{\beta_{0i}(1 - X)}{1 + \beta_{0i}(1 - X)} \tag{5.11}$$

Replacing Equations 5.10 and 5.11 into Equations 5.7 to 5.9 gives:

$$\text{BSTR} \quad \int \frac{2 \cdot \alpha \cdot \beta_{0i}}{1 + \alpha + \beta_{0i}(1 - X) - \sqrt{f(X)}} \, dX = \frac{k \cdot e}{K_M} \cdot t = \frac{m_{cat} \cdot a_{sp}}{K_M} \cdot t \tag{5.12}$$

$$\text{CPBR} \quad \int \frac{2 \cdot \alpha \cdot \beta_{0i}}{1 + \alpha + \beta_{0i}(1 - X) - \sqrt{f(X)}} \, dX = \frac{k \cdot e}{K_M} \cdot \tau = \frac{k \cdot E}{K_M \cdot F} = \frac{M_{cat} \cdot a_{sp}}{K_M \cdot F} \tag{5.13}$$

$$\text{CSTR} \quad \frac{2 \cdot \alpha \cdot \beta_{0i} \cdot X}{1 + \alpha + \beta_{0i}(1 - X) - \sqrt{f(X)}} = \frac{k \cdot e}{K_M} \cdot \tau = \frac{k \cdot E}{K_M \cdot F} = \frac{M_{cat} \cdot a_{sp}}{K_M \cdot F} \tag{5.14}$$

In the case of BSTR and CPBR, the integral can be solved analytically or numerically. Similar expressions can be derived for more complex kinetic mechanisms. In the case of

IDR, the expression of η_G will not be analytical as it is for EDR (see Equation 5.10) but instead a numerical correlation (see Equation 3.33), which, taking into consideration Equation 5.3, can be expressed as:

$$\eta_G = f(\beta_{0i}, X, \Phi) \tag{5.15}$$

As in the case of EDR, for so long as the enzyme remains fully active during reactor operation, Φ will be constant, since it will depend only on the kinetic parameters of the enzyme, the substrate diffusion coefficient, and the dimension and geometry of the catalyst particle (see Equation 3.23). Φ will thus remain constant during reactor operation and, for any given value of Φ, Equation 5.15 is merely:

$$\eta_G = f(\beta_{0i}, X) \tag{5.16}$$

Since this is a numerical correlation, it is convenient to transform it into an equation by fitting the values to a mathematical expression, so that, even though it does not have a clear physical meaning, it will be helpful for solving Equations 5.7 to 5.9. When considering enzyme inactivation during reactor operation (see Chapter 6), Φ will not be a constant, just like α in EDR.

Solved Problems

5.1 The dipeptide Z-Gly-Leu-NH$_2$ (GLA) is hydrolyzed with a neutral protease from *Bacillus subtilis* immobilized in a 2 mm-diameter spherical porous polyurethane particle whose density is $0.8\,g \cdot cm^{-3}$. At 35 °C, the intrinsic kinetic parameters of the immobilized protease are $V''_{max} = 2.5$ mmoles of hydrolyzed dipeptide per minute per gram of catalyst and $K_M = 70$ mM. The diffusion coefficient of the dipeptide in the support is $D_{eff} = 5 \cdot 10^{-5}\,cm^2 \cdot s^{-1}$. At such a temperature, hydrolysis is carried out in a CSTR fed with $600\,L \cdot h^{-1}$ of a 35 mM dipetide solution. Consider first-order kinetics applicable for the purpose of evaluating η_G only.

(a) Determine the catalyst mass required to obtain 75% conversion of the dipeptide into Z-glycine and leucine amide. Estimate the minimum reaction volume required.

(b) Recalculate the catalyst mass required if the reaction is carried out at 20 °C, at which temperature the D_{eff} for the dipeptide is reduced by 20%. Consider that the energy of activation of the reaction is $4\,Kcal \cdot mol^{-1}$ and the standard enthalpy change of dissociation of the enzyme–dipeptide complex is $1.35\,kcal\,mol^{-1}$; consider that $R = 1.98\,kcal \cdot mol^{-1} \cdot K^{-1}$ and the Arrhenius equation is applicable.

Resolution (a): In the case of a well-stirred CSTR, all enzyme particles in the reactor will be at the substrate concentration corresponding to the outlet concentration. η_G must thus be evaluated at the outlet value of X. From Equation 5.9:

$$\eta_G \frac{k \cdot e}{K_M} \tau = \eta_G \frac{k \cdot E}{K_M \cdot F} = \eta_G \frac{M_{cat} \cdot a_{sp}}{K_M \cdot F} = \frac{s_i}{K_M} X + \frac{X}{1 - X} \tag{5.17}$$

Data:

$a_{sp,35\,°C} = 2.5\,\text{mmol} \cdot \text{min}^{-1} \cdot \text{g}^{-1} = 2\,\text{mmol} \cdot \text{min}^{-1} \cdot \text{cm}^{-3}$
$K_{M,35\,°C} = 70\,\text{mM} = 0.07\,\text{mmol} \cdot \text{cm}^{-3}$
$D_{eff,35\,°C} = 5 \cdot 10^{-5}\,\text{cm}^2 \cdot \text{s}^{-1} = 3 \cdot 10^{-3}\,\text{cm}^2 \cdot \text{min}^{-1}$
$R_p = 0.1\,\text{cm}$
$L_{eq} = 0.0333\,\text{cm}$
$X = 0.75$
$F = 600\,\text{L} \cdot \text{h}^{-1} = 10\,\text{L} \cdot \text{min}^{-1}$
$[\text{GLA}]_i = 35\,\text{mM} = 0.035\,\text{mmol} \cdot \text{cm}^{-3}$

From Equation 3.23:

$$\Phi_{35\,°C} = 0.0333\sqrt{\frac{2}{0.07 \cdot 3 \cdot 10^{-3}}} = 3.25$$

From Equation 3.40:

$$\eta_{G,35\,°C} = \frac{1}{3.25}\left[\frac{1}{tanh\,9.75} - \frac{1}{9.75}\right] = 0.276$$

From Equation 5.17:

$$0.276 \cdot \frac{M_{cat,35\,°C} \cdot 2.5}{70 \cdot 10} = 0.5 \cdot 0.75 + 3$$

Answer (a):

$$M_{cat,35\,°C} = 3424\,\text{g}$$

Considering a value of $100\,\text{g} \cdot \text{L}^{-1}$ as the maximum allowable solid concentration in the reactor:

$$V_{eff} = 34.2\,\text{L}$$

Resolution (b):

$$E_a = 4000\,\text{cal} \cdot \text{mol}^{-1}; \Delta H^0 = 1350\,\text{cal} \cdot \text{mol}^{-1}$$

From Equation 2.35:

$$a_{sp,20\,°C} = a_{sp,35\,°C} \cdot exp\left[-\frac{4000}{1.98}\left(\frac{1}{293} - \frac{1}{308}\right)\right]$$

$$a_{sp,20\,°C} = 1.79\,\text{mmol} \cdot \text{min}^{-1} \cdot \text{g}^{-1} = 1.43\,\text{mmol} \cdot \text{min}^{-1} \cdot \text{cm}^{-3}$$

From Equation 2.32:

$$K_{M,20\,°C} = K_{M,35\,°C} \cdot exp\left[\frac{1350}{1.98}\left(\frac{1}{293} - \frac{1}{308}\right)\right]$$

$$K_{M,20°C} = 78.4 \, \text{mM}$$

$$D_{eff,20°C} = 2.4 \cdot 10^{-3} \, \text{cm}^2 \cdot \text{min}^{-1}$$

From Equation 3.23:

$$\Phi_{20°C} = 0.0333 \sqrt{\frac{1.43}{0.0784 \cdot 2.4 \cdot 10^{-3}}} = 2.90$$

From Equation 3.40:

$$\eta_{G,20°C} = \frac{1}{2.9} \left[\left[\frac{1}{tanh\, 8.7} - \frac{1}{8.7} \right] \right] = 0.305$$

From Equation 5.17:

$$0.305 \cdot \frac{M_{cat,20°C} \cdot 1.79}{78.4 \cdot 10} = 0.446 \cdot 0.75 + 3$$

Answer (b):

$$M_{cat,20°C} = 4788 \, \text{g}$$

At 20 °C the incidence of IDR is lower than at 35 °C (lower Φ and higher η_G) because temperature affects the catalytic potency of the enzyme ($V_{max} \cdot K_M^{-1}$) more strongly than the mass-transfer rate, so that at 20 °C the system moves away from the mass-transfer limitation. Despite this, reactor performance is better at 35 °C (a lower amount of catalyst is required for the same process task) because the improvement in kinetic parameters at this temperature (higher $V_{max} \cdot K_M^{-1}$) outweighs the higher incidence of IDR (lower η_G).

5.2 The continuous hydrolysis of a maltose (M) syrup at a concentration five times its Michaelis constant is conducted in a CSTR with cross-linked enzyme aggregates (CLEAs) of glucoamylase with Michaelis–Menten behavior. CLEAs are approximately spherical and are subjected to IDR revealed by a Thièle modulus of 3.8. A steady-state conversion of 0.65 is obtained. In order to minimize the impact of IDR, CLEAs of a lower size (10–90% of the original) are prepared. For the calculation of η_G, assume first-order kinetics with respect to maltose concentration.

Make a prediction of the steady-state conversion attainable with the different sizes of catalyst.

Resolution: From Equation 5.9:

$$\frac{[M]_i}{K_M} \cdot X + \frac{X}{1 - X} = \eta_G \frac{M_{cat} \cdot a_{sp}}{K_M \cdot F} \tag{5.18}$$

The right-hand side of the equation represents the process task and only η_G depends on particle size, so Equation 5.8 can be simplified to:

$$\frac{[M]_i}{K_M} \cdot X + \frac{X}{1-X} = \eta_G \cdot C \tag{5.19}$$

From Equation 3.40, for the original CLEAs:

$$\eta_{G,} = \frac{1}{3.8} \cdot \left[\left[\frac{1}{tanh\,11.4} - \frac{1}{11.4} \right] \right] = 0.24$$

So, from Equation 5.19:

$$C = \frac{5 \cdot 0.65 + \dfrac{0.65}{0.35}}{0.24} = 21.28$$

The reactor performance for a catalyst of any particle size is:

$$5X + \frac{X}{1-X} = \eta_G \cdot 21.28 \tag{5.20}$$

From Equation 3.23, the Thièle modulus at varying particle sizes (Φ_R) is calculated. Then, from Equation 3.40, the effectiveness factors are calculated. Finally, from Equation 5.20, the steady-state substrate conversion at different particle sizes is calculated.

Answer:

$R_P \cdot R_{P,0}^{-1} = \Phi_R \cdot \Phi_0^{-1}$	Φ_R	η_G	X
1	3.80	0.240	0.650
0.9	3.42	0.264	0.686
0.75	2.85	0.310	0.743
0.5	1.90	0.434	0.835
0.4	1.52	0.514	0.868
0.3	1.14	0.623	0.898
0.2	0.76	0.767	0.921
0.1	0.38	0.923	0.937

*Subscript 0 denotes the original CLEA

5.3 Tannase (tannin acyl esterase; EC 3.1.1.20) is a commercially important enzyme used in juice debittering and in the production of gallic acid. The catalyzed reaction can be represented as:

$$\text{Tannic acid} + 10\,H_2O \rightarrow 10\,\text{Gallic acid} + \text{Glucose}$$
$$\text{TA} \qquad\qquad\qquad \text{GA} \qquad\qquad \text{G}$$
$$C_{76}H_{52}O_{46} \qquad\qquad\qquad C_7H_6O_5 \qquad C_6H_{12}O_6$$

A CSTR is operated with 1 kg of *Aspergillus niger* tannase immobilized on the surface of chitin flakes fed $120\,L \cdot h^{-1}$ of a tannin solution with an equivalent concentration of $100\,g \cdot L^{-1}$ tannic acid. The enzyme has a specific activity of 500 IU per gram of catalyst (1 IU is the amount of tannase hydrolyzing 1 μmol of tannic acid per minute under the reaction conditions), its intrinsic K_M is 25 mM and its calculated Damkoehler number is 5.

(a) Determine the steady-state conversion of tannic acid into gallic acid at the reactor outlet. (b) Compare this conversion with that attainable under ideal conditions (no mass-transfer limitations). (c) What is the composition (in $g \cdot L^{-1}$) of the product stream in both cases?

Resolution: Data:

$\alpha = 5$

$s_{0i} = 100\,000 \cdot 1700^{-1} = 58.82 \text{ mM}$

$K_M = 25 \text{ mM}$

$\beta_{0i} = 2.35$

$M_{cat} = 1000 \text{ g}$

$a_{sp} = 0.5 \text{ mmol} \cdot \text{min}^{-1} \cdot \text{g}^{-1}$

$F = 2\,L \cdot \text{min}^{-1}$

From Equation 5.14:

$$\frac{23.5 \cdot X}{8.35 - 2.35 \cdot X - \sqrt{5.523 \cdot X^2 + 7.755 \cdot X + 22.723}} = \frac{1000 \cdot 0.5}{25 \cdot 2} = 10$$

Answer: (a) Solving this equation gives: $X = 0.618$. (b) From Equation 4.33 and Table 4.1: $X = 0.888$. Steady-state conversion is reduced by 30% as a consequence of EDR. (c) From Equation 4.1:

$$[GA] = 10 \cdot [TA]_0 \cdot X$$
$$[G] = [TA]_0 \cdot X$$
$$[TA] = [TA]_0 \cdot (1 - X)$$

	Product Composition ($g \cdot L^{-1}$)		
	[GA]	[G]	[TA]
EDR ($\alpha = 5$)	61.8	6.5	38.2
No EDR	88.8	9.4	11.2

5.4 Galactose oxidase (EC 1.1.3.9) catalyzes the oxidation of galactose by molecular oxygen. A set of experiments is conducted with galactose oxidase from *Gibberella fujikuroi* immobilized in a microporous cellophane membrane. The enzyme kinetics approach first-order kinetics with respect to galactose concentration and a value of 0.06 s^{-1} is determined for the first-order rate constant. Working under steady-sate conditions in CPBR with membranes of varying thickness and with a constant residence time of 10 seconds, the following results are obtained:

Membrane Thickness (mm)	0.1	0.2	0.4	0.6	0.8	1.0
Conversion	0.445	0.428	0.374	0.314	0.262	0.221

Estimate the approximate value of the diffusion coefficient of galactose in the cellophane membrane.

Resolution: From Equation 5.8:

$$\int \frac{dX}{v(X) \cdot \eta_G} = \frac{dX}{\eta_G \cdot k \cdot s_{0i}(1 - X)} = \frac{\tau}{s_{0i}}$$

Since the kinetics of reaction is first order, η_G does not depend on substrate concentration (see Equation 3.40):

$$\int \frac{dX}{(1 - X)} = \eta_G \cdot k \cdot \tau = 0.6 \cdot \eta_G \tag{5.21}$$

Solving Equation 5.21 gives:

$$\eta_G = \frac{-\ln(1 - X)}{0.6} = \frac{\tanh\left(\dfrac{\Phi}{2}\right)}{\dfrac{\Phi}{2}} \tag{5.22}$$

From Equation 3.23:

$$\Phi = L\sqrt{\frac{k}{D_{eff}}} \tag{5.23}$$

$$D_{eff} = k \cdot \left(\frac{L}{\Phi}\right)^2 \tag{5.24}$$

From Equations 5.22 to 5.24:

L (cm)	X	η_G	Φ	$D_{eff} \cdot 10^5$ (cm$^2 \cdot$ s^{-1})
0.01	0.445	0.981	0.48	2.60
0.02	0.428	0.931	0.95	2.66
0.04	0.374	0.781	1.89	2.69
0.06	0.314	0.628	2.83	2.70
0.08	0.262	0.506	3.78	2.69
0.10	0.221	0.416	4.72	2.69

Answer: The effective diffusion coefficient of galactose in the porous cellophane membrane is $2.7 \cdot 10^{-5}$ cm$^2 \cdot$ s^{-1}.

5.5 α-amylase derived from *Aspergillus niger* is immobilized in calcium alginate beads in order to hydrolyze cornstarch at the laboratory scale. The reaction is conducted in a

CPBR with a working volume of $50\,cm^3$ and the reactor is operated at a feed flow rate of $8.33\,cm^3 \cdot min^{-1}$. Under these conditions, EDR are negligible. Determine the conversion reached in this process, given that the enzyme follows Michaelis–Menten kinetics, the reactor bed has a void fraction (ε) of 0.4, the substrate concentration in the feed stream is $120\,g \cdot L^{-1}$ and the biocatalyst has a specific activity of $0.0258\,g$ hydrolyzed starch per gram of catalyst per minute, a density of 1.15 g catalyst per cubic centimetre, an intrinsic Michaelis constant of $40\,g \cdot L^{-1}$, and a Thièle modulus of 5. Compare the reactor performance with that under ideal conditions (no mass transfer limitations).

Resolution: A material balance on a differential element of the catalyst bed gives:

$$-F\frac{ds}{dl} = v(e,s) \cdot \eta_G(s,\Phi) \cdot A \cdot \varepsilon \tag{5.25}$$

where A represents the sectional area of the reactor and l its length. Writing the former equation in its dimensionless form gives:

$$\frac{dX}{d\lambda} = \eta_G(\beta_{0i} \cdot (1-X), \Phi) \cdot \frac{1-X}{1+\beta_{oi} \cdot (1-X)} \cdot \frac{M_{cat} \cdot a_{sp}}{F \cdot K_M} \tag{5.26}$$

where λ is the dimensionless length, defined as $l \cdot L_R^{-1}$, L_R representing in this case the total reactor bed length and l any position along that length.

Before solving Equation 5.26, the values of its unknown parameters must be calculated. M_{cat} is determined as:

$$M_{cat} = V_{bed} \cdot \rho_{app} \tag{5.27}$$

where V_{bed} represents the bed volume in the CPBR, ρ_{app} is the catalyst apparent density, and V_R is the reaction volume. So:

$$M_{cat} = 50 \cdot 1.15 = 57.5\,g$$

$$\beta_{0i} = \frac{s_{0i}}{K_M} = \frac{120}{40} = 3$$

$$K_M = 0.04\,g \cdot cm^{-3}$$

Replacing the values of these parameters in Equation 5.26 gives:

$$\frac{dX}{d\lambda} = \eta_G(3 \cdot (1-X), 5) \cdot \frac{1-X}{1+3 \cdot (1-X)} \cdot 4.45 \tag{5.28}$$

Equation 5.26 is a nonlinear differential equation, so it must be solved numerically. To this end, fourth-order Runge–Kutta will be employed here. According to Section A.4.2, the following terms are defined:

$$X(\lambda_i + 1) = X(\lambda_i) + \tfrac{1}{6} \cdot (k_1 + 2 \cdot k_2 + 2 \cdot k_3 + k_4) \cdot h$$

$$k_1 = 4.45 \cdot \eta_{G1} \cdot \frac{1 - X(\lambda_i)}{1 + 3 \cdot (1 - X(\lambda_i))} \qquad \eta_{G1} = \eta(3 \cdot (1 - X(\lambda_i)), 5)$$

$$k_2 = 4.45 \cdot \eta_{G2} \cdot \frac{1 - (X(\lambda_i) + 0.5 \cdot k_1 \cdot h)}{1 + 3 \cdot (1 - (X(\lambda_i) + 0.5 \cdot k_1 \cdot h))} \qquad \eta_{G2} = \eta(3 \cdot (1 - (X(\lambda_i) + 0.5 \cdot k_1 \cdot h)), 5)$$

$$k_3 = 4.45 \cdot \eta_{G3} \cdot \frac{1 - (X(\lambda_i) + 0.5 \cdot k_2 \cdot h)}{1 + 3 \cdot (1 - (X(\lambda_i) + 0.5 \cdot k_2 \cdot h))} \qquad \eta_{G3} = \eta(3 \cdot (1 - (X(\lambda_i) + 0.5 \cdot k_2 \cdot h)), 5)$$

$$k_3 = 4.45 \cdot \eta_{G4} \cdot \frac{1 - (X(\lambda_i) + k_3 \cdot h)}{1 + 3 \cdot (1 - (X(\lambda_i) + k_3 \cdot h))} \qquad \eta_{G4} = \eta(3 \cdot (1 - (X(\lambda_i) + k_3 \cdot h)), 5)$$

η_{Gi} can be estimated using Figure 3.7, which represents the solution of Equation 3.32 for Michaelis–Menten kinetics. In order to obtain more accurate values, η_{Gi} is calculated using the finite-difference method described in Problem 3.14. Assuming zero initial conversion and taking 10 intervals for λ ($h = 0.1$), the following results are obtained:

λ	$X(\lambda_i)$	η_1	k_1	η_2	k_2	η_3	k_3	η_4	k_4	$X(\lambda_{i+1})$
0	0	0.417	0.464	0.413	0.457	0.413	0.457	0.410	0.450	0.046
0.100	0.046	0.410	0.450	0.406	0.443	0.406	0.443	0.402	0.436	0.090
0.200	0.090	0.402	0.436	0.398	0.429	0.398	0.429	0.395	0.423	0.133
0.300	0.133	0.395	0.423	0.391	0.416	0.391	0.416	0.387	0.409	0.175
0.400	0.175	0.387	0.409	0.384	0.402	0.384	0.402	0.380	0.395	0.215
0.500	0.215	0.380	0.395	0.376	0.389	0.376	0.389	0.373	0.382	0.254
0.600	0.254	0.373	0.382	0.369	0.375	0.369	0.375	0.366	0.369	0.291
0.700	0.291	0.366	0.369	0.362	0.362	0.362	0.362	0.359	0.356	0.327
0.800	0.327	0.359	0.356	0.355	0.349	0.355	0.349	0.352	0.343	0.362
0.900	0.362	0.352	0.343	0.348	0.336	0.348	0.336	0.345	0.330	0.396
1.000	0.396									

Answer: Under the operation reactor conditions, $X = 0.396$.

In order to determine the performance of an ideal reactor (without IDR), the same procedure can be followed, now considering η_{Gi} to be equal to 1. In this manner, a value of $X = 0.85$ is obtained for ideal CPBR, which is 2.15 times higher than the conversion determined under IDR.

The reactor performance in both cases is represented in Figure 5.1. As can be seen, in order to achieve an X of 0.85 in CPBR under IDR, reactor length must be increased 3.1 times that of ideal CPBR. In other words, a τ 3.1 times higher is required, so if due to hydrodynamic considerations (i.e. a nonviable pressure drop) it is not possible to increase the reactor length, reducing F 3.1 times will allow $X = 0.85$ to be obtained.

5.6 In the previous problem, the reactor length was increased 3.1 times, but severe compaction of the alginate bed occurred, producing a high pressure drop. This makes this strategy impracticable as a means of increasing the substrate conversion to the desired level of 0.85. The feed flow rate could be decreased by 3.1 times in order to

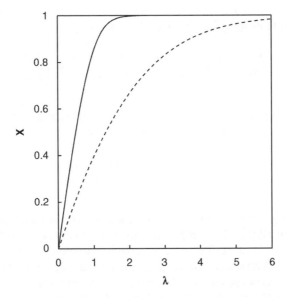

Figure 5.1 *Substrate conversion (X) along the CPBR dimensionless length (λ) for Michaelis–Menten kinetics. Solid line: ideal CPBR; dashed line: CPBR under internal diffusional restrictions.*

obtain the same result, but under such operating conditions EDR will no longer be negligible, with a Biot number of 5 being estimated. Determine the X profile across the reactor length under these operating conditions.

Resolution: Proceeding as in Problem 5.5, reactor behavior is given by Equation 5.29:

$$\frac{dX}{d\lambda} = \eta_G(\beta_{0i} \cdot (1-X), \Phi, Bi) \cdot \frac{1-X}{1+\beta_{oi} \cdot (1-X)} \cdot \frac{M_{cat} \cdot asp}{F \cdot K_M} \qquad (5.29)$$

Replacing the values of the parameters in Equation 5.29 as in Problem 5.5 gives:

$$\frac{dX}{d\lambda} = \eta_G(3 \cdot (1-X), 5, 5) \cdot \frac{1-X}{1+3 \cdot (1-X)} \cdot 13.786$$

Note that η_G now depends also on the Biot number. It is thus no longer possible to use Figure 3.7 for the estimation of η_G; in order to determine η_G, the substrate concentration profile inside the catalytic sphere must be determined by solving Equation 3.27. The finite-difference method will be used (see Problem 3.14 for details) taking nodes.

Using backwards finite differences and applying the boundary condition at $\rho = 0$ for node 1 gives:

$$\beta_1 = \beta_2$$

Using centered finite differences for nodes 2–20 gives:

$$\frac{\beta_{i+1} - 2 \cdot \beta_i + \beta_{i-1}}{h^2} + \frac{2}{\rho} \cdot \frac{(\beta_{i+1} - \beta_{i-1})}{2 \cdot h} - 9 \cdot \Phi^2 \cdot \frac{\beta_i}{1 + \beta_i} = 0$$

Using backwards finite differences and applying the boundary and continuity conditions at $\rho = 1$ (see Equations 3.22 and 3.42) for node 21 gives:

$$Bi \cdot (\beta_0 - \beta_{21}) = \frac{(\beta_{21} - \beta_{20})}{h}$$

The finite-difference scheme generates a nonlinear system of equations, which can be solved from a set of initial values for β_i using Newton's method (see Section A.4.1). As in Problem 3.14, the substrate profile inside the catalyst particle corresponding to first-order kinetics will be employed as a starting point. Such a profile is determined by Equations 3.36 using Equation 3.42 as boundary condition at the catalyst-liquid interface. These equations were not solved in Chapter 3, so they will be solved here. Section A.3.4 shows that Equation A.88 represents the general solution for Equation 3.36:

$$\beta = \left(\frac{1}{\rho}\right) \cdot \left(C_1 \cdot e^{-3 \cdot \Phi \cdot \rho} + C_2 \cdot e^{3 \cdot \Phi \cdot \rho}\right) \tag{A.88}$$

Equation A.88 must be evaluated in its boundary and continuity conditions for $\rho = 0$ and 1 (Equations 3.22 and 3.42, respectively) in order to determine the values of the constants C_1 and C_2. Note that both boundary and continuity conditions, expressed in terms of $d\beta/d\rho$, correspond to:

$$\frac{d\beta}{d\rho} = -\left(\frac{C_1 \cdot e^{-3 \cdot \Phi \cdot \rho} + C_2 \cdot e^{3 \cdot \Phi \cdot \rho}}{\rho^2}\right) + \left(\frac{-3 \cdot \Phi \cdot C_1 \cdot e^{-3 \cdot \Phi \cdot \rho} + 3 \cdot \Phi \cdot C_2 \cdot e^{3 \cdot \Phi \cdot \rho}}{\rho}\right)$$

$$\tag{5.30}$$

Equation 5.30 becomes undefined at $\rho = 0$, so L'Hospital's theorem must be used to calculate the limit of $d\beta/d\rho$ when ρ approaches 0^+:

$$\left.\frac{d\beta}{d\rho}\right|_{\rho=0} = Lim_{\rho \to 0^+} -\left(\frac{\frac{d^2}{d\rho^2}(C_1 \cdot e^{-3 \cdot \Phi \cdot \rho} + C_2 \cdot e^{3 \cdot \Phi \cdot \rho})}{\frac{d^2}{d\rho^2}(\rho^2)}\right)$$

$$+ \left(\frac{\frac{d}{d\rho}(-3 \cdot \Phi \cdot C_1 \cdot e^{-3 \cdot \Phi \cdot \rho} + 3 \cdot \Phi \cdot C_2 \cdot e^{3 \cdot \Phi \cdot \rho})}{\frac{d}{d\rho}(\rho)}\right)$$

$$\left.\frac{d\beta}{d\rho}\right|_{\rho=0} = \lim_{\rho\to 0^+} 9\Phi^2 \cdot \left(-\left(\frac{C_1 \cdot e^{-3\cdot\Phi\cdot\rho} + C_2 \cdot e^{3\cdot\Phi\cdot\rho}}{2}\right) + \left(C_1 \cdot e^{-3\cdot\Phi\cdot\rho} + C_2 \cdot e^{3\cdot\Phi\cdot\rho}\right)\right)$$

$$\left.\frac{d\beta}{d\rho}\right|_{\rho=0} = \frac{9\Phi^2}{2} \cdot (C_1 + C_2)$$

Evaluating the boundary condition at $\rho=0$ gives:

$$C_2 = -C_1 \tag{5.31}$$

Using Equation 5.31 to calculate the boundary condition at $\rho=1$ gives:

$$\beta_s = C_1 \cdot \left(e^{-3\cdot\Phi} - e^{3\cdot\Phi}\right) \tag{5.32}$$

Applying the continuity condition at $\rho=1$ (Equation 3.42) gives:

$$Bi \cdot (\beta_0 - \beta_s) = \left.\frac{d\beta}{d\rho}\right|_{\rho=1} = C_1 \cdot \left(\left(e^{3\cdot\Phi} - \cdot e^{-3\cdot\Phi}\right) - 3\cdot\Phi\cdot\left(e^{3\cdot\Phi} + \cdot e^{-3\cdot\Phi}\right)\right) \tag{5.33}$$

Replacing Equation 5.32 into Equation 5.33 and then solving for C_1 gives:

$$C_1 = \frac{Bi \cdot \beta_0}{(1 - Bi)\cdot\left(e^{3\cdot\Phi} - \cdot e^{-3\cdot\Phi}\right) - 3\cdot\Phi\cdot\left(e^{3\cdot\Phi} + \cdot e^{-3\cdot\Phi}\right)} \tag{5.34}$$

Replacing Equations 5.34 and 5.31 into Equation A.88 gives:

$$\beta = \frac{Bi \cdot \beta_0 \cdot \left(e^{-3\cdot\Phi\cdot\rho} - \cdot e^{3\cdot\Phi\cdot\rho}\right)}{\rho\cdot\left((1 - Bi)\cdot\left(e^{3\cdot\Phi} - \cdot e^{-3\cdot\Phi}\right) - 3\cdot\Phi\cdot\left(e^{3\cdot\Phi} + \cdot e^{-3\cdot\Phi}\right)\right)} \tag{5.35}$$

By reorganizing and using the identities $\sinh(x)=0.5\cdot(e^x - e^{-x})$ and $\cosh(x)=0.5\cdot(e^x + e^{-x})$, Equation 5.35 becomes:

$$\beta = \frac{Bi \cdot \beta_0 \cdot \sinh(3\cdot\Phi\cdot\rho)}{\rho\cdot\left((Bi - 1)\cdot\sinh(3\cdot\Phi\cdot) + 3\cdot\Phi\cdot\cosh(3\cdot\Phi)\right)} \tag{5.36}$$

By using Equation 5.36 to generate the initial values for β_i, the nonlinear system of equations formed by the finite differences is solved in a few iterations by employing Newton's method. It is important to recall that Newton's method has a fast convergence; however, convergence is not ensured, so the use of appropriate starting values is important, especially when the system of equations is highly nonlinear and has multiple solutions, as in this case. Once the substrate profile inside the sphere is determined by Newton's method, η_G is calculated by numerical integration (see Problem 3.14).

Now Equation 5.29 can be solved using fourth-order Runge–Kutta, as illustrated in Problem 5.5. Taking an integration step (h) equal to 0.1, the following results are obtained:

λ	$X(\lambda_i)$	η_1	k_1	η_2	k_2	η_3	k_3	η_4	k_4	$X(\lambda_{i+1})$
0.000	0.000	0.630	0.183	0.611	0.179	0.612	0.179	0.594	0.175	0.061
0.100	0.061	0.594	0.175	0.577	0.172	0.577	0.172	0.560	0.168	0.119
0.200	0.119	0.560	0.168	0.544	0.165	0.544	0.165	0.528	0.161	0.173
0.300	0.173	0.528	0.161	0.512	0.158	0.513	0.158	0.498	0.155	0.225
0.400	0.225	0.498	0.155	0.483	0.152	0.483	0.152	0.468	0.149	0.273
0.500	0.273	0.468	0.149	0.454	0.146	0.455	0.146	0.441	0.143	0.318
0.600	0.318	0.441	0.143	0.427	0.140	0.428	0.140	0.415	0.137	0.361
0.700	0.361	0.415	0.137	0.402	0.135	0.402	0.135	0.390	0.132	0.401
0.800	0.401	0.390	0.132	0.378	0.130	0.378	0.130	0.367	0.127	0.439
0.900	0.439	0.367	0.127	0.355	0.125	0.356	0.125	0.345	0.123	0.475
1.000	0.475									

Answer: The results indicate that $X = 0.475$ is reached in the CPBR, which is 44% lower than expected when considering only IDR ($X = 0.85$) and 52.5% lower than ideal ($X = 0.999$). A comparison of CPBR performance under no diffusional restrictions, under IDR and under IDR plus EDR is presented in Figure 5.2, which shows that, even with Biot numbers higher than 1, external mass-transfer limitations exert a great influence on η_G. In fact, an increase in reactor length of 2.85 times is required to obtain $X = 0.85$ under current conditions of operation.

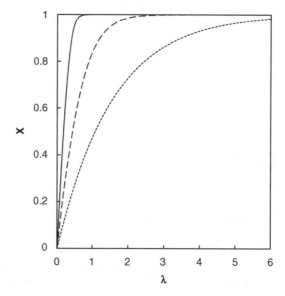

Figure 5.2 *Substrate conversion (X) along CPBR dimensionless length (λ) for Michaelis–Menten kinetics. Solid line: ideal CPBR; dashed line: CPBR given the presence of internal diffusional restrictions; dotted line: CPBR given the presence of external and internal diffusional restrictions.*

In Problem 5.5, $X = 0.396$ was obtained when the reactor was fed at a rate of $8.33\,cm^3\cdot min^{-1}$, which implies a volumetric productivity $(s_i \cdot X \cdot \tau^{-1})$ of $19.79\,g\cdot cm^{-3}\cdot min^{-1}$, while when the reactor was fed with $2.77\,cm^3\cdot min^{-1}$, $X = 0.475$, and a volumetric productivity of $7.91\,g\cdot cm^{-3}\cdot min^{-1}$ were attained. This analysis illustrates the interplay between diffusional restrictions and operation conditions. In the operation of reactors with immobilized enzymes, it is generally preferred that high feed flow rates be used, in order to reduce EDR, and that small catalyst particle sizes be used, in order to reduce IDR; however, under such conditions a high pressure drop will be produced in packed-bed reactors, compacting the catalyst bed and lowering the catalyst's lifespan, due to mechanical stress; besides, this will tend to produce channeling, so that the fluid flow regime strongly departs from ideal. The best conditions of operation in packed-bed rectors will result not just from a kinetic and mass-transfer analysis but also from the hydrodynamic conditions of the system.

5.7 Partially hydrolyzed lecithin has interesting applications in the food industry. Its production by enzymatic hydrolysis of soybean lecithin using a recombinant phospholipase A2 derived from porcine pancreas is under study. The enzyme is immobilized in spherical particles of DEAE-Sephadex with an average radius of $80\,\mu m$. The catalyst has a specific maximum reaction rate of $0.1\,\mu mol \cdot g^{-1} \cdot min^{-1}$, an intrinsic Michaelis constant of $40\,\mu M$, and a density of $1.5\,g\cdot cm^{-3}$. A value of $10^{-7}\,cm^2\cdot s^{-1}$ has been estimated for the effective diffusion coefficient of lecithin inside the catalyst particle. Previous studies show that the enzyme follows Michaelis–Menten kinetics and that EDR are negligible under the reaction conditions. Determine the reaction time needed to attain an X of 0.9 in a BSTR with an initial lecithin concentration of $0.8\,mM$ and a catalyst concentration of $10\,g\cdot L^{-1}$.

Resolution: From a material balance to the reactor:

$$-\frac{ds}{dt} = \eta_G \cdot k \cdot e \cdot \frac{s}{K_M + s}$$

which can be transformed into:

$$\frac{dX}{dt} = \eta_G(\beta_{oi} \cdot (1 - X), \Phi) \cdot \frac{(1 - X)}{1 + \beta_{oi} \cdot (1 - X)} \cdot \frac{m_{cat} \cdot a_{sp}}{K_M} \tag{5.37}$$

Equation 5.37 can be solved using the fourth-order Runge–Kutta numerical method. Since the enzyme has Michaelis–Menten kinetics, η_G can be estimated using Figure 3.7. The values of the parameters in Equation 5.37 are:

$$\beta_{oi} = 20$$

$$\frac{m_{cat} \cdot a_{sp}}{K_M} = \frac{10 \cdot 0.1}{40} = 0.025\,min^{-1}$$

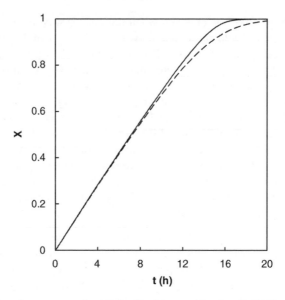

Figure 5.3 *BSTR performance under Michaelis–Menten kinetics. Solid line: BSTR under ideal conditions; dashed line: BSTR under IDR. X, substrate conversion; t, reaction time.*

Replacing these parameters in Equation 5.37 gives:

$$\frac{dX}{dt} = \eta_G(20 \cdot (1 - X), 2, 11) \cdot \frac{(1 - X)}{1 + 20 \cdot (1 - X)} \cdot 0.025 \qquad (5.38)$$

To apply the Runge–Kutta method, an integration step (h) must be chosen. Picking too large of a step (h) will reduce the calculation time but also lower the accuracy; the opposite will occur if h is too small. Under ideal conditions (in the absence of IDR), the reaction time will be 13.5 hours. Therefore, an h value of 30 minutes seems adequate, which gives almost 27 intervals of integration. By solving Equation 5.38 using Runge–Kutta (with h = 30 minutes), as shown in Problem 5.5, the profile of X versus time shown in Figure 5.3 is obtained. This figure indicates that a reaction time of 15 hours is required in order to reach X = 0.9, which is 11% higher than the value calculated for ideal BSTRs. Further-more, in Figure 5.3 the performances of ideal and real (subject to IDR) BSTRs are compared. Since the value of Φ is small and that of β_{oi} is high, the influence of IDR on the performance of the BSTR is only perceived at values of X greater than 50%.

Supplementary Problems

5.8 A device that removes CO_2 from blood is under development. The system consists of a CPBR with immobilized carbonic anhydrase (EC 4.2.1.1) in chitosan microspheres of 0.1 mm diameter. In preliminary tests at 37 °C, the reactor containing 10 µL of catalyst at $200\,\mathrm{IU} \cdot \mathrm{mL_{catalyst}}^{-1}$ (1 IU is the amount of enzyme that converts 1 µmol of CO_2 into bicarbonate each minute at 37 °C) was fed with $9\,\mathrm{mL} \cdot \mathrm{h}^{-1}$ of a saline solution containing $17\,\mathrm{mg} \cdot \mathrm{L}^{-1}$ dissolved CO_2. The diffusion coefficient of CO_2 within the catalyst matrix is $9 \cdot 10^{-9} \, m^2 \cdot s^{-1}$ and the Michaelis constant of CO_2 at $37 \, ^\circ C$ is 1.5 mM. Assume that first order kinetics with respect to CO_2 concentration is applicable.

(a) Calculate the level of CO_2 removal attained and the concentration of CO_2 at the device outlet. (b) If another device consisting of a polypropylene-membrane CSTR containing 100 mg of 1 mm-width catalyst membrane with $20 \, IU \cdot g^{-1}$ (density $= 0.9$ $g \cdot cm^{-3}$) works under the same conditions, calculate the level of CO_2 removal and the concentration of CO_2 at the device outlet. Compare the answer with the results obtained for (a). (c) What membrane width will make the performances of the two devices equal? (d) What amount of catalyst membrane will be required to make the performances of the two devices equal?

Answer: (a) 94.8%; $[CO_2] = 0.088 \, mg \cdot L^{-1}$. (b) 78.7%, $[CO_2] = 0.362 \, mg \cdot L^{-1}$.

The CPBR device performs better than membrane CSTR, since under almost equal operating conditions, CPBR is kinetically favored.

(c) It is not possible to make the performances of the two devices equal simply by reducing the membrane width. (d) $M_{cat} = 0.49 \, g$.

An amount of catalyst five times higher is required in order to obtian a conversion as high as that in the first device.

5.9 A CSTR bearing 1107 g of α-amylase immobilized in the surfaces of 2 mm-diameter beads of polyurethane-coated stainless steel (density $= 5.6 \, g \cdot cm^{-3}$) is fed $600 \, L \cdot h^{-1}$ of a maltopentaose syrup of $150 \, g \cdot L^{-1}$. The catalyzed reaction is:

$$C_{30}H_{52}O_{26} + 4H_2O \rightarrow 5C_6H_{12}O_6$$
$$\text{maltopentaose} \qquad\qquad \text{glucose}$$
$$\text{(M)} \qquad\qquad\qquad \text{(G)}$$

Initial reaction rates (v) at high agitation speed are obtained at various maltopentaose concentrations [M]:

[M]	5	12	20	30	50	100
v	58.1	113.2	155.3	190.8	233.4	280.3

v is in $mmol \cdot min^{-1} \cdot cm^{-2}$; [M] is in mM

Determine the steady-state conversion of the reactor as a function of the film mass-transfer coefficient (h') of maltopentaose and the concentration of glucose in the product stream. Consider a range of h' from 10 to $0.1 \, cm \cdot min^{-1}$.

Answer:

h' $(cm \cdot min^{-1})$	α	X	[G] $(g \cdot L^{-1})$
10	1.39	0.681	110.9
8	1.74	0.666	108.5
6	2.32	0.642	104.6
4	3.48	0.597	97.3
2	6.97	0.486	79.2
1	13.93	0.349	56.9
0.5	27.86	0.221	36.0
0.1	139.3	0.055	9.0

5.10 L-asparaginase (EC 3.5.1.1) is an important enzyme in cancer therapy. It catalyzes the hydrolysis of L-asparagine (AG; $C_4H_8N_2O_3$) into L-aspartic acid and ammonia and by doing so deprives fast-replicating cells of this essential amino acid. L-asparaginase is immobilized in a biocompatible epoxy resin for use in an extracorporeal shunt in order to remove L-asparagine from the patient bloodstream. The following kinetic measurements are made with 50 mg catalyst per mL:

t (mins)	0	1	2	5	10	15	20	25
[AG] (mM)	125	116.2	107.5	83.0	47.9	22.7	8.6	2.8

Design the column enzyme module to remove 90% of the blood L-asparagine from a patient whose level prior to treatment is $550 \, \mu g \cdot mL^{-1}$ serum. The treatment consists in practicing a venous–arterial shunt through the enzyme module, assisted by a peristaltic pump that delivers $10 \, mL \cdot min^{-1}$. The catalyst particles are 0.1 mm in diameter, with an effective density of $0.7 \, g \cdot mL^{-1}$ and an apparent density (under the packing conditions in the column) of $0.2 \, g \cdot mL^{-1}$. The diffusion coefficient of L-asparagine in water at the operation temperature is $2 \cdot 10^{-5} \, cm^2 \cdot s^{-1}$. Assume a diameter-to-catalyst-bed-height ratio of 0.2.

Answer: Volume of catalyst bed $= 23.5 \, cm^3$; reactor diameter $= 1.82 \, cm$; catalyst bed height $= 9.07 \, cm$.

5.11 Jack bean urease (urea amidohydrolase, EC 3.5.1.5) is immobilized in flat laminar polyvinyl alcohol gel particles of $1 \times 0.5 \times 0.01$ cm. The intrinsic kinetic parameters of the enzyme are $V_{max} = 0.32 \, \mu mol \cdot mL^{-1} \cdot s^{-1}$ and $K_M = 3.7 \, mM$. The diffusion coefficient for urea in water at 25 °C is $1.43 \cdot 10^{-5} \, cm^2 \cdot s^{-1}$. The void fraction in the gel matrix is estimated at 0.6 and the pore tortuosity at 2.5.

(a) Assuming first-order kinetics for urea concentration, find the performance (conversion versus flow rate) of CSTR and CPBR with 100 g of gel catalyst (density very close to 1) in the treatment of a waste stream containing $1.25 \, g \cdot L^{-1}$ of urea (OCN_2H_4). (b) If the maximum allowable level of disposal of urea is 100 ppm, what will the maximum allowable flow rate in each type of reactor be? (c) If the CSTR has to process the same flow rate as the CPBR, what conversion will be obtained in each case?

Answer: The results are summarized in the following table, with the answer to (a) given in the second and third rows, the answer to (b) denoted by the asterisks in the second and third rows and the answer to (c) given in the fourth row.

X	$F_{CPBR} \, L \cdot min^{-1}$	$F_{CSTR} \, L \cdot min^{-1}$	$X_{F,CSTR \, = \, F,CPBR}$
0.1	4.7	4.46	0.095
0.2	2.22	1.98	0.182
0.3	1.39	1.16	0.263
0.4	0.97	0.74	0.338
0.5	0.71	0.50	0.411
0.6	0.54	0.33	0.478

(*Continued*)

X	F_{CPBR} L·min^{-1}	F_{CSTR} L·min^{-1}	$X_{F,CSTR\,=\,F,CPBR}$
0.7	0.41	0.21	0.547
0.8	0.31	0.12	0.615
0.9	0.22	0.055	0.692
0.92*	0.20*	0.043*	
0.95	0.17	0.026	0.744
0.99	0.11	0.005	0.818

5.12 A CSTR operates with 3 kg of β-amylase (EC 3.2.1.2) immobilized in the surface of stainless-steel particles covered with a polyurethane film, with a specific activity of 300 IU · $g_{catalyst}^{-1}$. The following initial rate data are obtained in the hydrolysis of maltotetraose (MT) at varying initial concentrations:

[MT] (mM)	5	12	20	30	50	100
v	58.1	113.3	155.3	190.8	233.4	280.4

v is in μmoles of MT hydrolyzed per minute per gram of catalyst

The reactor is fed with $90\,L \cdot h^{-1}$ of a $335.7\,g \cdot L^{-1}$ maltotetraose ($C_{24}H_{42}O_{21}$) solution. Determine the steady-state conversion obtained.

Answer:

$$X = 0.798$$

5.13 β-galactosidase from *Kluyveromyces lactis* is immobilized in chitosan spherical beads to hydrolyze the lactose present in sweet whey produced in the manufacture of Cheddar cheese. The catalyst has a specific activity of $1200\,IU \cdot g^{-1}$, a density of $0.9\,g \cdot cm^{-3}$, a Michaelis constant for lactose of 34 mM, and a galactose competitive inhibition constant of 93 mM. The reaction is performed in a CPBR, which has to process $450\,L \cdot h^{-1}$ of whey with 5% (w/v) of lactose. A substrate conversion of 0.8 is mandatory for this use of the hydrolyzed whey. The catalyst beads have an average radius of 0.1 mm, EDR are negligible, and the effective diffusion coefficients of galactose and lactose inside the chitosan particles are $4.3 \cdot 10^{-6}$ and $2.8 \cdot 10^{-6}$ $cm^2 \cdot s^{-1}$, respectively.

How much catalyst mass is required in this process, compared to the amount theoretically needed in the absence of a mass-transfer limitation?

Answer: A catalyst mass 5% higher than ideal (no mass-transfer limitations) is required. The presence of product inhibition decreases the impact of diffusional restrictions on reactor performance.

6

Enzyme Reactor Design and Operation under Biocatalyst Inactivation

Enzyme reactor design and operation under ideal conditions and in the presence of mass-transfer limitations were presented in Chapters 4 and 5, respectively. In this chapter, the effect of enzyme inactivation on enzyme reactor design and operation will be analyzed. Enzymes are labile catalysts, so enzyme inactivation during reactor operation is inherent to their nature. Substantial improvements in enzyme catalyst robustness have been attained in recent decades, making more stable enzymes that allow prolonged and more efficient use available; however, no matter how stable an enzyme is, it will eventually be forced to a point of substantial decay, and the moment in which an enzyme catalyst must be discarded and replaced is a major issue in biocatalysis, being a crucial decision in process economics. The evaluation of enzyme decay during reactor operation is thus an aspect of major concern. Enzymes that are dissolved in the reaction medium tend to be poorly stable and their recovery is impractical, so their efficiency of use is modest and considerable inactivation occurs during reactor operation. With immobilized enzymes, stability is supposed to be high enough to allow prolonged use, either in continuous processes or in sequential batch processes with catalyst conventional recovery after each batch.

6.1 Mechanistically Based Mathematical Models of Enzyme Inactivation

Enzyme inactivation is the reflection of alterations in protein structure and as such is quite a complex molecular process. Strangely enough (or perhaps not so strangely), enzyme inactivation has been modeled according to very simple mechanisms, mostly involving single-stage first-order kinetics, as presented in Section 2.7.2 (see Equation 2.38). Single-stage first-order kinetics assumes that the enzyme suffers a single-step transition from a native fully active conformation to a completely inactive form. This is clearly an over-simplification of a process in which multiple intermediate, progressively less active

Problem Solving in Enzyme Biocatalysis, First Edition. Andrés Illanes, Lorena Wilson and Carlos Vera.
© 2014 John Wiley & Sons, Ltd. Published 2014 by John Wiley & Sons, Ltd. Companion Website:
http://www.wiley.com/go/illanes-problem-solving

enzyme species can occur (series-type enzyme inactivation) and enzyme heterogeneity (especially in the case of immobilized enzymes) can involve isoenzymes with different patterns of inactivation (parallel-type enzyme inactivation).

Based on these considerations, Henley and Sadana [1] proposed an elegant matricial view of enzyme inactivation, taking into account both series and parallel inactivation. Unfortunately, the number of parameters (transition-rate constants and relative specific activities of intermediate species or of initial isoenzymes) that must be fitted becomes very high as the mechanism under consideration gains in complexity, so in practice rather simple mechanisms (mostly two-stage series type) have been considered, though in many cases this is a significant step forward in describing enzyme inactivation over the conventional single-stage first-order classical mechanism [2]. Further, enzyme inactivation is usually modeled based on data obtained under nonreactive conditions, and this may lead to considerable error if used to predict the actual behavior of enzymes during reactor operation [3], such as when reaction substrates and/or products affect enzyme stability, as has been seen on many occasions [4–6]. The former aspect will be dealt with in this section, while the modulation effect on enzyme inactivation will be analyzed in Section 6.2.

Single-stage first-order inactivation mechanism can be represented as:

$$E_{active} \xrightarrow{k_D} E_{inactive}$$

Equation 2.38 describes such a mechanism, in which k_D (the first-order inactivation rate constant) represents the structural transition rate constant from a totally active enzyme species to a totally inactive one. Under nonreactive conditions, such a rate is a function of operational parameters (temperature, pH, ionic strength) but, as has been mentioned, it may also be affected by the concentrations of the catalytic modulators (substrates and/or products).

A variation on such a mechanism considers that the final enzyme species is not fully inactive; in such a case, the corresponding model is:

$$\frac{e}{e_0} = (1 - A') \cdot \exp(-k_D \cdot t) + A' \tag{6.1}$$

A two-stage series mechanism has proven to be quite useful in describing enzyme inactivation. This can be represented by the following scheme:

$$E_{active} \xrightarrow{k_{D1}} E_{intermediate} \xrightarrow{k_{D2}} E_{inactive}$$

Many experimental behaviors have been well modeled according to such mechanisms [7]. Models for enzyme inactivation according to two-stage series mechanisms without and with residual enzyme activity are represented by Equations 6.2 and 6.3, respectively:

$$\frac{e}{e_0} = (1 + A) \cdot \frac{k_{D1}}{k_{D2} - k_{D1}} \cdot \exp(-k_{D1} \cdot t) - A \cdot \frac{k_{D1}}{k_{D2} - k_{D1}} \cdot \exp(-k_{D2} \cdot t) \tag{6.2}$$

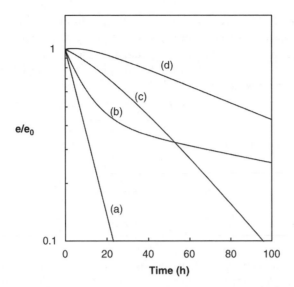

Figure 6.1 *Profiles of enzymatic inactivation according to different mechanisms. (a) Single-stage first-order kinetics without residual activity and $k_D = 0.1\,h^{-1}$. (b) Two-stage kinetics with convex profile of inactivation and $A = 0.4$, $k_{D1} = 0.1\,h^{-1}$ and $k_{D2} = 0.005\,h^{-1}$. (c) Two-stage kinetics with a concave profile of inactivation and $A = 0.6$, $k_{D1} = 0.3\,h^{-1}$, and $k_{D2} = 0.05\,h^{-1}$. (d) Two-stage kinetics with a grace period and $A = 1.05$, $k_{D1} = 0.1\,h^{-1}$, and $k_{D2} = 0.01\,h^{-1}$. e: enzyme activity at any time; e_0: initial enzyme activity.*

$$\frac{e}{e_0} = A' + \left(1 + \frac{A \cdot k_{D1} - A' \cdot k_{D2}}{k_{D2} - k_{D1}}\right) \cdot \exp(-k_{D1} \cdot t)$$

$$- \left(\frac{A \cdot k_{D1} - A' \cdot k_{D2}}{k_{D2} - k_{D1}}\right) \cdot \exp(-k_{D2} \cdot t) \tag{6.3}$$

Single-stage first-order mechanisms can easily be identified experimentally in a residual-activity-versus-inactivation-time semilog plot, through which a straight line is obtained whose slope is $-k_D$. Departure from this linearity is quite frequent, especially with immobilized enzymes, for which several nonlinear patterns of inactivation may be observed, most of which are well modeled by two-stage series mechanisms [8–10]. These behaviors are represented in Figure 6.1.

6.2 Effect of Catalytic Modulators on Enzyme Inactivation

Enzyme stability has customarily been determined under nonreactive conditions, even though these may not reflect the actual behavior of the catalyst during reactor operation. Modulation by catalytic effectors (substrates and products of reaction) can be significant and must be taken into account when describing the operational stability of an enzyme. Although it was originally proposed as a substrate protection effect [4], the concept has been generalized to allow the modulation effect to be either favorable (protection) or unfavorable (deprotection) and products, as well as substrates, to exert such kinds of modulation; in fact, any substance interacting with the enzyme structure during catalysis is a potential modulator of its stability [11]. Therefore, the free enzyme and each of the enzyme complexes inactivate at different rates, so, given single-stage first order kinetics, Equation 2.38 can be rewritten as:

$$-\frac{dem}{dt} = k_{D,M} \cdot em \tag{6.4}$$

If $k_{D,M}$ is the first-order inactivation rate constant of the enzyme–modulator M complex:

$$k_{D,M} = k_D \cdot (1 - n_M) \tag{6.5}$$

then, by definition, the modulation factor (n_M) of modulator M on enzyme stability is:

$$n_M = 1 - \frac{k_{D,M,}}{k_D} \tag{6.6}$$

A positive modulator (protectant) is one in which n_M is positive (greater than 0); a negative modulator (deprotectant) is one in which n_M is negative (less than 0). If n_M is 0, no modulation effect is exerted by M. This analysis can be extended to any other mechanism of enzyme inactivation. In the case of a series-type inactivation mechanism, there will be a modulation factor for each stage of enzyme inactivation. We have developed a model based on a two-stage series inactivation mechanism for the hydrolysis of penicillin G with penicillin G acylase, in which the substrate and both products of reaction are inhibitors, making them potential modulators of enzyme activity [6]. A similar analysis has been carried out for the hydrolysis of lactose with β-galactosidase, in which the product galactose, being a competitive inhibitor, is a modulator of enzyme stability, while glucose, having no inhibition effect, is not; in both stages of enzyme inactivation, galactose modulation factors are dependent on temperature, allowing the reactor temperature to be optimized for the hydrolysis of lactose in whey permeate [12].

6.3   Mathematical Models for Different Enzyme Kinetics, Modes of Operation, and Reactor Configurations under Biocatalyst Inactivation

6.3.1   Nonmodulated Enzyme Inactivation

A simple model for a single-stage first-order inactivation mechanism will be considered in this section. This model corresponds to Equation 2.38.

6.3.1.1   Batch Reactors

Given that the enzyme inactivates during operation, from Equations 2.38 and 4.4:

$$\int_0^X \frac{dX}{v(e,X)} = \int_0^t \frac{k \cdot e_0 \exp(-k_D \cdot t)}{s_i} dt \tag{6.7}$$

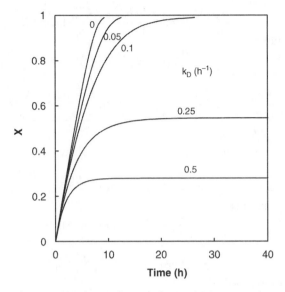

Figure 6.2 *BSTR performance under single-stage first-order kinetics of enzyme inactivation for Michaelis–Menten kinetics with $s_i = 100\,mM$, $K_M = 10\,mM$, and different values of k_D (h^{-1}). The reactor was designed to obtain a conversion of 95% after 8 hours of reaction under zero enzyme inactivation. X: substrate conversion; k_D: first-order inactivation rate constant.*

By integration, the left-hand side of this equation corresponds to those in Table 4.1 for the different kinetic models of single-substrate reactions. Denoting this expressions as $f(s_i, X)$, the equation describing batch stirred-tank reactor (BSTR) performance is:

$$f(s_i, X) = \frac{k \cdot e_0[1 - \exp(-k_D \cdot t)]}{K_M \cdot k_D} = \frac{m_{cat} \cdot a_{sp}[1 - \exp(-k_D \cdot t)]}{K_M \cdot k_D} \qquad (6.8)$$

For simple Michaelis–Menten kinetics, Equation 6.8 becomes:

$$\frac{s_i}{K_M} \cdot X - \ln(1 - X) = \frac{k \cdot e_0[1 - \exp(-k_D \cdot t)]}{K_M \cdot k_D} = \frac{m_{cat} \cdot a_{sp}[1 - \exp(-k_D \cdot t)]}{K_M \cdot k_D} \qquad (6.9)$$

Figure 6.2 shows the results of Equation 6.9 under different levels of enzyme inactivation.

For reactions involving two substrates, the procedure is similar. Under enzyme inactivation, Equations 4.16 to 4.20 now become:

OS (A limiting)

$$\frac{K_{MA}K'_{MB}}{a_i(b_i - a_i)} \ln\left|\frac{b_i - a_iX}{b_i(1 - X)}\right| - \frac{K'_{MB}}{a_i} \ln\left|\frac{b_i - a_iX}{b_i}\right| + X =$$
$$= \frac{m_{cat} \cdot a_{sp}}{a_i \cdot k_D}[1 - \exp(-k_D \cdot t)] \qquad (6.10)$$

OS (B limiting)
$$\frac{K_{MA}K'_{MB}}{b_i(a_i - b_i)} \ln\left|\frac{a_i - b_i X}{a_i(1 - X)}\right| - \frac{K'_{MB}}{b_i} \ln(1 - X) + X$$
$$= \frac{m_{cat} \cdot a_{sp}}{b_i \cdot k_D}[1 - \exp(-k_D \cdot t)] \tag{6.11}$$

RS (A limiting)
$$\frac{K_{MA}K'_{MB}}{a_i(b_i - a_i)} \ln\left|\frac{b_i - a_i X}{b_i(1 - X)}\right| - \frac{K'_{MB}}{a_i} \ln\left|\frac{b_i - a_i X}{b_i}\right| - \frac{K'_{MA}}{a_i} \ln(1 - X) + X =$$
$$= \frac{m_{cat} \cdot a_{sp}}{a_i \cdot k_D}[1 - \exp(-k_D \cdot t)] \tag{6.12}$$

Ping-Pong (A limiting)
$$X - \frac{\alpha \cdot K_{MA}}{a_i} \ln(1 - X) - \frac{K''_{MB}}{a_i} \ln(b_i - a_i \cdot X) =$$
$$= \frac{m_{cat} \cdot a_{sp}}{a_i \cdot k_D}[1 - \exp(-k_D \cdot t)] \tag{6.13}$$

Ping-Pong (B limiting)
$$X - \frac{K''_{MB}}{b_i} \ln(1 - X) - \alpha \frac{K_{MA}}{b_i} \ln(a_i - b_i \cdot X) =$$
$$= \frac{m_{cat} \cdot a_{sp}[1 - \exp(-k_D \cdot t)]}{b_i \cdot k_D} \tag{6.14}$$

(The nomenclature is the same as in Chapter 4.)

6.3.1.2 Continuous Reactors

The performance of continuous reactors under ideal conditions was analyzed in Section 4.4.2 for steady-state operation, but this analysis does not apply in the presence of enzyme inactivation. However, pseudo-steady-state conditions can be reasonably assumed so long as the catalyst is stable enough to produce a smooth decay of activity during reactor operation (this should be the case if the enzyme is to be used continuously). Initial steady-state conversion (X_0) can be obtained as in Section 4.4.2. After a certain reactor operation time t, enzyme activity decays to a certain value E(t) and pseudo-steady-state conversion becomes X(t). The reactor performance equations at times 0 and t are then:

$$\text{At } t = 0 \qquad \frac{k \cdot E_0}{K_M \cdot F} = \frac{M_{cat} \cdot a_{sp,0}}{K_M \cdot F} = f(s_i, X_0) \tag{6.15}$$

$$\text{At } t = t \qquad \frac{k \cdot E(t)}{K_M \cdot F} = \frac{M_{cat} \cdot a_{sp}(t)}{K_M \cdot F} = f(s_i, X(t)) \tag{6.16}$$

The corresponding kinetic expressions for single-substrate reactions are those given in Table 4.1.

For multiple-substrate reactions, the procedure is the same but Equations 6.15 and 6.16 become:

$$\text{At } t = 0 \qquad \frac{k \cdot E_0}{K_M \cdot F} = \frac{M_{cat} \cdot a_{sp,0}}{K_M \cdot F} = f(a_i, b_i, X_0) \tag{6.17}$$

$$\text{At } t = t \qquad \frac{k \cdot E(t)}{K_M \cdot F} = \frac{M_{cat} \cdot a_{sp}(t)}{K_M \cdot F} = f(a_i, b_i, X(t)) \tag{6.18}$$

If enzyme decay is represented by a certain function of time $\varphi(t)$, from Equations 6.15 to 6.18:

$$\text{One-substrate} \quad \varphi(t) = \frac{E(t)}{E_0} = \frac{a_{sp}(t)}{a_{sp,0}} = \frac{f(s_i, X(t))}{f(s_i, X_0)} \tag{6.19}$$

$$\text{Two-substrate} \quad \varphi(t) = \frac{E(t)}{E_0} = \frac{a_{sp}(t)}{a_{sp,0}} = \frac{f(a_i, b_i, X(t))}{f(a_i, b_i, X_0)} \tag{6.20}$$

Equations 6.15 to 6.20 allow the enzyme reactor performance ($X(t)$ versus t) to be determined for any kinetic model and for a continuous mode of reactor operation. For simple Michaelis–Menten kinetics and single-stage first-order kinetics of enzyme inactivation:

$$\text{CPBR} \quad \ln\frac{E(t)}{E_0} = \ln\frac{a_{sp}(t)}{a_{sp,0}} = \ln\frac{\dfrac{s_i}{K_M}X(t) - \ln[1 - X(t)]}{\dfrac{s_i}{K_M}X_0 - \ln[1 - X_0]} = -k_D \cdot t \tag{6.21}$$

$$\text{CSTR} \quad \ln\frac{E(t)}{E_0} = \ln\frac{a_{sp}(t)}{a_{sp,0}} = \ln\frac{\dfrac{s_i}{K_M}X(t) + \dfrac{X(t)}{1 - X(t)}}{\dfrac{s_i}{K_M}X_0 + \dfrac{X_0}{1 - X_0}} = -k_D \cdot t \tag{6.22}$$

Equations 6.21 and 6.22 allow the curve of operation of the reactors (evolution of substrate conversion with time) to be determined and can be used as design equations, allowing determination of the reactor volume required for certain processing tasks (see Section 4.4.2). A graphical representation of Equations 6.21 and 6.22 is given in Figure 6.3.

As can be seen, the behaviors of the two reactors tend to approach each other as the operation time increases, meaning that the greatest difference occurs under steady-state conditions (time 0 in Figure 6.3). A sound comparison between continuous packed-bed reactor (CPBR) and continuous stirred-tank reactor (CSTR) performance must then be based on the ratio of the areas under the whole operation curve, rather than that of the initial values of conversion. In the case illustrated in Figure 6.3, the ratio of X_0 is 1.17 while the ratio of areas is only 1.06. The situation is analogous to that in which there is more than one substrate reaction.

6.3.2 Modulated Enzyme Inactivation

With modulated enzyme inactivation, the concentrations of the modulators (substrates and products of reaction) change with time, and as a consequence the inactivation rates also change. A rather simple case can be used to illustrate reactor design under enzyme modulation. Consider an immobilized enzyme, free of diffusional restrictions, catalyzing the hydrolysis of a substrate into two products, one of which is a competitive inhibitor. Then:

$$v(X) = \frac{v(X)}{k \cdot e} = \frac{s_i(1 - X)}{K_M\left[1 + \dfrac{s_i X}{K_{IC}}\right] + s_i(1 - X)} \tag{6.23}$$

(a)

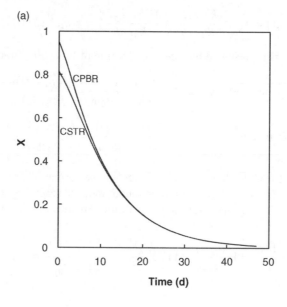

(b)

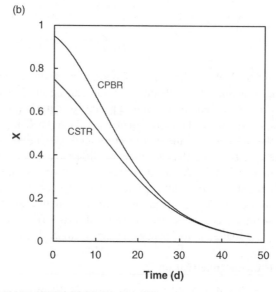

Figure 6.3 *CPBR and CSTR reactor performance under single-stage first-order enzyme inactivation kinetics ($k_D = 0.1\,h^{-1}$) for: (a) Michaelis–Menten kinetics; (b) competitive inhibition by product; (c) total noncompetitive inhibition by product; and (d) total uncompetitive inhibition by substrate. $s_i = 100\,mM$, $K_M = K_{IC} = K_{INC} = 10\,mM$, $K_{IUC} = 100\,mM$. The enzyme load was calculated to obtain a steady-state substrate conversion of 95% in CPBR. X: substrate conversion.*

(c)

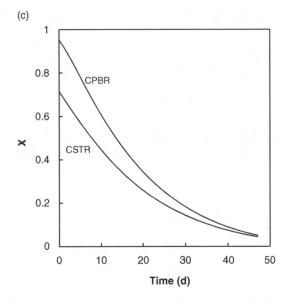

(d)

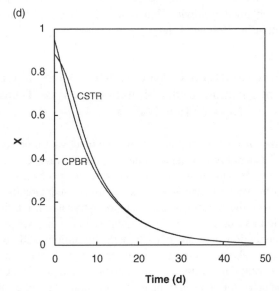

Figure 6.3   *(Continued)*

   Given a two-stage series mechanism of inactivation and assuming that each enzyme species (free enzyme, enzyme–substrate complex, and enzyme–inhibitory product complex) deactivates at a different rate, by introducing the concept of a modulation factor for each enzyme species at each stage of enzyme inactivation (Equation 6.6), the following equation can be derived for the modulated enzyme inactivation rate [2]:

$$-\frac{de}{dt} = e \cdot k \left[\frac{(1 - A) \cdot \exp(-k_{D1} \cdot t) \cdot [1 - \upsilon(X) \cdot N_1(X)]}{\exp(-k_{D1} \cdot t) + \frac{k_{D1} \cdot A}{(k_{D2} - k_{D1})} [\exp(-k_{D1} \cdot t) - \exp(-k_{D2} \cdot t)]} \right.$$

$$\left. + \frac{A \cdot k_{D2} \cdot [\exp(-k_{D1} \cdot t) - \exp(-k_{D2} \cdot t)] \cdot [1 - \upsilon(X) \cdot N_2(X)]}{(k_{D2} - k_{D1})\exp(-k_{D1} \cdot t) + k_{D1} \cdot A \cdot [\exp(-k_{D1} \cdot t) - \exp(-k_{D2} \cdot t)]} \right]$$

(6.24)

where:

$$N_1(X) = n_{1S} + n_{1P} \cdot \frac{K_M \cdot X}{K_{IC} \cdot (1 - X)}$$

(6.25)

$$N_2(X) = n_{2S} + n_{2P} \cdot \frac{K_M \cdot X}{K_{IC} \cdot (1 - X)}$$

(6.26)

are the lumped modulation factors for stages 1 and 2 of enzyme inactivation. BSTR performance can be obtained by solving Equations 6.23 and 6.24. The interested reader can find more information in reference [3].

6.4 Mathematical Models for Enzyme Kinetics, Modes of Operation, and Reactor Configurations under Simultaneous Mass-Transfer Limitations and Enzyme Inactivation

The effects of mass-transfer limitations and enzyme inactivation on reactor performance have been analyzed separately in Chapter 5 and Section 6.3, respectively. However, both phenomena can be present simultaneously in heterogeneous enzyme catalysis. Moreover, there is an interplay between the two, since inactivation, by reducing the catalytic potential of the enzyme, moves the system away from mass-transfer limitations; therefore, an apparent stabilization can be observed in immobilized enzymes subjected to diffusional restrictions. The interaction between mass-transfer limitations and enzyme inactivation provokes a complex behavior during reactor operation, because substrate depletion magnifies the impact of the mass-transfer limitations and enzyme inactivation counteracts it. That is to say, the behavior of an enzyme reactor under the influence of both phenomena deserves attention.

Under mass-transfer limitations and enzyme inactivation, Equation 4.4 can be rewritten as:

$$\frac{dX}{dt} = \eta(X, t) \frac{k \cdot e(t) \cdot \upsilon(X, t)}{K_M \cdot \beta_i}$$

(6.27)

The differential Equation 6.27 can be solved for any type of enzyme kinetics, mass-transfer limitation, and enzyme inactivation pattern, provided that sound expressions are available for the effectiveness factor $\eta(X,t)$ (or $\eta_G(X,t)$), the dimensionless reaction rate $\upsilon(X,t)$, and the enzyme inactivation rate $e(t)$. Note that mass-transfer moduli (Damkoehler number or Thièle modulus) depend on the concentration of active enzyme, so that in this case they are no longer constant and will vary according to the enzyme inactivation during reactor operation. The magnitude of these moduli will decrease with reaction time, so the impact of mass-transfer limitations will be reduced.

To illustrate reactor performance under simultaneous mass-transfer limitations and enzyme inactivation, consider the simple case of a BSTR working with a moderately labile immobilized enzyme with Michaelis–Menten intrinsic kinetics, subjected to nonmodulated single-stage first-order kinetics of enzyme inactivation under external diffusional restrictions (EDR). In such a case, the intrinsic rate equation is:

$$\upsilon(X, t) = \frac{\beta_i(1 - X)}{1 + \beta_i(1 - X)} \tag{6.28}$$

The expression for the effectiveness factor (see Equation 5.10) is:

$$\eta(X,t) = \frac{[(1 + \beta_i(1 - X)]\left[1 + \alpha(t) + \beta_i(1 - X)) - \sqrt{[1 + \alpha(t) - \beta_i(1 - X)]^2 + 4\beta_i(1 - X)}\right]}{2\beta_i(1 - X) \cdot \alpha(t)} \tag{6.29}$$

where:

$$\alpha(t) = \frac{k \cdot e(t)}{K_M \cdot h} \tag{6.30}$$

and the kinetics of enzyme inactivation is represented by Equation 2.38.

By replacing Equations 6.28 to 6.30 and Equation 2.38 into Equation 6.27, $X = f(\beta_i,t)$ is obtained through the numerical solution of the system of differential Equations 6.27 and 2.38. If all kinetic, inactivation, and mass-transfer parameters are evaluated as functions of temperature, the optimal temperature for reactor operation can be determined [12]. More complex reaction kinetics, enzyme inactivation kinetics, and mass-transfer limitations (i.e. internal diffusional restrictions) will add to the mathematical complexity, but the methodology used to tackle the problem will be the same.

Didactic software has been constructed by which to simulate enzyme reactor behavior under different scenarios of enzyme kinetics, mass-transfer limitations, and enzyme inactivation. The program is fed all of the kinetic, mass-transfer, and inactivation parameters and initial conditions and delivers a graphical representation of enzyme performance. This allows the different variables involved to be played with and the interactions between them to be appreciated. A demo of this software is presented in Epsilon Software Information.

6.5 Strategies for Reactor Operation under Biocatalyst Inactivation

With conventional batch processes in which the enzyme is dissolved in the reaction medium, the enzyme is usually discarded after reaching the desired conversion. Enzyme recovery is seldom practiced, because the enzyme is usually poorly stable and has rather low activity remaining at the end of the batch; besides, recovery in an active form is not a simple task. The situation is quite different in the case of continuous reactor operation with immobilized enzymes; here, the enzyme remains in the reactor while the substrate(s) flow through to be converted into product(s), and the retention time of the catalyst is much greater than the fluid retention time. Under enzyme inactivation, the challenge is to produce a reactor output of uniform quality; that is, to obtain a fairly constant conversion throughout reactor operation. Several strategies have been considered in order to meet this goal. Obviously, enzyme decay can be counteracted by fresh catalyst make-up, but this may be cumbersome in the case of CPBR and limited by fluid mechanic considerations in the case of CSTR. In both cases, the catalyst decay profile must be well described if it is to be compensated for by the catalyst make-up rate; continuous addition of solid catalyst is hard to perform, so intermittent additions that closely match the continuous enzyme decay must be made. Profiling of the temperature during reactor operation has been proposed based on the idea that a change in temperature may produce an imbalance between enzyme activity and stability (due to the different magnitudes of the respective energies of activation), with a stronger effect on the reaction rate than on the inactivation rate; this strategy has some rather obvious drawbacks, since it cannot be used for long and will likely lead to operational instability. Substrate concentration can also be profiled in order to compensate for enzyme decay, although the response is not easy to predict, especially if enzyme stability is modulated by the reaction substrates and products; another drawback is that, although the quality of the product will remain constant (constant X), the product concentration will vary, which may complicate downstream processing.

A very sound strategy for maintaining a constant quality output is to compensate for enzyme inactivation by means of a corresponding decrease in flow rate, since X will remain constant as long as the $E \cdot F^{-1}$ ratio is unaltered (see, for example, Equations 4.23 and 4.33). So, the condition for maintaining constant X will be:

$$\frac{E}{F} = \frac{E_0}{F_0} = constant \tag{6.31}$$

If considering an exponential decay of enzyme activity with time, for example:

$$E = E_0 \cdot \exp(-k_D \cdot t)$$

and the condition that satisfies Equation 6.31 is:

$$F = F_0 \cdot \exp(-k_D \cdot t)$$

The obvious drawback of this strategy is that it produces a variable output for downstream operations. However, fluctuations in flow rate can be reduced if multiple staggered reactors are used. For an allowable output flow rate fluctuation (ratio of minimum to maximum allowable flow rate to downstream operations; R_F) and a catalyst replacement criterion (number of half-lives of catalyst use before the catalyst is discarded; $N_{t1/2}$), the number of reactors required (N_R) can be calculated as [13]:

$$N_R = -N_{t1/2} \frac{\ln 2}{\ln R_F} \tag{6.32}$$

The reactor staggering time (time interval between each reactor startup) is then:

$$t_s = \frac{t_C}{N_R} = \frac{t_{1/2} \cdot N_{t1/2}}{N_R} \tag{6.33}$$

Equation 6.32 allows the number of reactors required to produce a reactor output with the predetermined allowable flow fluctuation and catalyst replacement criterion to be determined. In practice, one extra reactor will be required to allow for discharge, cleaning, and filling of the reactor after completion of its operation round. For example, if the replacement criterion is two half-lives, the allowable flow fluctuation is $\pm 10\%$ and the operational half-life of the enzyme is 60 days, from Equations 6.32 and 6.33:

$$N_R = -2 \cdot \frac{\ln 2}{\ln\left(\dfrac{0.9}{1.1}\right)} = 6.9$$

$$t_s = \frac{60 \cdot 2}{8} = 15 \text{ days}$$

Under such conditions the required number of reactors is $N_R + 1 = 8$ and the staggering time is 15 days. The number of reactors required will increase with the reduction in the allowable flow rate fluctuation and the increase in the duration of enzyme use before replacement.

From the enzyme inactivation pattern, whatever it may be, constant quality can be obtained by following the same pattern of reactor feed flow rate. Making use of a control valve and suitable software for simulating enzyme decay, it is possible to operate the reactor to ensure constant quality.

Solved Problems

6.1 A 60 L working-volume BSTR is used for the hydrolysis of cheese whey permeate ($40 \text{ g} \cdot \text{L}^{-1}$ of lactose) and concentrated cheese whey permeate ($150 \text{ g} \cdot \text{L}^{-1}$ of lactose) with a fungal β-galactosidase whose Michaelis constant is 90 mM and whose competitive inhibition constant for galactose is 9 mM. Lactose concentration must be reduced by 80% after 5 hours of reaction.

   (a) Determine the operation curve for the reactor working with both permeates, given that under the conditions of reaction the enzyme half-life is 2 days. (b) Compare the results obtained with those under ideal conditions (no enzyme inactivation).

Resolution (a): Data:

$K_M = 90 \text{ mM}$; $K_{IC} = 9 \text{ mM}$

$s_{i,1} = 117 \text{ mM}$; $s_{i,2} = 439 \text{ mM}$

$V_{eff} = 60 \text{ L}$

$X = 0.8$

$t = 300 \text{ minutes}$

$k_D = 0.347 \text{ days}^{-1} = 2.41 \cdot 10^{-4} \text{ min}^{-1}$

Given competitive inhibition by galactose and single-stage first-order enzyme kinetics of inactivation, from Equation 6.7 and Table 4.1:

$$s_i \cdot X \cdot \left(\frac{1}{K_M} - \frac{1}{K_{IC}}\right) - \left(1 + \frac{s_i}{K_{IC}}\right) \cdot \ln(1 - X) = \frac{k \cdot E_0 [1 - \exp(-k_D \cdot t)]}{V_{eff} \cdot K_M \cdot k_D} \tag{6.34}$$

From the data given, under initial conditions:

At $s_i = 117$ mM:

$$117 \cdot 0.8 \cdot \left(\frac{1}{90} - \frac{1}{9}\right) - \left(1 + \frac{117}{9}\right) \cdot \ln 0.2 = \frac{k \cdot E_0 [1 - \exp(-2.41 \cdot 10^{-4} \cdot 300)]}{60 \cdot 90 \cdot 2.41 \cdot 10^{-4}}$$

At $s_i = 439$ mM:

$$439 \cdot 0.8 \cdot \left(\frac{1}{90} - \frac{1}{9}\right) - \left(1 + \frac{439}{9}\right) \cdot \ln 0.2 = \frac{k \cdot E_0 [1 - \exp(-2.41 \cdot 10^{-4} \cdot 300)]}{60 \cdot 90 \cdot 2.41 \cdot 10^{-4}}$$

Then:

$$\text{At } s_i = 117 \text{ mM} \quad k \cdot E_0 = 245.77 \text{ mmoles} \cdot \text{min}^{-1} = 245,770 \text{ IU}$$
$$\text{At } s_i = 439 \text{ mM} \quad k \cdot E_0 = 839.53 \text{ mmoles} \cdot \text{min}^{-1} = 839,530 \text{ IU}$$

From Equation 6.34, with varying t:

$$\text{At } s_i = 117 \text{ mM} \quad -11.7 \cdot X - 14 \cdot \ln(1 - X) = \frac{245.77 \cdot [1 - \exp(-2.41 \cdot 10^{-4} \cdot t)]}{60 \cdot 90 \cdot 2.41 \cdot 10^{-4}}$$

$$\tag{6.35}$$

$$\text{At } s_i = 439 \text{ mM} \quad -43.9 \cdot X - 49.8 \cdot \ln(1 - X) = \frac{839.53 [1 - \exp(-2.41 \cdot 10^{-4} \cdot t)]}{60 \cdot 90 \cdot 2.41 \cdot 10^{-4}}$$

$$\tag{6.36}$$

Answer (a):

From Equations 6.35 and 6.36:

t (minutes)	X_{117mM}	X_{439mM}
10	0.136 (0.136)*	0.153 (0.153)
25	0.255 (0.255)	0.272 (0.273)
50	0.381 (0.383)	0.396 (0.397)
100	0.537 (0.540)	0.546 (0.549)
150	0.636 (0.640)	0.641 (0.646)
200	0.706 (0.712)	0.709 (0.715)
250	0.759 (0.766)	0.760 (0.767)
300	0.800 (0.808)	0.800 (0.808)

*Values in parentheses correspond to solution to (b)

Resolution (b): Given no enzyme inactivation, from Equation 4.5 and Table 4.1:

$$s_i \cdot X \cdot \left(\frac{1}{K_M} - \frac{1}{K_{IC}} \right) - \left(1 + \frac{s_i}{K_{IC}} \right) \cdot \ln(1-X) = \frac{k \cdot E_0}{V_{eff} \cdot K_M} \cdot t \qquad (6.37)$$

From the data provided:

At $s_i = 117\,\text{mM}$:

$$-11.7 \cdot X - 14 \cdot \ln(1-X) = \frac{245.77}{60 \cdot 90} \cdot t \qquad (6.38)$$

At $s_i = 439\,\text{mM}$:

$$-43.9 \cdot X - 49.8 \cdot \ln(1-X) = \frac{839.53}{60 \cdot 90} \cdot t \qquad (6.39)$$

Answer (b):

The results of Equations 6.38 and 6.39 correspond to the values in parentheses in the above table. As can be appreciated, the differences are not significant. This is because immobilized enzymes are usually stable, so the activity lost in the first batch is very low (in this case, the enzyme half-life was 48 hours and the batch duration was only 3 hours), so given single-stage first-order inactivation kinetics, the residual activity after the first batch is: exp $(-2.41 \cdot 10^{-4} \cdot 300) = 0.93$. In this situation, enzyme loss is only 7%, so the enzyme will be reused many times before it reaches a significant level of inactivation.

6.2 A yeast invertase inactivates thermally according to second-order kinetics, with a second-order inactivation rate constant at reactor operating temperature $= 1.76 \cdot 10^{-6}$ $\text{IU}^{-1}\text{h}^{-1}$. Construct the operation curve (X versus t) of a CPBR with $1000\,\text{g}$ of chitin-immobilized invertase ($100\,\text{IU} \cdot \text{g}^{-1}$) fed with a sucrose concentration of 10 times the Michaelis constant whose steady-state conversion under such conditions is 0.9

Resolution: The mechanism of second-order inactivation is represented by:

$$-\frac{dE}{dt} = k_D \cdot E^2 \qquad (6.40)$$

Integrating Equation 6.40 gives:

$$\frac{1}{E(t)} - \frac{1}{E_0} = k_D \cdot t \qquad (6.41)$$

Given Michaelis–Menten kinetics, from Equations 6.15 and 6.16:

$$E_0 = \frac{K_M \cdot F}{k} \cdot [10 \cdot X_0 - \ln(1 - X_0)] \qquad (6.42)$$

$$E(t) = \frac{K_M \cdot F}{k} \cdot [10 \cdot X(t) - \ln(1 - X(t))] \qquad (6.43)$$

From Equations 6.41 to 6.43:

$$t = \frac{k}{K_M \cdot F \cdot k_D} \cdot \left[\frac{1}{10 \cdot X(t) - \ln(1 - X(t))} - \frac{1}{10 \cdot X_0 - \ln(1 - X_0)} \right] \tag{6.44}$$

Data:

$$E_0 = 100,000 \text{ IU}$$
$$k_D = 1.76 \cdot 10^{-6} \text{ IU}^{-1} \cdot h^{-1}$$
$$X_0 = 0.9$$

From Equation 6.42:

$$\frac{k \cdot E_0}{K_M \cdot F} = 11.3 \tag{6.45}$$

From Equation 6.44:

$$t = \frac{11.3}{100,000 \cdot 1.76 \cdot 10^{-6}} \cdot \left[\frac{1}{10 \cdot X(t) - \ln(1 - X(t))} - \frac{1}{10 \cdot X_0 - \ln(1 - X_0)} \right] \tag{6.46}$$

$$t = 642.05 \cdot \left[\frac{1}{10 \cdot X(t) - \ln(1 - X(t))} \right] - 56.82$$

Answer: From Equation 6.46:

t (h)	0	5	10	20	50	75	100	200
X(t)	0.900	0.849	0.800	0.711	0.526	0.431	0.364	0.225

6.3 A certain penicillin acylase is inhibited under high concentrations of penicillin G, with a substrate inhibition constant of 54 mM and a Michaelis constant of 27 mM. The reaction is carried out with immobilized penicillin acylase in both CPBR and CSTR fed with 135 mM penicillin G potassium salt, under such conditions that the steady-state performance of both reactors is the same.

   If the enzyme is inactivated according to single-stage first-order kinetics with a half-life of 20 days, obtain the performance (X versus t) of both reactors.

Resolution: Data:

$$K_M = 27 \text{ mM}; \quad K_S = 54 \text{ mM}, s_i = 135 \text{ mM}$$
$$t_{1/2} = 20 \text{ days} \quad k_D = 0.0347 \text{ days}^{-1}$$

At steady state, from Equation 4.23 and Table 4.1:

$$\frac{k \cdot E}{K_M \cdot F} = 5X_0 - \ln(1 - X_0) + 12.5X_0(1 - 0.5X_0) = 5X_0 + \frac{X_0}{1 - X_0} + 12.5X_0(1 - X_0) \tag{6.47}$$

Solving Equation 6.47:

$$X_0 = 0.872$$

$$\frac{k \cdot E}{K_M \cdot F} = 12.565$$

Dividing Equation 6.15 by Equation 6.16, and given single-stage first-order kinetics of enzyme inactivation (Equation 2.39):

$$t = \frac{\ln\frac{E_0}{E(t)}}{k_D} = \frac{\ln\frac{f(s_i,X_0)}{f(s_i,X(t))}}{k_D} \tag{6.48}$$

For both CPBR and CSTR, Equation 6.48 becomes:

$$\text{CPBR} \quad t = 28.82 \cdot \ln\left[\frac{12.565}{5X(t) - \ln(1 - X(t)) + 12.5X(t)(1 - 0.5X(t))}\right] \tag{6.49}$$

$$\text{CSTR} \quad t = 28.82 \cdot \ln\left[\frac{12.565}{5X(t) + \frac{X(t)}{1 - X(t)} + 12.5X(t)(1 - X(t))}\right] \tag{6.50}$$

Answer: From Equations 6.49 and 6.50:

t (days)	0	1	5	7.5	10	15	20	25	30	40	50
X_{CPBR}	0.872	0.841	0.714	0.642	0.579	0.469	0.384	0.315	0.261	0.179	0.125
X_{CSTR}	0.872	0.864	0.822	0.785	0.735	0.569	0.462	0.363	0.290	0.192	0.130

As can be seen, the decay in conversion is faster in CPBR than in CSTR, so under enzyme inactivation a proper comparison between the two must be made over the whole operation time and not just for the steady-state initial conversion.

6.4 Stachyose (S) is a tetrasaccharide ($C_{24}H_{42}O_{21}$) which upon hydrolysis produces different mixtures of monosaccharides. Stachyose is poorly fermented and is present in significant quantities in certain legumes, so its ingestion may produce abdominal discomfort. A certain α-galactosidase catalyzes the hydrolysis of stachyose to produce 1 mol of glucose, 1 mol of fructose, and 2 mol of galactose per 1 mol of stachyose. The following initial rate data have been obtained at low stachyose concentrations with an immobilized α-galactosidase under conditions in which mass-transfer limitations are negligible:

[S]	0.15	0.30	0.45	0.60	0.75	1.00
$v \cdot 10^3$	13.1	25.5	37.2	48.2	58.7	75.0

[S] is in mM; v is in $mmol \cdot min^{-1} \cdot g_{cat}^{-1}$

However, the enzyme is subjected to uncompetitive inhibition at high stachyose concentrations, with an inhibition constant 50% greater than the Michaelis constant. (a) Determine the steady-state operation curve (flow rate versus conversion) of CPBR and CSTR loaded with 1 kg of immobilized enzyme where a lupin hydrolysate containing $20\,g \cdot L^{-1}$ of stachyose is being treated. At what conversion do both reactors have the same performance? (b) What will the operation curve (X versus t) be at this conversion, if the enzyme is inactivated according to single-stage first-order kinetics and with a half-life of 25 days? (c) What will the product composition (in $g \cdot L^{-1}$) be after 5 days of operation?

Resolution (a): From the initial rate data in the table and using any method to determine the kinetic parameters (see Section 2.6):

$V_{max} = 0.45\,mmol \cdot min^{-1} \cdot g_{catalyst}^{-1}$
$K_M = 5\,mM$
$K_S = 7.5\,mM$
$s_i = 20\,g \cdot L^{-1} = 30\,mM$
$M_{cat} = 1000\,g$

From Equations 4.23 and 4.33 and Table 4.1:

$$CPBR \quad 6 \cdot X - \ln(1 - X) + 24 \cdot X \cdot (1 - 0.5 \cdot X) = \frac{1000 \cdot 0.45}{5 \cdot F} \tag{6.51}$$

$$CSTR \quad 6 \cdot X + \frac{X}{1 - X} + 24 \cdot X \cdot (1 - X) = \frac{1000 \cdot 0.45}{5 \cdot F} \tag{6.52}$$

So:

$$F_{CPBR} = \frac{90}{6 \cdot X - \ln(1 - X) + 24 \cdot X \cdot (1 - 0.5 \cdot X)} \tag{6.53}$$

$$F_{CSTR} = \frac{90}{6 \cdot X + \frac{X}{1 - X} + 24 \cdot X \cdot (1 - X)} \tag{6.54}$$

Answer (a): From Equations 6.53 and 6.54:

X	0.3	0.5	0.7	0.9	0.91	0.92	0.93	0.94	0.95
F_{CPBR}	10.87	7.09	5.51	4.60	4.55	4.41	4.46	4.41	4.36
F_{CSTR}	12.38	9.00	7.78	5.44	5.13	4.79	4.41	3.97	3.48

F: in $L \cdot min^{-1}$

At $X = 0.929$, both reactors have the same performance. The inversion point now occurs at a high substrate conversion, so across most of the operation range CSTR is a better option than CPBR.

Resolution (b): Data:

$X = 0.929$
$k_D = \ln 2 \cdot 25^{-1} = 0.0277\,days^{-1}$

From Equations 6.15, 6.16, and 6.19 (considering single-stage first-order kinetics) and Table 4.1:

$$\text{CPBR} \quad t = \frac{1}{k_D} \cdot \ln \frac{\frac{s_i}{K_M} X_0 - \ln[1 - X_0] + \frac{s_i^2 X_0}{K_M K_S}(1 - 0.5 \cdot X_0)}{\frac{s_i}{K_M} X(t) - \ln[1 - X(t)] + \frac{s_i^2 X(t)}{K_M K_S}(1 - 0.5 \cdot X(t))} \tag{6.55}$$

$$\text{CSTR} \quad t = \frac{1}{k_D} \cdot \ln \frac{\frac{s_i}{K_M} X_0 + \frac{X_0}{1 - X_0} + \frac{s_i^2 X_0}{K_M K_S}(1 - X_0)}{\frac{s_i}{K_M} + \frac{X(t)}{1 - X(t)} + \frac{s_i^2 X(t)}{K_M K_S}(1 - X(t))} \tag{6.56}$$

From Equations 6.55 and 6.56:

$$t_{CPBR} = 36.067 \cdot \ln \frac{20.16}{6 \cdot X(t) - \ln[1 - X(t)] + 24 \cdot X(t) - 12 \cdot X(t)^2} \tag{6.57}$$

$$t_{CSTR} = 36.067 \cdot \ln \frac{20.24}{6 \cdot X(t) + \frac{X(t)}{1 - X(t)} + 24 \cdot X(t)(1 - X(t))} \tag{6.58}$$

Answer (b): From Equations 6.57 and 6.58:

t (days)	0	1	2	3	4	5	10	15	20	25
X_{CPBR}	0.929	0.901	0.871	0.839	0.807	0.776	0.638	0.531	0.446	0.377
X_{CSTR}	0.929	0.926	0.922	0.919	0.915	0.911	0.883	0.835	0.707	0.512

As can be seen, the decay in performance is more pronounced in CPBR than in CSTR.

Resolution (c): After 5 days, the molar stachyose conversion is 0.776 for CPBR and 0.911 for CSTR:

	[mM]		$[g \cdot L^{-1}]$	
	CPBR	CSTR	CPBR	CSTR
residual stachyose	$30 \cdot 0.224 = 6.72$	$30 \cdot 0.089 = 2.67$	4.48	1.78
glucose	$30 \cdot 0.776 = 23.28$	$30 \cdot 0.911 = 27.33$	4.19	4.92
fructose	$30 \cdot 0.776 = 23.28$	$30 \cdot 0.911 = 27.33$	4.19	4.92
galactose	$60 \cdot 0.776 = 46.56$	$60 \cdot 0.911 = 54.66$	8.38	9.84

Answer (c): The product composition for CPBR and CSTR is given in the fourth and fifth columns of the table. Products of a higher quality are obtained with CSTR because the residual stachyose is at a much lower concentration than it is in CPBR.

6.5 A metagenomic phenol hydroxylase (EC 1.14.1) has been identified that is very active over recalcitrant phenolic compounds. However, the enzyme is strongly inhibited by one of the products of reaction in a noncompetitive mode. The product noncompetitive inhibition constant is 5 mM, while the Michaelis constant for the phenolic compound being degraded is 25 mM. A CPBR is operated for 10 days to process an effluent with $5000\,mg \cdot L^{-1}$ of this compound (average molecular mass $= 750\,Da$) and the flow rate is adjusted to deliver an effluent of constant quality (70% removal of the phenolic compound). The catalyst loaded in the reactor is 20 kg and its specific activity is $320\,IU \cdot g^{-1}$.

(a) If the enzyme inactivates according to single-stage first-order kinetics and the inactivation rate constant is $5.25 \cdot 10^{-3}\,h^{-1}$, calculate the total volume processed during reactor operation and the amount of phenolic compound discharged. (b) Recalculate for a CSTR operating under similar conditions and compare with the results for (a). (c) If the inhibition is competitive (the value of the inhibition constant is the same), how does this modify the results obtained in (a) and (b) for CPBR and CSTR, respectively? Discuss.

Resolution (a): Data:

$K_M = 25\,mM$
$K_{INC} = 5\,mM$
$a_{sp} = 0.32\,mmol \cdot min^{-1} \cdot g^{-1}$
$s_i = 5000\,mg \cdot L^{-1} = 6.667\,mM$
$M_{cat} = 20\,000\,g$
$k_D = 5.25 \cdot 10^{-3}\,h^{-1}$
$X = 0.7$
$t_{op} = 240\,h$

The first step is to calculate the initial flow rate (F_0). From Equation 6.15 and Table 4.1:

$$\frac{M_{cat} \cdot a_{sp}}{K_M \cdot F_0} = s_i X \left[\left(\frac{1}{K_M} - \frac{1}{K_{INC}} \right) \right] - \left[\left(1 + \frac{s_i}{K_{INC}} \right) \right] \cdot \ln(1 - X) + \frac{s_i^2 X^2}{2 K_M K_{INC}} \quad (6.59)$$

From Equation 6.59:

$$\frac{20,000 \cdot 0.32}{25 \cdot F_0} = -0.7467 + 2.8094 + 0.08722$$

$$F_0 = 119.1\,L \cdot min^{-1} = 7146\,L \cdot h^{-1}$$

To keep X(t) constant, the flow rate profile F(t) should follow the enzyme inactivation profile E(t) (see Equations 6.15 and 6.16). So (E(t)/F(t) = constant). Then:

$$\frac{F(t)}{F_0} = \frac{E(t)}{E_0} = \exp[-k_D \cdot t] \quad (6.60)$$

and the total processed volume (V_{proc}) is:

$$V_{proc} = \int_0^t F(t)dt = \int_0^t F_0 \cdot \exp[-k_D \cdot t]dt = F_0 \cdot \frac{[1 - \exp(-k_D \cdot t)]}{k_D} \quad (6.61)$$

$$V_{proc} = 7146 \cdot \frac{[1 - \exp(-5.25 \cdot 10^{-3} \cdot 240)]}{5.25 \cdot 10^{-3}} = 975049 \, \text{L}$$

$$M_{phenolic} = 975049 \cdot 6.667 \cdot 0.3 \cdot \frac{750}{10^6} = 1463 \, \text{kg}$$

Answer (a): A total of $975\,049\,\text{L}$ is processed. The amount of phenolic compound discharged during operation is $1463\,\text{kg}$.

Resolution (b): From Equation 6.15 and Table 4.1:

$$\frac{M_{cat} \cdot a_{sp}}{K_M \cdot F_0} = \frac{s_i \cdot X}{K_M} + \frac{X}{1 - X} + \frac{s_i \cdot X^2}{K_{INC}(1 - X)} + \frac{s_i^2 \cdot X^2}{K_M \cdot K_{INC}} \tag{6.62}$$

From Equation 6.62:

$$\frac{20,000 \cdot 0.32}{25 \cdot F_0} = 0.1867 + 2.3333 + 2.1779 + 0.1742$$

$$F_0 = 52.54 \, \text{L} \cdot \text{min}^{-1} = 3152.4 \, \text{L} \cdot \text{h}^{-1}$$

$$V_{proc} = 3152.4 \cdot \frac{[1 - \exp(-5.25 \cdot 10^{-3} \cdot 240)]}{5.25 \cdot 10^{-3}} = 430135$$

$$M_{phenolic} = 430135 \cdot 6.667 \cdot 0.3 \cdot \frac{750}{10^6} = 645 \, \text{kg}$$

Answer (b): A total of $430\,135\,\text{L}$ can be processed in a CSTR. The amount of phenolic compound discharged during operation will be $645\,\text{kg}$. The processing capacity of CSTR is 41% lower than that of CPBR.

Resolution (c): If the product of phenolic degradation is a competitive inhibitor then from Equation 6.15 and Table 4.1:

$$\text{CPBR} \quad \frac{M_{cat} \cdot a_{sp}}{K_M \cdot F_0} = s_i \cdot X \cdot \left(\frac{1}{K_M} - \frac{1}{K_{IC}} \right) - \left(\frac{1}{K_M} - \frac{1}{K_{IC}} \right) \cdot \ln(1 - X) \tag{6.63}$$

$$\text{CSTR} \quad \frac{M_{cat} \cdot a_{sp}}{K_M \cdot F_0} = \frac{s_i}{K_M} \cdot X + \frac{X}{1 - X} + \frac{s_i}{K_{NC}} \cdot \frac{X^2}{1 - X} \tag{6.64}$$

From Equations 6.63 and 6.64:

$$\text{CPBR} \quad \frac{20,000 \cdot 0.32}{25 \cdot F_0} = s_i \cdot X \cdot \left(\frac{1}{K_M} - \frac{1}{K_{IC}} \right) - \left(1 + \frac{s_i}{K_{IC}} \right) \cdot \ln(1 - X) \tag{6.65}$$

$$\text{CSTR} \quad \frac{20,000 \cdot 0.32}{25 \cdot F_0} = \frac{s_i}{K_M} \cdot X + \frac{X}{1 - X} + \frac{s_i}{K_{NC}} \cdot \frac{X^2}{1 - X} \tag{6.66}$$

From Equations 6.65 and 6.66:

$$\text{CPBR} \quad \frac{20,000 \cdot 0.32}{25 \cdot F_0} = -0.7467 + 2.8094 \tag{6.67}$$

$$\text{CSTR} \quad \frac{20,000 \cdot 0.32}{25 \cdot F_0} = 0.18667 + 2.3333 + 2.1779 \tag{6.68}$$

Answer (c):

$$F_{0,CPBR} = 124.1\,\text{L} \cdot \text{min}^{-1} = 7446\,\text{L} \cdot \text{h}^{-1}$$
$$F_{0,CSTR} = 54.5\,\text{L} \cdot \text{min}^{-1} = 3270\,\text{L} \cdot \text{h}^{-1}$$
$$V_{proc,CPBR} = 7446 \cdot \frac{[1 - \exp(-5.25 \cdot 10^{-3} \cdot 240)]}{5.25 \cdot 10^{-3}} = 1,015,983\,\text{L}$$
$$V_{proc,CSTR} = 3270 \cdot \frac{[1 - \exp(-5.25 \cdot 10^{-3} \cdot 240)]}{5.25 \cdot 10^{-3}} = 446,181\,\text{L}$$

6.6 A column reactor packed with 550 g of chitin-immobilized fungal β-galactosidase (apparent density $\rho_{app} = 0.28\,\text{g} \cdot \text{mL}^{-1}$) is fed with 21 bv · h^{-1} (catalyst bed volumes per hour) of demineralized concentrated cheese whey permeate with 60.3 g · L^{-1} lactose. The reactor operates at 53 °C, at which temperature the following parameters are obtained:

Maximum reaction rate $= 480\,\mu\text{mol}$ of glucose per minute per gram of catalyst
Michaelis contant for lactose $= 100\,\text{mM}$
Competitive inhibition constant for galactose $= 75\,\text{mM}$
Half-life $= 15$ days

(a) Determine the operation curve of the reactor. Makes any assumptions explicit.
(b) Solve (a) but considering now that the enzyme is also inhibited by the product glucose in a noncompetitive mode, with an inhibition constant of 90 mM. (c) Solve (b) for a CSTR operating under the same conditions. Compare your results.

Resolution (a): Data:

$K_M = 100\,\text{mM}$
$K_{IC} = 75\,\text{mM}$
$a_{sp} = 0.48\,\text{mmol} \cdot \text{min}^{-1} \cdot \text{g}_{catalyst}^{-1}$
$s_i = 60.3\,\text{g} \cdot \text{L}^{-1} = 176.3\,\text{mM}$ lactose
$M_{cat} = 550\,\text{g}$
$\rho_{app} = 0.28\,\text{g} \cdot \text{mL}^{-1}$
$t_{1/2} = 15$ days

Given single-stage first-order kinetics of enzyme inactivation:

$k_D = \ln 2/15 = 4.62 \cdot 10^{-2}\,\text{days}^{-1}$
$F = 21\,\text{bv} \cdot \text{h}^{-1} = \dfrac{21 \cdot 550}{0.28 \cdot 1000} = 41.25\,\text{L} \cdot \text{h}^{-1} = 0.6875\,\text{L} \cdot \text{min}^{-1}$

From Equation 4.23 and Table 4.1:

$$\frac{M_{cat} \cdot a_{sp}}{K_M \cdot F} = s_i \cdot X(t) \cdot \left(\frac{1}{K_M} - \frac{1}{K_{IC}}\right) - \left(1 + \frac{s_i}{K_{IC}}\right) \cdot \ln(1 - X(t))$$

and under steady-state initial conditions:

$$\frac{M_{cat} \cdot a_{sp}}{K_M \cdot F} = s_i \cdot X_0 \left(\frac{1}{K_M} - \frac{1}{K_{IC}}\right) - \left(1 + \frac{s_i}{K_{IC}}\right) \cdot \ln(1 - X_0)$$

$$\frac{550 \cdot 0.48}{100 \cdot 0.6875} = 176.3 \cdot X_0 \left(\frac{1}{100} - \frac{1}{75}\right) - \left(1 + \frac{176.3}{75}\right) \ln(1 - X_0)$$

$$3.84 = -0.588 \cdot X_0 - 3.351 \cdot \ln(1 - X_0) \tag{6.69}$$

From Equation 6.19, given single-stage first-order kinetics of enzyme inactivation:

$$t = \frac{1}{k_D} \cdot \ln \frac{s_i \cdot X_0 \cdot \left(\frac{1}{K_M} - \frac{1}{K_{IC}}\right) - \left(1 + \frac{s_i}{K_{IC}}\right) \cdot \ln(1 - X_0)}{s_i \cdot X(t) \cdot \left(\frac{1}{K_M} - \frac{1}{K_{IC}}\right) - \left(1 + \frac{s_i}{K_{IC}}\right) \cdot \ln(1 - X(t))}$$

$$t = \frac{1}{4.62 \cdot 10^{-2}} \cdot \ln \frac{3.84}{-0.588 \cdot X(t) - 3.351 \cdot \ln(1 - X(t))} \tag{6.70}$$

***Resolution (b)*:** Data:

$K_{INC} = 90 \, \text{mM}$

From Equation 4.23 and Table 4.1:

$$\frac{M_{cat} \cdot a_{sp}}{K_M \cdot F} = s_i X \left[\frac{1}{K_M} - \frac{1}{K_{IC}} - \frac{1}{K_{INC}}\right] - \left[1 + \frac{s_i}{K_{IC}} + \frac{s_i}{K_{INC}}\right] \ln(1 - X) + \frac{s_i^2 \cdot X^2}{2 \cdot K_M \cdot K_{INC}}$$

and under steady-state initial conditions:

$$\frac{M_{cat} \cdot a_{sp}}{K_M \cdot F} = s_i X_0 \left[\frac{1}{K_M} - \frac{1}{K_{IC}} - \frac{1}{K_{INC}}\right] - \left[1 + \frac{s_i}{K_{IC}} + \frac{s_i}{K_{INC}}\right] \ln(1 - X_0) + \frac{s_i^2 \cdot X_0^2}{2 \cdot K_M \cdot K_{INC}}$$

$$\frac{550 \cdot 0.48}{100 \cdot 0.6875} = 176.3 \cdot X_0 \left(\frac{1}{100} - \frac{1}{75} - \frac{1}{90}\right) - \left(1 + \frac{176.3}{75} + \frac{176.3}{90}\right) \cdot \ln(1 - X_0)$$

$$+ \frac{176.3^2 \cdot X_0^2}{2 \cdot 100 \cdot 90}$$

$$3.84 = -2.547 \cdot X_0 - 5.31 \cdot \ln(1 - X_0) + 1.727 \cdot X_0^2 \tag{6.71}$$

From Equation 6.19, given single-stage first-order kinetics of enzyme inactivation:

$$t = \frac{1}{k_D} \cdot \ln \frac{s_i \cdot X_0 \left[\frac{1}{K_M} - \frac{1}{K_{IC}} - \frac{1}{K_{INC}} \right) \right] - \left[1 + \frac{s_i}{K_{IC}} + \frac{s_i}{K_{INC}} \right] \cdot \ln(1 - X_0) + \frac{s_i^2 \cdot X_0^2}{2 \cdot K_M \cdot K_{INC}}}{s_i \cdot X(t) \left[\frac{1}{K_M} - \frac{1}{K_{IC}} - \frac{1}{K_{INC}} \right) \right] - \left[1 + \frac{s_i}{K_{IC}} + \frac{s_i}{K_{INC}} \right] \cdot \ln(1 - X(t)) + \frac{s_i^2 \cdot X(t)^2}{2 \cdot K_M \cdot K_{INC}}}$$

$$t = \frac{1}{4.62 \cdot 10^{-2}} \cdot \ln \frac{3.84}{-2.547 \cdot X(t) - 5.31 \cdot \ln(1 - X(t)) + 1.727 \cdot X(t)^2} \tag{6.72}$$

Resolution (c): From Equation 4.33 and Table 4.1:

$$\text{CSTR} \quad \frac{M_{cat} \cdot a_{sp}}{K_M \cdot F} = \frac{X}{1 - X} + \frac{s_i \cdot X^2}{1 - X} \cdot \left(\frac{1}{K_{IC}} + \frac{1}{K_{INC}} \right) + \frac{s_i X}{K_M} \cdot \left(1 + \frac{s_i \cdot X}{K_{INC}} \right)$$

and under steady-state initial conditions:

$$\frac{M_{cat} \cdot a_{sp}}{K_M \cdot F} = \frac{X_0}{1 - X_0} + \frac{s_i \cdot X_0^2}{1 - X_0} \cdot \left(\frac{1}{K_{IC}} + \frac{1}{K_{INC}} \right) + \frac{s_i X_0}{K_M} \cdot \left(1 + \frac{s_i \cdot X_0}{K_{INC}} \right)$$

$$\frac{550 \cdot 0.48}{100 \cdot 0.6875} = \frac{X_0}{1 - X_0} + \frac{176.3 \cdot X_0^2}{1 - X_0} \cdot \left(\frac{1}{75} + \frac{1}{90} \right) + \frac{176.3 \cdot X_0}{100} \cdot \left(1 + \frac{176.3 \cdot X_0}{100} \right)$$

$$3.84 = \frac{X_0}{1 - X_0} + 4.31 \cdot \frac{X_0^2}{1 - X_0} + 1.763 \cdot X_0 \cdot (1 + 1.96 \cdot X_0) \tag{6.73}$$

From Equation 6.19, given single-stage first-order kinetics of enzyme inactivation:

$$t = \frac{1}{k_D} \cdot \ln \frac{3.84}{\frac{X(t)}{1 - X(t)} + \frac{s_i \cdot X(t)^2}{1 - X(t)} \cdot \left(\frac{1}{K_{IC}} + \frac{1}{K_{INC}} \right) + \frac{s_i X(t)}{K_M} \cdot \left(1 + \frac{s_i \cdot X(t)}{K_{INC}} \right)}$$

$$t = \frac{1}{4.62 \cdot 10^{-2}} \cdot \ln \frac{3.84}{\frac{X(t)}{1 - X(t)} + 4.31 \cdot \frac{X(t)^2}{1 - X(t)} + 1.763 \cdot X(t) \cdot (1 + 1.96 \cdot X(t))} \tag{6.74}$$

Answer (a): Solving Equation 6.69:

$$X_0 = 0.72$$

Solving Equation 6.70:

t (days)	1	2	3	4	5	7.5	10	15	20	25
X(t)	0.704	0.688	0.672	0.656	0.640	0.600	0.559	0.482	0.409	0.344

Answer (b): Solving Equation 6.71:

$$X_0 = 0.591$$

Solving Equation 6.72:

t (days)	1	2	3	4	5	7.5	10	15	20	25
X(t)	0.577	0.563	0.549	0.535	0.521	0.488	0.455	0.394	0.338	0.288

Answer (c): Solving Equation 6.73:

$$X_0 = 0.447$$

Solving Equation 6.74:

t (days)	1	2	3	4	5	7.5	10	15	20	25
X(t)	0.437	0.427	0.417	0.408	0.398	0.375	0.352	0.309	0.269	0.233

Performance of CBTR at steady state is much better than that of CSTR because the enzyme in this case is inhibited by both products. Differences tend to be reduced during reactor operation as the enzyme becomes progressively more inactivated.

6.7 *Aspergillus niger* glucose oxidase is immobilized in spherical chitosan particles of average diameter (d_p) in a range from 0.1 to 2.0 mm. The effective diffusion coefficient of glucose (D_{eff}) in the chitosan matrix is estimated at $2.4 \cdot 10^{-5}$ $cm^2 \cdot min^{-1}$. Under the reaction conditions, the enzyme has a specific activity of $330\ IU \cdot g_{catalyst}^{-1}$.

(a) Determine the range of global effectiveness factors of such catalyst particles, indicating, according to your judgment, what catalyst particle size can be considered free of mass-transfer limitations. Consider in this case that under excess oxygen, the reaction of glucose oxidation is well described by first-order kinetics with respect to glucose concentration, with the first-order rate constant being $2 \cdot 10^{-3}\ s^{-1}$.

(b) Determine the course of reaction (X versus t) at 0.2 M initial glucose concentration for a range of catalyst particle sizes and at a concentration of 10 g of catalyst per liter of reaction volume. Assume in this case that the reaction follows Michaelis–Menten kinetics, with a K_M for glucose of 5 mM and a half-life of 5 hours.

Resolution (a): From Equation 3.23:

$$\Phi = \frac{d_p}{6} \cdot \sqrt{\frac{k}{D_{eff}}} = 11.785 \cdot d_p$$

Given first-order kinetics, values of η_G for different values of Φ are obtained from Equation 3.40.

Answer (a):

d_p (cm)	0.01	0.015	0.05	0.1	0.15	0.20	0.25	0.30
Φ	0.118	0.177	0.589	1.179	1.768	2.357	2.946	3.536
η_G	0.992	0.982	0.839	0.610	0.459	0.364	0.301	0.256

Catalysts can be considered essentially free of mass-transfer limitations at a d_p below 0.15 mm.

Resolution (b): Combining Equations 5.7 and 6.7 gives:

$$\int_0^X \frac{dX}{v(e,X)\cdot\eta_G(X)} = \int_0^t \frac{k\cdot e_0 \exp(-k_D\cdot t)}{s_i}\,dt \tag{6.75}$$

In this case, for first-order reaction kinetics and single-stage first-order inactivation kinetics, η_G can be considered constant (independent of X), so Equation 6.75 can be integrated to obtain:

$$\frac{s_i}{K_M}X - \ln(1-X) = \eta_G\frac{k\cdot e_0}{K_M\cdot k_D}[1-\exp(-k_D\cdot t)] = \eta_G\frac{m_{cat}\cdot a_{sp}}{K_M\cdot k_D}[1-\exp(-k_D\cdot t)] \tag{6.76}$$

Data:

$s_i = 200\,\text{mM}$
$K_M = 5\,\text{mM}$
$k_D = 2.3\cdot10^{-3}\,\text{min}^{-1}$
$m_{cat} = 10\,\text{g}\cdot\text{L}^{-1}$
$a_{sp} = 330\,\text{IU}\cdot\text{g}^{-1} = 0.33\,\text{mmol}\cdot\text{min}^{-1}\cdot\text{g}^{-1}$

Using these data in Equation 6.76 gives:

$$40X - \ln(1-X) = 287\cdot\eta_G\cdot[1-\exp(-2.3\cdot10^{-3}\cdot t)] \tag{6.77}$$

By solving Equation 6.77 for different values of η_G (different values of d_p), the corresponding operation curves can be obtained.

Answer (b): For $d_p = 0.1\,\text{mm}$ ($\eta_G = 0.992$):

t (min)	10	20	30	40	50	60	70	80
X	0.158	0.311	0.459	0.603	0.740	0.867	0.970	≈ 1

For $d_p = 1$ mm ($\eta_G = 0.610$):

t (min)	10	20	40	60	80	100	120	150
X	0.097	0.191	0.373	0.545	0.705	0.852	0.969	≈ 1

For $d_p = 3$ mm ($\eta_G = 0.256$):

t (min)	10	20	50	100	150	250	350	400	450
X	0.041	0.080	0.157	0.366	0.518	0.767	0.944	0.990	≈ 1

As can be appreciated, catalyst particle size has a strong effect on reactor performance when internal diffusional restrictions are present.

6.8 An enzyme immobilized in the outer surface of an impervious support is used as a catalyst for a reaction conducted at an initial dimensionless substrate concentration of 6. With the active enzyme, a Damkoehler number of 5 is determined. The enzyme inactivates according to first-order kinetics, with a half-life of 1 day. Evaluate the effectiveness factor as a function of substrate conversion at different levels of enzyme inactivation (assume that first-order inactivation kinetics apply) and discuss the results.

Resolution: Data:

$\alpha_i = 5$; the subscript i indicates the initial value (i.e. when the enzyme has not yet suffered inactivation)

$\beta_{0i} = 6$; the subscript 0 indicates the bulk reaction medium

$t_{1/2} = 24$ hours

$k_D = \ln 2 \cdot 24^{-1} = 0.0289\,h^{-1}$

From Equation 6.29:

$$\eta(X,t) = \frac{[1 + 6(1 - X)] \cdot [(1 + \alpha(t) + 6(1 - X) - \sqrt{(1 + \alpha(t) - 6(1 - X))^2 + 24(1 - X)}]}{12 \cdot \alpha(t) \cdot (1 - X)}$$

$$(6.78)$$

where from Equation 6.30, given first-order enzyme inactivation kinetics:

$$\alpha(t) = \alpha_i \cdot \exp(-0.0289 \cdot t) = 5 \cdot \exp(-0.0289 \cdot t) \qquad (6.79)$$

By replacing Equation 6.79 into Equation 6.78, $\eta(X,t)$ can be obtained for different values of $\alpha(t)$, corresponding to different reaction times.

Answer:

X	$\eta(X,t)$				
	t = 0 hours	t = 10 hours	t = 30 hours	t = 50 hours	t = 100 hours
0.1	0.796	0.862	0.935	0.967	0.993
0.2	0.756	0.833	0.921	0.960	0.991
0.3	0.709	0.795	0.901	0.950	0.989
0.4	0.653	0.747	0.874	0.936	0.986
0.5	0.589	0.688	0.836	0.915	0.982
0.6	0.517	0.616	0.784	0.885	0.975
0.7	0.438	0.531	0.711	0.837	0.963
0.8	0.353	0.435	0.611	0.761	0.942
0.9	0.262	0.327	0.482	0.641	0.896

The value of the effectiveness factor decreases with substrate conversion because the substrate concentration is reduced and so the relative magnitude of mass-transfer limitations is increased. The value of the effectiveness factor increases with time because the enzyme becomes progressively more inactivated and the system becomes less sensitive to mass-transfer limitations.

6.9 The synthesis of cephalexin (Cx) from phenylglycine methyl ester (PGME) and 7-amino-3-deacetoxy cephalosporanic acid (7ADCA) catalyzed by penicillin acylase is a kinetically controlled reaction that can be described by an ordered sequential mechanism in which the enzyme binds first to the acyl donor (PGME) and then to the nucleophile (7ADCA) to yield the antibiotic. The corresponding dissociation constants for the enzyme-PGME secondary complex and the enzyme PGME-7ADCA tertiary complex are 5 and 20 mM, respectively. Synthesis is usually conducted under nucleophile limitation.

(a) Determine the operation curve (X versus t) of a BSTR operating with 90 mM PGME and 30 mM 7ADCA initial concentration and $2\,g\cdot L^{-1}$ of penicillin acylase immobilized in glyoxyl-agarose (GAPA) with a specific activity of $250\,IU\cdot g^{-1}$. (b) If GAPA is inactivated during the course of the reaction, so that its operational half-life is 24 hours, analyze the new results and compare them with those for (a). (c) What will the increase in operation time be in sequential batches with catalyst retention in the reactor if the final conversion is to be maintained at 0.95? (d) If, besides inactivation, GAPA is subjected to external diffusional restrictions with respect to 7ADCA and if under operating conditions the Damkoehler number is 3 (for the determination of the effectiveness factor, assume the kinetics of reaction of 7ADCA is of the Michaelis–Menten type), analyze the new results and compare them with those for (a) and (b).

Resolution (a): The synthesis of Cx with penicillin acylase can be represented as:

$$PGME + 7ADCA \rightarrow C_X + MeOH$$
$$\quad(A) \qquad\quad (B)$$

In this case, the limiting substrate is 7ADCA (B). So Equation 4.17 applies:

$$\frac{K_{MA}K'_{MB}}{b_i(a_i - b_i)} \ln\left|\frac{a_i - b_i X}{a_i(1 - X)}\right| - \frac{K'_{MB}}{b_i} \ln(1 - X) + X = \frac{k \cdot e}{b_i} t = \frac{m_{cat} \cdot a_{sp}}{b_i} t \quad (4.17)$$

Data:

$K_{MA} = 5\,mM$
$K'_{MB} = 20\,mM$
$a_{sp} = 0.25\,mmol \cdot min^{-1} \cdot L^{-1}$
$m_{cat} = 2\,g \cdot L^{-1}$
$b_i = 30\,mM$
$a_i = 90\,mM$

Using these in Equation 4.17 gives:

$$0.0556 \ln\left|\frac{3 - X}{3(1 - X)}\right| - 0.667 \cdot \ln(1 - X) + X = 0.0167 \cdot t \quad (6.80)$$

Resolution (b): Data:

$t_{1/2} = 24\,hours$
$k_D = 4.81 \cdot 10^{-4}\,min^{-1}$

From Equation 6.11:

$$f(X) = 0.0556 \ln\left|\frac{3 - X}{3(1 - X)}\right| - 0.667 \cdot \ln(1 - X) + X = 34.65 \cdot [1 - \exp(-k_D \cdot t)] \quad (6.81)$$

$$t = -2079 \cdot \ln(1 - 0.0289 \cdot f(X)) \quad (6.82)$$

Answer (a) and (b): Solving Equation 6.80 gives the results for (a) as t_a in the following table. Replacing Equation 6.81 into Equation 6.82 and solving it gives the results for (b) as t_b:

X	0.1	0.2	0.3	0.4	0.5	0.6	0.7	0.8	0.9	0.95
t_a (min)	10.41	21.38	33.05	45.57	59.34	74.85	93.11	116.5	152.3	185.3
t_b (min)	10.48	21.56	33.43	46.23	60.41	76.50	95.59	120.3	158.8	194.7

t_a: time under no enzyme inactivation ($k_D = 0$); t_b: time under enzyme inactivation ($k_D = 4.81 \cdot 10^{-4}\,min^{-1}$)

The time required to achieve the same X increases as a consequence of enzyme inactivation, but the differences are rather small because the stability of the enzyme is high, so the effect in one batch is mild. This difference increases with the number of batches.

Resolution (c): At the end of the first batch, $X = 0.95$ after 194.7 minutes. At the end of the second batch, from Equation 2.39:

$e_2 = e \cdot \exp(-4.81 \cdot 10^{-4} \cdot 194.7) = 0.91 \cdot e$

$f(X) = 34.65 \cdot 0.91 \cdot (1 - \exp(-4.81 \cdot 10^{-4} \cdot t_2)) = 31.53 \cdot (1 - \exp(-4.81 \cdot 10^{-4} \cdot t_2)$

$t_2 = -2079 \cdot \ln(1 - 0.0317 \cdot f(X))$

For $X = 0.95$: $t_2 = 214.6$

At the end of the third batch, from Equation 2.39:

$e_3 = e_2 \cdot \exp(-4.81 \cdot 10^{-4} \cdot 214.6) = 0.9 \cdot e_2$

$f(X) = 28.38 \cdot (1 - \exp(-4.81 \cdot 10^{-4} \cdot t_3))$

$t_2 = -2079 \cdot \ln(1 - 0.0352 \cdot f(X))$

For $X = 0.95$: $t_3 = 239.7$

At the end of the fourth batch, from Equation 2.39:

$e_4 = e_3 \cdot \exp(-4.81 \cdot 10^{-4} \cdot 239.7) = 0.89 \cdot e_3$

$f(X) = 25.26 \cdot (1 - \exp(-4.81 \cdot 10^{-4} \cdot t_4))$

$t_4 = -2079 \cdot \ln(1 - 0.0396 \cdot f(X))$

For $X = 0.95$: $t_4 = 271.7$

And so on ...

Answer (c):

Batch	$t_{X=0.95}$	Δt_1 (minutes)	Δt_2 (minutes)
1	194.7		
2	214.6	19.9	19.9
3	239.7	25.1	45.0
4	271.7	32.0	77.0

Δt_1: time increase from batch to batch; Δt_2: time increase with respect to first batch

Resolution (d): Data:

$\alpha_i = 3$

If the enzyme kinetics is controlled by 7ADCA and if the mass-transfer limitations mostly affect 7ADCA, since its concentration is lower than that of PGME, the effectiveness factor can be approximated by Equation 6.29:

$$\eta(X,t) = \frac{[(1 + \beta_i(1-X)]\left[1 + \alpha(t) + \beta_i(1-X) - \sqrt{[1 + \alpha(t) - \beta_i(1-X)]^2 + 4\beta_i(1-X)}\right]}{2\beta_i(1-X) \cdot \alpha(t)}$$

where, in this particular case:

$$\alpha(t) = \frac{k \cdot e(t)}{K'_{MB} \cdot h} = \frac{m_{cat} \cdot a_{sp} \cdot \exp(-k_D \cdot t)}{K'_{MB} \cdot h} = \alpha_i \cdot \exp(-k_D \cdot t) = 3 \cdot \exp(-k_D \cdot t)$$

$$\beta_i = \frac{b_i}{K'_{MB}} = 1.5$$

From Equations 4.8 and 4.12:

$$\frac{dX}{dt} = \frac{\eta(X,t)}{b_i} \cdot \frac{m_{cat} \cdot a_{sp} \cdot \exp(-k_D \cdot t) \cdot (a_i - b_i \cdot X) \cdot b_i(1 - X)}{(a_i - b_i \cdot X)(b_i(1-X) + K'_{MB}) + K_{MA} \cdot K'_{MB}} \tag{6.83}$$

Equation 6.83 must be solved using numerical methods, such as Euler or Runge–Kutta. Here the explicit Euler method (Section 4.1) will be used. Given Equation A.90, Equation 6.83 becomes:

$$X(t_{i+1}) = X(t_i)$$
$$+ \Delta t \cdot \left(\frac{\eta(X(t_i), t_i)}{b_i} \cdot \frac{m_{cat} \cdot a_{sp} \cdot \exp(-k_D \cdot t_i) \cdot (a_i - b_i \cdot X(t_i)) \cdot b_i(1 - X(t_i))}{(a_i - b_i \cdot X(t_i))(b_i(1-X(t_i)) + K'_{MB}) + K_{MA} \cdot K'_{MB}} \right) \tag{6.84}$$

The explicit Euler method is the simplest way of solving Equation 6.84. However, this procedure requires a small integration step (Δt) if it is to be stable and offer an adequate accuracy of results, necessitating much calculation work. Fortunately, this method can be easily implemented in any spreadsheet software, such as MS Excel. Replacing the values of the parameters in Equation 6.84 and taking a Δt to be equal to 1 minute gives:

$$X(t_{i+1}) = X(t_i)$$
$$+ 1 \cdot \left(\frac{\eta(X(t_i), t_i)}{30} \cdot \frac{2 \cdot 0,25 \cdot \exp(-k_D \cdot t_i) \cdot (90 - 30 \cdot X(t_i)) \cdot 30(1 - X(t_i))}{(90 - 30 \cdot X(t_i))(30(1 - X(t_i)) + 20) + 5 \cdot 20} \right) \tag{6.85}$$

The results in the following table were obtained by calculating X iteratively from Equation 6.85. These results show that in the presence of EDR the reaction time required to attain X = 0.95 is 2.81 times that in the absence of inactivation ($521 \cdot 185.3^{-1}$) and 2.78 times that in its presence ($542 \cdot 194.7^{-1}$).

X	0.1	0.2	0.3	0.4	0.5	0.6	0.7	0.8	0.9	0.95
t'_a (minutes)	20	41	65	92	124	163	213	282	402	521
t'_b (minutes)	20	41	65	93	125	165	216	289	414	542

t'_a: time under no enzyme inactivation ($k_D = 0$); t'_b: time under enzyme inactivation ($k_D = 4.81 \cdot 10^{-4} \, min^{-1}$)

Supplementary Problems

6.10 In Problem 4.18, steady-state conversions of acrylonitrile into acrylamide with immobilized nitrile hydratase in CPBR and CSTR configurations were calculated, giving 0.897 and 0.687, respectively. If the immobilized enzyme under such conditions has a half-life of 100 hours, the inactivation kinetics is single-stage first-order and pseudo-steady state can be assumed throughout the reactor operation, determine the performance equations (X versus t) of both reactors. Comment on the results obtained.

Answer:

t (minutes)	0	20	40	80	160	240	300	360	420	480
X_{CPBR}		0.897	0.852	0.801	0.693	0.491	0.331	0.240	0.170	0.119	0.082
X_{CSTR}		0.687	0.650	0.611	0.534	0.391	0.274	0.204	0.150	0.107	0.076

As can be seen, the differences between the two reactors tend to decrease with increasing operation time. Under steady state, X_{CPBR} is 30% higher than X_{CSTR}; after 8 hours, it is only 8% higher.

6.11 In Problem 5.12, the steady-state conversion of maltotetraose (MT) into maltose in CSTR was determined to be 0.798. If the immobilized β-amylase under such conditions has a half-life of 10 days, the inactivation kinetics is single-stage first-order and pseudo-steady state can be assumed throughout the reactor operation, determine the operation curve of the reactor (X versus t) up to 50 days. If after one half-life of operating time the temperature is reduced so that the inactivation rate constant is reduced to one half, what will the operation curve look like from then on (assume that enzyme activity is not affected by this temperature change)? What will the result be if the activity of the enzyme is reduced by 30% with the temperature shift?

Answer:

t (days)	0	1	2	3	5	10	20	25	30	35	40	50
X_a	0.798	0.778	0.756	0.731	0.676	0.521	0.274	0.196	0.139	0.099	0.070	0.035
X_b						0.521	0.382	0.324	0.274	0.232	0.196	0.139
X_c						0.378	0.272	0.229	0.194	0.163	0.137	0.098

X_a: conversion; X_b: conversion after temperature shift; X_c: conversion after temperature shift considering reduction in activity

6.12 A BSTR is working with a novel enzyme and the operator is only allowed to measure the conversion of the substrate into product. She assumes that the enzyme follows Michaelis–Menten kinetics and some laboratory data allow her to estimate that the initial substrate concentration is six times the Michaelis constant. She calculates that the conversion attained after one shift (8 hours) will meet the requirement of 90%. However, as the plant supervisor you know:

- that the magnitude of the Michaelis constant was underestimated by 20%;
- that the enzyme is competitively inhibited by the product of reaction with an inhibition constant four times the magnitude of the Michaelis constant;
- that the enzyme is inactivated during the operation, with its half-life under such conditions being 35 hours.

(a) Calculate the magnitude of the error in the estimation of the conversion made by the plant operator.

(b) Later, more data from the laboratory indicate that product inhibition is noncompetitive, although the magnitude of the inhibition constant is the same as that previously reported. What is the magnitude of the error now?

Answer:

(a) Given competitive inhibition and enzyme inactivation:

$$X = 0.75; \text{error} = 20\%$$

(b) Given noncompetitive inhibition and enzyme inactivation:

$$X = 0.637; \text{error} = 41.3\%$$

6.13 (a) Calculate the CPBR–CSTR catalyst mass ratio under different steady-state conversions if the kinetics of the enzymatic reaction is first-order with respect to substrate concentration. (b) How will this ratio vary under enzyme inactivation when the reactor operates for one, two, and three half-lives and the steady-state conversion is 0.9? (c) Answer (a) and (b) again but now assuming Michaelis–Menten kinetics under a dimensionless initial substrate concentration of 10. (d) Answer (a) and (b) again but now assuming that the enzyme is competitively inhibited by one of the products of reaction and that the inhibition constant is twice the Michaelis constant.

Answer (a):

X	0.1	0.3	0.5	0.7	0.9	0.95	0.99
M_{CPBR}/M_{CSTR}	0.948	0.832	0.693	0.516	0.256	0.158	0.047

The differences are strongly dependent on conversion, being quite significantly in favor of CPBR under high conversion (which is what is to be expected in an industrial operation).

Answer (b): At $X_0 = 0.9$:

	X	
	CPBR	CSTR
$t = t_{1/2}$	0.683	0.818
$t = 2\, t_{1/2}$	0.437	0.692
$t = 3\, t_{1/2}$	0.250	0.529

Decay, as for most kinetic mechanisms, is slower in CSTR than in CPBR.

Answer (c):

X	0.1	0.3	0.5	0.7	0.9	0.95	0.99
M_{CPBR}/M_{CSTR}	0.995	0.979	0.949	0.879	0.630	0.439	0.133

The trend is the same as in (a), but differences are significantly attenuated, as substrate concentration approaches saturation, where zero-order kinetics prevail and no difference between configurations occur.

At $X_0 = 0.9$:

	X	
	CPBR	CSTR
$t = t_{1/2}$	0.496	0.684
$t = 2\,t_{1/2}$	0.253	0.387
$t = 3\,t_{1/2}$	0.128	0.200

Decay is again slower in CSTR than in CPBR. Decay is more pronounced for both reactor configurations in the case of Michaelis–Menten kinetics than it is with first-order kinetics.

Answer (d):

X	0.1	0.3	0.5	0.7	0.9	0.95	0.99
M_{CPBR}/M_{CSTR}	0.970	0.894	0.783	0.613	0.313	0.191	0.054

Differences are smaller than in first-order kinetics but higher than in Michaelis–Menten kinetics.

6.14 A BSTR is operated with $300\,mg \cdot L^{-1}$ of immobilized β-glucosidase to remove the cellobiose content in a wood hydrolysate. The reactor is loaded with cellobiose to an initial concentration of $6\,g \cdot L^{-1}$. The process operates at $25\,°C$, but it has been proposed that the temperature be increased to $40\,°C$. The following data have been obtained with the immobilized enzyme:

- Michaelis constant for cellobiose at $25\,°C = 14\,mM$
- competitive inhibition constant for glucose at $25\,°C = 70\,mM$
- reaction rate constant at $25\,°C = 0.05\,mmol \cdot min^{-1} \cdot mg_{catalyst}^{-1}$
- energy of activation of the reaction of cellobiose hydrolysis $= 15\,kcal \cdot mol^{-1}$
- energy of activation of β-glucosidase inactivation $= 20\,kcal \cdot mol^{-1}$
- enzyme half-life at $25\,°C = 2.4$ hour

In the range from 25 to $40\,°C$, the Michaelis constant and the glucose inhibition constant do not vary significantly. Compare the reactor performance at both temperatures. Is it reasonable to increase the temperature?

Answer:

t (hours)	0.25	0.5	1	2	3	5	10	30	50	90	100
$X_{25\,°C}$	0.012	0.023	0.046	0.090	0.131	0.208	0.364	0.689	0.803	0.871	0.871
$X_{40\,°C}$	0.304	0.476	0.636	0.725	0.742	0.747	0.747	0.747			

An increase in temperature from 25 to $40\,°C$ produces a significant increase in productivity but the conversion attained is lower than at $25\,°C$. Both parameters must be properly weighed in economic terms, but in this case the decrease in conversion at $40\,°C$ (12 percentage points) is not as significant as the increase in productivity (about 37 times), so the suggestion that the temperature be increased is reasonable.

6.15 A *Rhodococcus* sp. nitrilase (EC 3.5.5.1) catalyzes the hydrolysis of phenylacetonitrile to phenylacetic acid and ammonia. Ammonia is a noncompetitive inhibitor
of the enzyme, with an inhibition constant of 25 mM, while the Michaelis constant
for phenylacetonitrile is 7.5 mM. The second-order thermal inactivation rate constant
of the nitrilase at the reactor operating temperature is $2.15 \cdot 10^{-6} \, \text{IU}^{-1} \cdot \text{h}^{-1}$. A CPBR
operates with 50 000 IU of nitrilase fed with 120 mM substrate concentration, obtaining
a steady-state substrate conversion of 0.75. A CSTR, operating at the same flow rate
and feed substrate concentration, is operated with the amount of enzyme necessary to
obtain the same steady-state substrate conversion than in the CPBR.

(a) Determine the evolution of substrate conversion with time for both reactors. (b)
How much more catalyst mass should be used in CPBR than in CSTR?

Answer:

(a)

t (hours)	0	1	2	3	5	10	15	20	25	50
X_{CPBR}	0.75	0.700	0.664	0.633	0.579	0.482	0.416	0.368	0.331	0.225
X_{CSTR}	0.75	0.718	0.688	0.660	0.613	0.523	0.461	0.414	0.378	0.271

(b) 81.4% more.

6.16 The enzyme fumarate hydratase (EC 4.2.1.2) catalyzes the hydration of fumarate
to produce S-malate. An immobilized fumarate hydratase is used under conditions
at which the equilibrium of reaction is well displaced to the formation of S-malate.
A CSTR is operated under conditions at which the initial steady-state conversion
of fumarate into S-malate is 96%. After 48 hours of operation at a dimensionless
inlet substrate concentration (ratio of inlet substrate concentration to Michaelis
constant) of 5, the conversion decays to 90%, and a temperature shift of $+10\,°C$ is
effected; at the new temperature, the Michaelis constant is 20% lower and the
enzyme specific activity is doubled, but the first-order inactivation rate constant is
increased by 50%.

(a) Determine the first-order inactivation rate constant at the initial temperature. (b)
Determine the fumarate–malate conversion immediately following the temperature
shift. (c) Determine the reactor operation curve at the second temperature.

Answer: (a) $0.01579 \, \text{h}^{-1}$. (b) 0.965.

(c)

t (hours)*	0	1	2	3	4	5	6
X	0.965	0.918	0.855	0.781	0.703	0.629	0.564
t (hours)*	7	8	9	10	15	20	30
X	0.509	0.463	0.423	0.389	0.277	0.214	0.147

*Additional time over the 48 hours at the lower temperature

6.17 Pyranose oxidase (glucose 2-oxidase, EC 1.1.3.10) catalyzes the conversion of glucose and molecular oxygen into 2-dehydro-glucose and hydrogen peroxide. Pyranose oxidase from *Phanerocheate chrysosporium* is immobilized in micro-porous glass beads of 600 μm average diameter and used in a CPBR at 37 °C. The reactor, packed with 100 g of immobilized enzyme, is fed with $90 \, L \cdot h^{-1}$ of a $1.8 \, g \cdot L^{-1}$ glucose solution saturated with oxygen (solubility of oxygen in the reaction medium $= 7 \, mg \cdot L^{-1}$). The specific activity of the catalyst is $1000 \, IU \cdot g^{-1}$ (1 IU is defined as the amount of pyranose oxidase that converts 1 μmol of glucose per minute under excess glucose and oxygen saturation at 37 °C). The intrinsic Michaelis constant for oxygen at this temperature is 15 mM and the operational half-life of the catalyst is 150 hours. The diffusion coefficient of oxygen is estimated at $3 \cdot 10^{-5} \, cm^{-2} \cdot s^{-1}$ and the density of the porous glass is $0.6 \, g \cdot cm^{-3}$. Determine the operation curve of the reactor, making any assumptions explicit.

Answer:

First-order kinetics with respect to oxygen concentration can be assumed since $[O_2] \ll K_{MO2}$. The glucose concentration should not affect the reaction rate since it is in very high excess.

t (hours)	0	10	50	100	200	300	400	500	750
X		0.90	0.89	0.840	0.77	0.60	0.44	0.31	0.21	0.07

6.18 The synthesis of ascorbyl palmitate from ascorbic acid (A) and palmitic acid (B) is carried out in an organic medium at 60 °C with a lipase from *Pseudomonas stutzeri* immobilized in octyl agarose with a specific activity of $225 \, IU \cdot g_{catalyst}^{-1}$. A kinetic model has been developed for a sequential ordered mechanism in which first ascorbic acid binds to the enzyme to produce the ascorbyl–enzyme complex and then this is attacked by palmitic acid to yield the ester. The kinetic parameters of the enzyme are: $K_{MA} = 2.4 \, mM$; $K_{MB} = 3.6 \, mM$; $K'_{MA} = 6.6 \, mM$ (see nomenclature in Chapter 4). The production team is discussing whether the best strategy for limiting the reaction is to use ascorbic or palmitic acid. You propose solving this dispute by carrying out a continuous run in a laboratory column reactor. The test is carried out by loading the column with 1.4 g of catalyst and operating it at a flow rate of $20 \, mL \cdot min^{-1}$ using A : B ratios of 2.0 and 0.5 and a limiting substrate concentration of 10 mM.

(a) Determine the initial steady-state conversions that should be attained if ascorbic acid (A) and palmitic acid (B) are the limiting substrates. (b) How will the conversion vary with time if the half-life of the immobilized enzyme is 3 hours under reaction conditions?

Answer (a):

$X_{0,Alim} = 0.803$

$X_{0,Blim} = 0.583$

At steady state there is a definite advantage to limiting the reaction by ascorbic acid (the substrate that first binds to the enzyme).

Answer (b):

t (hours)	0	0.25	0.5	1	2	3	4	5	7.5	10
$X_{0,Alim}$	0.803	0.770	0.738	0.674	0.555	0.452	0.365	0.294	0.168	0.095
$X_{0,Blim}$	0.583	0.559	0.536	0.489	0.406	0.333	0.271	0.219	0.127	0.073
$r_{A/B}$ *	0.726	0.726	0.726	0.726	0.732	0.737	0.742	0.745	0.756	0.768

* Ratio $X_{0,Blim}/X_{0,Alim}$

With enzyme inactivation, the differences in reactor performance with A limiting versus B limiting are reduced with time, but only slightly, so that running the reactor under limitation of ascorbic acid is the best option.

6.19 Different surface-immobilized enzyme catalysts are prepared by contacting varying amounts of protein per gram of support. At a certain concentration of immobilized protein the Damkoehler number of the catalyst (α_0) is determined to be 1.5. The activity of the immobilized enzyme increases linearly with the amount of immobilized protein up to four times this level of immobilized protein and the catalysts are equally stable at the different protein loadings, enzyme inactivation being well described by first-order kinetics. Determine the variation of the effectiveness factor with substrate conversion in a BSTR for the different catalysts at initial dimensionless substrate concentrations (β_i) of 1 and 10 (a) at the beginning of the reaction and (b) after 0.53 half-lives of operating time. Comment on the results obtained.

Answer (a):

X	$\eta(X)$							
	$\alpha_0=1.5$ $\beta_i=1$	$\alpha_0=1.5$ $\beta_i=10$	$\alpha_0=3$ $\beta_i=1$	$\alpha_0=3$ $\beta_i=10$	$\alpha_0=4.5$ $\beta_i=1$	$\alpha_0=4.5$ $\beta_i=10$	$\alpha_0=6$ $\beta_i=1$	$\alpha_0=6$ $\beta_i=10$
0.1	0.646	0.983	0.445	0.961	0.334	0.933	0.266	0.896
0.25	0.613	0.976	0.415	0.944	0.310	0.903	0.246	0.850
0.5	0.551	0.951	0.363	0.883	0.268	0.800	0.212	0.710
0.75	0.480	0.858	0.307	0.700	0.225	0.566	0.178	0.467
0.9	0.433	0.667	0.273	0.465	0.199	0.350	0.157	0.279
0.95	0.417	0.551	0.262	0.363	0.191	0.268	0.150	0.212
0.99	0.403	0.433	0.252	0.273	0.184	0.199	0.144	0.157

The effectiveness factor decreases with the increase in Damkoehler number, which is expected because of the increase in mass-transfer limitation. The effectiveness factor increases with substrate concentration, also as expected, since the substrate transport rate across the stagnant liquid layer surrounding the catalyst is increased. The effectiveness factor decreases with increasing substrate conversion because the substrate concentration is lower and so is the rate of substrate transport across the stagnant liquid layer.

Answer (b): After 0.53 half-lives of operation time, α_0 decreases exponentially, giving rise to the corresponding α after this time:

X	$\eta(X)$							
	$\alpha_0 = 1.5$ $\alpha = 1.04$ $\beta_i = 1$	$\alpha_0 = 1.5$ $\alpha = 1.04$ $\beta_i = 10$	$\alpha_0 = 3$ $\alpha = 2.08$ $\beta_i = 1$	$\alpha_0 = 3$ $\alpha = 2.08$ $\beta_i = 10$	$\alpha_0 = 4.5$ $\alpha = 3.12$ $\beta_i = 1$	$\alpha_0 = 4.5$ $\alpha = 3.12$ $\beta_i = 10$	$\alpha_0 = 6$ $\alpha = 4.16$ $\beta_i = 1$	$\alpha_0 = 6$ $\alpha = 4.16$ $\beta_i = 10$
0.1	0.737	0.989	0.553	0.975	0.434	0.959	0.354	0.940
0.25	0.707	0.984	0.512	0.965	0.404	0.942	0.329	0.913
0.5	0.648	0.968	0.460	0.927	0.353	0.877	0.285	0.820
0.75	0.576	0.905	0.395	0.796	0.299	0.688	0.240	0.594
0.9	0.526	0.756	0.353	0.574	0.265	0.453	0.212	0.371
0.95	0.509	0.648	0.339	0.460	0.254	0.353	0.203	0.285
0.99	0.494	0.526	0.328	0.353	0.245	0.265	0.196	0.212

Comments made for (a) also apply when enzyme inactivation occurs, but in this case the effectiveness factors are higher than in (a) for any condition, because the lower activity of the catalysts moves the system away from mass-transfer limitations; at a sufficiently high substrate concentration and with significant enzyme inactivation the system will not be restricted by the substrate mass-transfer rate and the effectiveness factor will approach a value of 1.

References

1. Henley, J. and Sadana, A. (1986) Deactivation theory. *Biotechnology and Bioengineering*, **28**, 1277–1285.
2. Illanes, A. and Wilson, L. (2003) Enzyme reactor design under thermal inactivation. *Critical Reviews in Biotechnology*, **23**, 61–93.
3. Illanes, A., Wilson, L., and Tomasello, G. (2001) Effect of modulation of enzyme inactivation on temperature optimization for reactor operation with chitin-immobilized lactase. *Journal of Molecular Catalysis B:Enzymatic*, **1**, 531–540.
4. Chen, K. and Wu, J. (1987) Substrate protection of immobilized glucose isomerase. *Biotechnology and Bioengineering*, **30**, 817–824.
5. Villaume, I., Thomas, D., and Legoy, M. (1990) Catalysis may increase the stability of an enzyme: the example of horse liver alcohol dehydrogenase. *Enzyme and Microbial Technology*, **12**, 506–509.
6. Illanes, A., Altamirano, C., and Zuñiga, M.E. (1996) Thermal inactivation of penicillin acylase in the presence of substrate and products. *Biotechnology and Bioengineering*, **50**, 609–616.
7. Polaković, M. and Vrábel, P. (1996) Analysis of the mechanism and kinetics of thermal inactivation of enzymes: Critical assessment of isothermal inactivation experiments. *Process Biochemistry*, **31**, 787–800.
8. Aymard, C. and Belarbi, A. (2000) Kinetics of thermal deactivation of enzymes: a simple three parameters phenomenological model can describe the decay of enzyme activity, irrespectively of the mechanism. *Enzyme and Microbial Technology*, **27**, 612–618.
9. Longo, M. and Combes, D. (1999) Thermostability of modified enzymes: a detailed study. *Journal of Chemical Technology and Biotechnology*, **74**, 25–32.

10. Illanes, A., Altamirano, C., Aillapán, A. *et al.* (1998) Packed-bed reactor performance with immobilised lactase under thermal inactivation. *Enzyme and Microbial Technology*, **23**, 3–9.
11. Illanes, A., Altamirano, C., and Cartagena, O. (1994) Enzyme reactor performance under thermal inactivation, in *Advances in Bioprocess Engineering* (eds E. Galindo and O. Ramírez), Kluwer Acad Publ., Dordrecht, The Netherlands, pp. 467–472.
12. Illanes, A., Wilson, L., and Tomasello, G. (2000) Temperature optimization for reactor operation with chitin-immobilized lactase under modulated inactivation. *Enzyme and Microbial Technology*, **27**, 270–278.
13. Pitcher, W. (1978) Design and operation of immobilized enzyme reactors, in *Advances in Biochemical Engineering*, vol. **10** (eds T. Ghose, A. Fiechter, and A. Blakebrough), Springer Verlag, Berlin, pp 1–26.

7

Optimization of Enzyme Reactor Operation

Conejeros

Research involves purposely modifying an input variable (x) and then observing any changes produced in the system in order to determine its effects (y) (see Scheme 7.1). Generally only a few inputs and responses are considered relevant *a priori*, usually based on the judgment and previous knowledge of the researcher. In this way, as shown in Scheme 7.1, not all the inputs are quantified and controlled and not all responses are quantified.

From an engineering perspective, the aims of such experiments are first to determine the input variables that have a significant influence on a given response and second to determine the magnitude of their effects. This allows establishing set points for the inputs, in order to maximize or minimize a response or help set up input conditions capable of producing a homogeneity in that response (e.g. a product) within tolerance margins for the input variables. These goals can be achieved by quantitatively predicting process performance by means of an appropriate model; moreover, they can be mathematically represented by an objective function that is dependent on how input variables affect the process state variables given a set of system parameters.

This chapter addresses the optimization of enzyme reactors. Two modelling approaches are considered: mechanistic and statistical. The former is grounded in physics, chemistry, and mathematics. The field of mathematics devoted to the solution of optimization problems is called mathematical programming, which we can consider a set of techniques used to seek the optimization (maximization or minimization) of a single objective function, given a mathematical model describing the behavior of a biocatalytic system. The latter modelling approach considers the representation of the relationship between inputs and responses by means of empirical mathematical expressions, obtained by regression of experimental observations. Statistical models are preferred when mechanistic

Problem Solving in Enzyme Biocatalysis, First Edition. Andrés Illanes, Lorena Wilson and Carlos Vera.
© 2014 John Wiley & Sons, Ltd. Published 2014 by John Wiley & Sons, Ltd. Companion Website:
http://www.wiley.com/go/illanes-problem-solving

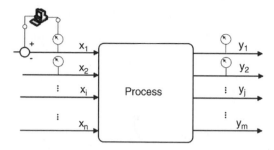

Scheme 7.1 *Structure of an experiment.*

models are too complex or when parameter estimation is too difficult to allow practical application. Nevertheless, statistical models shed no light on process phenomena.

7.1 Strategy for the Optimization of Enzyme Reactor Performance

7.1.1 Objective Function

The first studies of the optimization of enzyme reactor performance were related to batch operation, but with the growing use of immobilized enzymes as process catalysts, continuous modes of operation, whether packed-bed (CPBR) or stirred-tank (CSTR), have increasingly been examined.

The objective function that is to be optimized can depend on many factors, and its complexity will depend on how many of these are considered. Various criteria can be used for optimization, but product quality has first to be defined. This means that the conversion at reactor outlet must be established as a sound point of comparison for the conditions under study. Product quality must remain constant throughout the production campaign, so under conditions of enzyme inactivation various strategies for coping with this task can be envisaged.

In the case of sequential batch operation (usually with immobilized enzyme), fresh catalyst can be made up after each batch to compensate for the activity lost, or else reaction time can be progressively increased batch after batch until the desired conversion is reached. Both strategies have limitations (see Section 6.5), so the moment will inevitably come when the spent catalyst must be replaced by fresh one. For continuously operated reactors with immobilized enzymes, different strategies can also be envisaged for the delivery of a product of constant quality under conditions of enzyme inactivation (see Section 6.5); again, the moment will come when the spent catalyst must be wasted and replaced by fresh one. The criterion for catalyst replacement is undoubtedly critical to optimizing reactor operation, since it will determine the efficiency of catalyst use.

Some of the objectives that can be used to optimize reactor operation are presented in this section.

7.1.1.1 Volumetric Productivity

Volumetric productivity (the mass of product produced per unit time and unit reaction volume) has frequently been considered an adequate parameter for optimization. It is best suited to the case of batch reactors with soluble enzymes in which the catalyst is not recovered after the reaction. Volumetric productivity is described by Equation 4.45.

7.1.1.2 Specific Productivity

Specific productivity (the mass of product produced per unit time and unit mass of catalyst) is a better choice as it relates to the amount of catalyst used, which is more cost-effective. It is well suited to optimization with immobilized enzymes, which allow the catalyst to be recovered in a sequential batch operation or used for prolonged periods of time under continuous operation. Its drawback is that it weighs the enzyme catalyst mass and operation time equally, but their impacts on operation cost might well be of a different magnitude. Specific productivity is described by Equation 4.46.

7.1.1.3 Cost-Based Objective Function

A cost-based objective function is a still better approach as from a process perspective the goal is to minimize costs. The optimization time span may depend on the characteristics of the production system (for example, the raw material, or even the enzyme itself, can be seasonal), but it is advisable to carry out the evaluation on an annual basis, defining annual cost as the objective function. The main components of such cost-based objective function are as follows:

Enzyme cost: This considers the amount of enzyme catalyst expenditure during the period of analysis (e.g. one year). The required data are the catalyst mass required and the unit cost (or price) of the catalyst. Catalyst mass is calculated to fulfill the production task (see Chapters 4–6) and unit cost corresponds to the price at which the supplier delivers the enzyme (whether in soluble or immobilized format) or the cost of producing it domestically. The production of the catalyst will include fermentation (or extraction), purification to the required level and, eventually, immobilization. In the case of immobilized enzymes, the cost of the enzyme and of the support come into play and the method of immobilization will have complex effects (see Section 3.2). For example, if the enzyme is immobilized by covalent attachment, the biocatalyst is likely to be robust, highly stable, and amenable to reactivation but will also be expensive as a result of the materials and complex procedures involved, the rather high activity losses and the impossibility of recovering the support after enzyme exhaustion. By contrast, if the enzyme is immobilized by adsorption, the procedure will be simpler and cheaper and the support may be easy to recover after enzyme exhaustion, but the catalyst will be less stable and less specific with respect to the enzyme. In the case of carrier-free immobilized enzymes, there will obviously be no cost of support and the specific activity may be quite high, but handling of the catalyst in the reactor might be cumbersome.

Energy cost: This can be a significant component of the total operation cost. The main sources of energy consumption are maintenance of temperature during reactor operation, the energy balance of which will result from the heat of reaction; heat exchange with the surroundings and heat of agitation; heating (or in a few cases cooling) of the reactor feed stream to operation temperature; and agitation and pumping energy.

Reactor cost: This arises from the reactor design. Its depreciation value must be considered in the cost-based objective function.

A cost-based objective function can thus be expressed as:

$$CA = CA_{enzyme} + CA_{energy} + CA_{depreciation} \qquad (7.1)$$

The parameters of reactor operation and design will have various impacts on the cost components of the cost-based objective function [1].

7.1.2 Variables for Optimization of Enzyme Reactor Performance

Although the structure of the objective function can be quite simple, assigning the proper values can be complex. This is because, beyond the equations describing reactor operation (see Chapters 4–6), the critical variables for optimization must be considered, which are difficult to establish *a priori* and frequently arise from industry experience. Some of the variables identified as critical in terms of the optimization of enzyme reactor operation are as follows:

- *Temperature*: Whatever the objective function, temperature will always be a critical variable in any enzymatic process. It is particularly sensitive to optimization, having opposite effects on enzyme activity and stability (see Section 2.7.2). Temperature optimization is thus a major issue in enzyme biocatalysis and will be analyzed in Section 7.1.3.
- *Enzyme – substrate mass ratio*: This reactor operation variable will have a strong impact on the various components of the cost-based objective function. Increasing it leads to a higher enzyme cost but lower energy costs and, probably, a lower cost of reactor depreciation.
- *Substrate – substrate molar ratio*: For reactions involving more than one substrate, this parameter may have a strong incidence on reactor operation (see Section 4.4). Beyond its impact on conversion and productivity, it may be a key variable in the selectivity of the reaction and in substrate inhibition or poor substrate solubility. The strategy of feeding substrate to the reactor will affect the objective function, allowing its optimization.
- *Catalyst replacement*: In the case of immobilized enzyme reactors, in which the catalyst is supposed to be recovered and used for prolonged periods of time, the time at which the spent catalyst must be replaced by fresh one is a critical issue in terms of reactor optimization. The impact of this variable on processing cost is well known by industrial users of enzymes and accumulated experience indicates that optimal catalyst replacement time lies somewhere between one and three half-lives; that is, between times corresponding to 50 and 12.5% of residual activity. This is certainly a rule of thumb, but it establishes some boundaries that are worthwhile for consideration; however, the optimal point of replacement will vary from case to case and will be dictated by the balance between enzyme cost and reactor operation costs. For rather cheap enzymes, replacement will move to the upper limit of that range (higher value of residual activity at replacement), while for rather expensive enzymes it will move to the lower limit (lower value of residual activity at replacement). In other words, the more expensive the enzyme, the more extended its utilization. Although critical in terms of its impact on cost, approximation is mostly empirical and to the best of our knowledge, only preliminary studies on the effect of this parameter in a cost-based objective function have been reported [1].

7.1.3 Determination of Optimum Temperature

The optimization of reactor operation temperature requires that all kinetic and mass-transfer parameters involved in the corresponding equations (see Chapters 4–6) be expressed as validated temperature-explicit functions. An objective function must then be developed to determine the temperature that maximizes or minimizes such functions. In the case of a cost-based objective function, optimum temperature will be the one minimizing it.

The thermal dependence of kinetic parameters can be established based on thermodynamic correlations and Arrhenius expressions (see Section 2.7.2). Temperature-explicit functions for kinetic parameters are described by Equations 2.32, 2.33, and 2.34. Parameters of enzyme inactivation can also be determined as Arrhenius-type functions of temperature, as represented by Equation 2.40. In the case of modulated enzyme inactivation (see Section 6.2), modulation factors (n), as described by Equation 6.6, can be expressed as Arrhenius functions of temperature:

$$n_M = 1 - \frac{k_{D,M,0,}}{k_{D,0}} \cdot exp\left[-\frac{(E_{ia,M} - E_{ia})}{R \cdot T}\right] \tag{7.2}$$

These temperature-explicit functions can be experimentally determined according to the methodologies described in Section 2.6. A complete study of the temperature optimization of reactor operation based on temperature-explicit functions has been reported for the case of a batch stirred-tank reactor (BSTR) with chitin-immobilized β-galactosidase [1].

In the case of heterogeneous catalysis (see Section 3.4), mass-transfer parameters ought to be expressed as temperature-explicit functions as well. This implies determination of the temperature dependence of film mass-transfer coefficients or diffusion coefficients, according to the case.

7.2 Mathematical Programming for Static Optimization

Static optimization involves optimization of problems whose variables and objective functions are time-independent. An example of this type of problem is the maximization of the volumetric productivity in continuous reactors (CPBR and CSTR) in steady state, under ideal operation conditions (see Section 4.2).

Considering a mathematical function J(x) representing the objective function (see Section 7.1.1), the static optimization problem can be posed as determining vector x ∈ to R^n (see Section 7.1.2) such that it maximizes (or minimizes) J(x), that is:

$$max\, J(x)$$
$$x \in R^n \tag{7.3}$$

where J(x) is a continuous function that must be at least twice differentiable at the optimum point. Assuming that there exists a maximum for J(x), denoted as J(x*), the optimal point x* must satisfy both necessary and sufficient conditions for optimality. The necessary condition is that the gradient of J(x*), $\nabla J(x^*)$, be equal to zero:

$$\nabla J(x^*) = \left[\frac{dJ(x)}{dx_1}\bigg|_{x^*} \quad \frac{dJ(x)}{dx_2}\bigg|_{x^*} \quad \cdots \quad \frac{dJ(x)}{dx_{n-1}}\bigg|_{x^*} \quad \frac{dJ(x)}{dx_n}\bigg|_{x^*} \right] = 0 \qquad (7.4)$$

However, the solution satisfying Equation 7.4 may correspond to a maximum, a minimum, or a saddle point. To determine whether x* is maximum or not, the sufficiency conditions must be examined. These are related to the Hessian of J(x), denoted here as $\nabla^2 J$ (x) (see Equation 7.5). If $\nabla^2 J(x^*)$ is a negative-definite matrix or alternatively if its eigenvalues are all negative, x* is a maximum; on the other hand, if $\nabla^2 J(x^*)$ is a positive-definite matrix, or alternatively if its eigenvalues are all positive, x* is a minimum. If the eigenvalues of $\nabla^2 J(x^*)$ have different signs, x* is a saddle point.

$$\nabla^2 J(x^*) = \begin{bmatrix} \frac{\partial^2 J(x)}{\partial x_1^2}\bigg|_{x^*} & \frac{\partial^2 J(x)}{\partial x_1 \partial x_2}\bigg|_{x^*} & \cdots & \frac{\partial^2 J(x)}{\partial x_1 \partial x_{n-1}}\bigg|_{x^*} & \frac{\partial^2 J(x)}{\partial x_1 \partial x_n}\bigg|_{x^*} \\ \frac{\partial^2 J(x)}{\partial x_2 \partial x_1}\bigg|_{x^*} & & & & \\ \vdots & & \ddots & & \vdots \\ \frac{\partial^2 J(x)}{\partial x_{n-1} \partial x_1}\bigg|_{x^*} & & & & \\ \frac{\partial^2 J(x)}{\partial x_n \partial x_1}\bigg|_{x^*} & \frac{\partial^2 J(x)}{\partial x_n \partial x_2}\bigg|_{x^*} & \cdots & \frac{\partial^2 J(x)}{\partial x_n \partial x_{n-1}}\bigg|_{x^*} & \frac{\partial^2 J(x)}{\partial x_n^2}\bigg|_{x^*} \end{bmatrix} \qquad (7.5)$$

In this section, only unconstrained optimization has been described; the reader interested in constrained optimization is directed to the textbooks by Rao [2] and Edgar *et al.* [3].

7.3   Dynamic Programming

Dynamic programming deals with optimization problems whose objectives change with time. This situation can typically be associated with the operation of enzyme reactors, where catalyst inactivation with temperature or catalyst inhibition will affect the total amount of product produced across a given time of operation. Thus, the purpose of dynamic programming is to drive the process to an increase in the objective, either by setting a fixed value on a controlled variable or by finding its time-varying profile; for example, maximizing volumetric productivity for a given production time by changing the operation temperature or the feed flow rate. This is not an easy task as reaction kinetics can be highly nonlinear. Furthermore, operational variables may affect kinetic parameters, which will have an effect on process evolution and performance and will reflect on the objective value at end of the reaction time.

Modeling based on differential equations becomes a strong tool not only for the evaluation of operational conditions but also for determining the optimal profile of controlled variables with the aim of optimizing process performance.

Consider any process that can be modeled by the differential equation (e.g. rate equations):

$$\frac{dx}{dt} = f(x, v, u, t) \tag{7.6}$$

where x represents state variables, v model parameters, and u is a variable chosen as the control variable. By considering a final-time objective, such as maximizing the value of $J(x)$ at final time t_f, the optimization problem can be formally written as:

$$\max_{x,v,u,t} J(x)\big|_{t_f}$$

subject to:

$$\frac{dx}{dt} = f(x\big|_{t_0,} v, u, t) \tag{7.7}$$

$$x\big|_{t_0} = x_0 \tag{7.8}$$

$$u_{lo} \leq u \leq u_{up} \tag{7.9}$$

$$t_o \leq t \leq t_f \tag{7.10}$$

where x_0 is the initial conditions for the problem and u_{lo} and u_{up} are the allowed lower and upper bounds for the control variable. For example, if the controlled variable is the feed flow rate, its lower and upper bounds may be given by the pump capacity. Thus, in order to solve the dynamic optimization problem, Equation 7.6 can be discretized following any ordinary differential equation discretization scheme (e.g. Euler's method). As an illustration, if 10 time discretization intervals are considered, between 0 and t_f, this will generate 10 unknown values x_1 to x_{10} and 10 unknown values for controls u_0 to u_9. The initial value x_0 is assumed to be a given initial condition and x_{10} corresponds to the end value $x(t_f)$. The simplest implementation can be made using Euler's method, although any other may be used, and the recursion obtained from the discretized version of Equation 7.7 is:

$$x_{i+1} = x_i + \Delta t \cdot f(x_i, v, u_i, t) \tag{7.11}$$

where x_i values are linked through Equation 7.11 and values of u_i must be chosen within such bounds that they maximize $J(x)\big|_{t_f}$. The 10 discretized versions of Equation 7.11 yield a system of nonlinear algebraic equations that can be solved using Newton's method in order to maximize $J(x)\big|_{t_f}$ by changing the values of u_i.

7.4 Statistical Optimization by Surface Response Methodology

Surface response methodology (SRM) is perfectly suited to the modeling and optimization of complex processes where a mechanistic model is too difficult to use. These models usually involve several parameters, many of which are quite difficult to determine or estimate. As an example, consider the effect of pH on enzymatic reactions, as described in Section 2.7.1. Even for an enzymatic reaction with one nonionizable substrate, difficult experimental work is required to determine the six kinetic constants considered in the Michaelis – Davidsohn hypothesis. The complexity of the model and the determination of its parameters increase significantly if ionizable reactive species are involved in the reaction. Thus, modeling the effect of pH on reaction kinetics and determining the optimum pH might be addressed by means of SRM, as shown in Fajardo and Illanes [4] for the kinetically controlled synthesis of ampicillin with immobilized penicillin acylase.

It is important to bear in mind that one of the major goals of SRM is to provide by regression a model for process response that determines the influence of input variables on the response and consequently optimizes it. The most common models used to fit the response in SRM are linear ones of first $(\hat{y} = b_0 + \sum b_i \cdot x_i)$ or second order $(\hat{y} = b_0 + \sum b_i \cdot x_i + \sum b_{ii} \cdot x_i^2 + \sum \sum b_{ij} \cdot x_i \cdot x_i)$; higher-order models are seldom needed.

Experimental design is the first aspect that must be considered when applying SRM. Errors in experimental design occur when observations are made in an arbitrary way; for example, if factor A is considered much important than B, more experimental data for A will be collected. Moreover, replicates of the experimental observations are usually not considered and sampling may be unequally spaced.

Appropriate experimental design techniques provide desirable properties such as orthogonality, rotatability, and estimation of experimental error with a minimum of experimental observations. Thus, orthogonality ensures a minimization of the variance in regression of first-order coefficients. So, if a first-order model is used to fit the experimental observations obtained in an orthogonal experimental design, the regression coefficients associated with each factor are mutually independent. This is important to accurately quantifying the influence of each factor. Rotatability implies that the variance is the same for all points located at the same distance from the center, so the model provides a good prediction in all directions. When the response is fitted to a first-order model, the experimental design should be orthogonal, while if fitted to a second-order model it should be at least rotatable, because not all experimental designs for second-order models are orthogonal. Classic designs for second-order models are central composite and Box – Behnken; meanwhile, for first-order models the use of 2^k design with replicates of the central point and of simplex design is customary. The use of statistical package software is advised for this purpose.

Once the experimental design has been selected and the experiments carried out, the response is analyzed by fitting it to a linear model. Any linear model can be written as shown in Section A.15 (for details, see Section A.2). Thus, the value of the regression coefficients can be estimated using the least-square method (see Equation A.20).

The use of coded variables (x_c) is recommended because they allow the influence of each variable in the response function of the process to be determined by comparing the magnitudes of the regression coefficients. When noncoded variables are employed, the magnitude of the regression coefficient is dependent on the units of the input variables; for example, the coefficient estimated for the effect of temperature in a reaction will clearly not be the same when expressed in °F, °C, or K. Furthermore, the use of coded inputs helps avoid the ill conditioning of the $X^T \cdot X$ matrix when the magnitude of noncoded variable is very small, meaning that its inverse can be calculated more accurately and so too the model coefficient. Input variables are usually coded between $+1$ and -1 as follows:

$$x_c = \frac{x - a}{b} \tag{7.12}$$

with:

$$a = \frac{x_h + x_l}{2} \tag{7.13}$$

$$b = \frac{x_h - x_l}{2} \tag{7.14}$$

where x_l and x_h are the lowest and highest values considered for the input variable. Note that x_c is equal to 1 for $x = x_h$ and equal to -1 for $x = x_l$.

7.4.1 Assessing the Quality of SRM and its Parameters

Some questions that arise when building a model for a process response are: Is the model good enough to fit experimental observations? Are all parameters significant or does it contain dummy coefficients? What is the confidence level for the model prediction? This section presents the most common statistical tests used to answer these questions.

For a first approach to evaluating the existence of a linear relationship between response and any of the regression variables (q), two hypotheses are posed: the null hypothesis, which implies that all regression coefficients are equal to zero, and the alternative hypothesis, in which at least one of them is other than zero. A rejection of the null hypothesis means that at least one of the regression variables contributes to explaining the response. The null hypothesis is rejected if the value of the statistic F for the null hypothesis (F_0) is higher than the value of Fisher $-$ Snedecor distribution ($F_{\alpha,q,n-q-1}$) for a confidence level of $1 - \alpha$ with q degrees of freedom in the numerator and $n - q - 1$ degrees of freedom in the denominator. In this case, F_0 is defined as:

$$F_0 = \frac{SS_R \cdot q^{-1}}{SS_E \cdot (n - q - 1)^{-1}} \tag{7.15}$$

with:

$$SS_R = \sum (\hat{y}_i - \bar{Y})^2 = B^{*^T} \cdot X^T \cdot Y - \frac{\sum (y_i)^2}{n} \tag{7.16}$$

$$SS_E = \sum (y_i - \hat{y}_i)^2 = Y^T \cdot Y - B^{*^T} \cdot X^T \cdot Y \tag{7.17}$$

If the null hypothesis is rejected then we should discover whether all regression coefficients are significant or not. The simplest test in this case is to determine the confidence interval (CI) for each parameter according to Equation A.24. If the CI exceeds the estimated value for the regression coefficients, the variable associated with that parameter should be removed from the model.

A second approach consists in using the extra sum-of-squares analysis, which allows the question of whether a variable or group of variables provides a significant contribution to explaining the model's response to be determined. In order to do so, the full model containing q regressor variables is considered and its sum of squares due to the regression (SS_R) is computed, then the k variables studied are removed from the original q variables present in the model and the sum of squares due to the regression of the reduced model containing q-k variables $(SS_{R,q-k})$ is calculated. The regression sum of squares due to the k variables, given that the q-k variables were already present in the model $(SS_{R,k/q-k})$, is:

$$SS_{R,k/q-k} = SS_R - SS_{R,q-k} \qquad (7.18)$$

The k variables are significant if their regression coefficients are other than zero. We thus pose as a null hypothesis that these regression coefficients are equal to zero and as an alternative hypothesis that at least one of them is other than zero. The null hypothesis is rejected if F_0 is higher than $F_{\alpha,k,n-p}$. In this case F_0 corresponds to:

$$F_0 = \frac{(SS_{R,k/q-k}) \cdot k^{-1}}{SS_E \cdot (n-p)^{-1}} \qquad (7.19)$$

Although this procedure can be applied to a group of variables, it is recommended that it be applied to each variable individually in order to provide a deeper analysis of its contribution.

A third approach is to analyze the behavior of the adjusted determination coefficient (R_{adj}), since its value only improves when a significant variable is added to the model. The value of R_{adj} can be calculated using Equation A.27.

So far, only the quality of the model and of its parameters have been studied, but another interesting aspect of analysis is the quality of the predictions. The predicted response $(\hat{y})$ for a particular set of values of the model variables (x_g) is given by:

$$\hat{y} = x_g^T \cdot B^* \qquad (7.20)$$

where x_g^T is equal to $[1 \ x_{g,1} \ x_{g,2} \ \dots \ x_{g,p-1} \ x_{g,q}]$ and $x_{g,1}$ is the value of the ith variable in x_g. The 1-α CI for the predicted response at point x_g can then be calculated as follows:

$$\hat{y} \pm t(n-p, a/2) \cdot \sqrt{\frac{SS_E}{n-p} \cdot x_g^T (X^T \cdot X)^{-1} \cdot x_g} \qquad (7.21)$$

where $t(n-p, \alpha/2)$ represents the upper $\alpha/2$ quantile of Student's t-distribution for $n-p$ degrees of freedom.

7.4.2   Process Optimization by SRM

In order to maximize or minimize the process response (considering only SRM), a second-order model must be selected to represent the response; first-order models do not offer the possibility of appropriately representing the curvature in the response. For the purposes of this section, the second-order model can conveniently be represented as:

$$\hat{y} = b_o^* + x^T \cdot B_1^* + x^T \cdot B_2^* \cdot x \tag{7.22}$$

where x^T is equal to $[x_1 \ x_2 \ \dots \ x_{f-1} \ x_f]$ and represents any point in the process operability region, f represents the number of factors considered in the response, B_1^* corresponds to the vector containing the first-order regression coefficient and B_2^* is the estimated parameters for the interaction between variables and the quadratic terms. B_1^* is a column vector as B_2^* corresponds to a symmetric matrix of the following form:

$$B_1^* = \begin{bmatrix} b_1^* \\ b_2^* \\ \vdots \\ b_{f-1}^* \\ b_f^* \end{bmatrix} \tag{7.23}$$

$$B_2^* = \begin{bmatrix} b_{11}^* & 0.5 \cdot b_{12}^* & \cdots & 0.5 \cdot b_{1f-1}^* & 0.5 \cdot b_{1f}^* \\ & b_{22}^* & \cdots & 0.5 \cdot b_{2f-1}^* & 0.5 \cdot b_{2f-1}^* \\ & & & & \vdots \\ & & & \ddots & 0.5 \cdot b_{ff-1}^* \\ \text{sym} & & \cdots & & b_{ff}^* \end{bmatrix} \tag{7.24}$$

The maximum, minimum, or saddle point of Equation 7.22 is determined by taking its derivative with respect to x (see Appendix A and Section 2.1):

$$\frac{\partial \hat{y}}{\partial x} = B_1^* + 2 \cdot B_2^* \cdot x \tag{7.25}$$

Setting Equation 7.25 to zero and solving for x gives:

$$x = -\frac{(B_2^*)^{-1} \cdot B_1^*}{2} \tag{7.26}$$

In order to determine whether $\hat{y}(x)$ is a maximum, a minimum, or a saddle point in the process response, the Hessian of y evaluated at x ($\nabla^2 \hat{y}(x)$) should be determined as a positive- or negative-definite matrix. For those who are not familiar with this technique and regression analysis, textbooks on the topic by Montgomery [5] and Draper and Smith [6] are recommend.

Finally, another situation of interest arises when process variables are subject to constrains. In this case the process optimization can be conducted using a first- or second-order model. The use of Lagrange's multiplier methods is quite helpful in solving this problem. For the interested reader, textbooks by Rao [2], Edgar *et al.* [3] and Draper and Smith [6] are recommended.

Solved Problems

7.1 A BSTR with immobilized β-galactosidase is used for the hydrolysis of concentrated whey permeate with a lactose concentration of 285.71 mM in order to produce glucose syrup. The enzyme presents competitive inhibition by the product galactose and the reaction of hydrolysis is conducted to a final conversion of 85%. The reactor, which has a working volume of 1000 L, is loaded with $60\,g\cdot L^{-1}$ of biocatalyst with a specific activity of 368.5 IU/g. It operates in repeated batch mode, with catalyst recovery after each batch down to 25% residual activity, at which point the biocatalyst is discarded to initiate another cycle of production. The values of the kinetic and inactivation parameters of the biocatalyst at various temperatures are presented in Table 7.1.

Determine the reactor operation temperature which maximizes the specific productivity (mass of glucose produced per unit time and unit of biocatalyst mass) of one cycle of reactor operation (one cycle of reactor operation considers all batches until catalyst replacement), the temperature-explicit expressions for all parameters within the temperature range under study, and a temperature-explicit equation describing reactor operation, assuming that mass-transfer limitations are negligible in all cases.

Resolution: The equation describing the performance of a BSTR with an enzyme subjected to competitive inhibition by the product and with first-order kinetics of enzyme inactivation (see Table 4.1 and Equation 6.7) is:

$$s_i \cdot X \cdot \left[\frac{1}{K_M} - \frac{1}{K_{IC}}\right] - \left[1 + \frac{s_i}{K_{IC}}\right] \ln(1 - X) = \frac{k \cdot e_0}{k_D \cdot K_M} \cdot (1 - \exp(-k_D \cdot t)) \quad (7.27)$$

For each of the temperatures in Table 7.1, the reaction time required to obtain 85% conversion at the end of each batch is obtained from Equation 7.27, assuming sequential batch operation until the end of the operation cycle (25% residual activity of the biocatalyst). To illustrate this, results are presented for sequential batch operation at 45 °C. The values of K_M, K_{IC}, k, and k_D are obtained from Table 7.1 at 45 °C. The values of s_i (285.71 mM) and X (0.85) are given data.

$$e_0 = m_{cat} \cdot a_{sp} = 60\,g \cdot L^{-1} \cdot 368.5\,IU \cdot g^{-1} = 22,110\,IU \cdot L^{-1} = 1326.6\,mM \cdot h^{-1}$$

Table 7.1   *Kinetic and inactivation parameters of immobilized β-galactosidase.*

	5 °C	15 °C	25 °C	30 °C	35 °C	40 °C	45 °C
k (−)	0.121	0.233	0.429	0.525	0.700	0.724	0.400
K_M (mM)	12.53	18.13	25.59	30.14	35.31	41.16	47.75
K_{IC} (mM)	16.59	16.65	16.62	16.64	16.67	16.64	16.65
k_D (h^{-1})	0.010	0.016	0.027	0.031	0.040	0.048	0.056

Replacing these values in Equation 7.27 gives:

$$285.71(mM) \cdot 0.85 \cdot \left[\frac{1}{47.75(mM)} - \frac{1}{16.65(mM)}\right] - \left[1 + \frac{285.71(mM)}{16.65(mM)}\right] \ln(1 - 0.85) =$$

$$= \frac{0.4 \cdot 1326.6(mM \cdot h^{-1})}{0.056(h^{-1}) \cdot 47.75(mM)} \cdot \left(1 - \exp(-0.056(h^{-1}) \cdot t)\right)$$

$$t = 2.40 \text{ (h)}$$

Fractional residual activity after the first batch can be obtained from Equation 2.39:

$$\frac{e}{e_0} = \exp(-k_D \cdot t) = \exp(-0.056 \cdot 2.4(h)) = 0.874 \tag{2.39}$$

The initial activity of the second batch is then: $e_0 \cdot 0.874 = 1326.6\,(mM \cdot h^{-1}) \cdot 0.874 = 1159.5\,(mM \cdot h^{-1})$.

Proceeding in the same way for the following batches up to the point of catalyst replacement (residual activity $= 0.25$), the values in Table 7.2 are obtained.

After six batches the residual activity of the biocatalyst drops to 25%, so a new production cycle with fresh biocatalyst must begin.

Specific productivity is calculated by summing the amount of glucose produced in each batch up to the moment of catalyst replacement, which defines one cycle of reactor operation.

amount of glucose produced per batch $= s_i \cdot X \cdot V_{eff} \cdot MW_{glucose} = 0.28571\,M \cdot 0.85 \cdot 1000$
    $L \cdot 180\,g\,gmol^{-1} = 43\,713.6\,g = 43.7\,kg$
glucose produced per cycle $= 43.7\,kg \cdot batch^{-1}$ per six batches $= 262.2\,kg_{glucose}$
time of cycle (see Table 7.2) $= 25.07$ hours
mass of catalyst used $(M_{cat}) = 60\,g \cdot L^{-1} \cdot 1000\,L = 60\,000\,g = 60\,kg_{catalyst}$
specific productivity of cycle $= 262.2 \cdot 25.07^{-1} \cdot 60^{-1} = 0.174\,kg_{glucose} \cdot kg_{catalyst}^{-1} \cdot h^{-1}$
product-to-catalyst mass ratio $= 262.2 \cdot 60^{-1} = 4.374\,kg_{glucose} \cdot kg_{catalyst}^{-1}$

Proceeding in the same way for the other temperatures, results are obtained as summarized in Table 7.3.

Table 7.2 *Sequential BSTR performance at 45°C.*

Batch #	Initial Activity $(mM \cdot h^{-1})$	Time of Batch (h)	Total Operation Time (h)	Residual Activity (fraction of initial)
1	1326.6	2.40	2.40	0.874
2	1159.5	2.77	5.17	0.749
3	993.0	3.28	8.46	0.623
4	826.2	4.03	12.48	0.497
5	659.4	5.21	17.69	0.371
6	492.6	7.38	25.07	0.246

Table 7.3 *Sequential BSTR performance in the range from 5 to 45°C.*

Temperature (°C)	Number of Batches per Cycle	Total Time of Cycle (h)	Residual Activity at the End of Cycle (% of initial)	Specific Productivity $(kg_{glucose} \cdot kg_{catalyst}^{-1} \cdot h^{-1})$	Product – Catalyst Ratio $(kg_{glucose} \cdot kg_{catalyst}^{-1})$
5	24	134.15	26.15	0.130	17.48
15	24	86.42	25.09	0.202	17.48
25	21	51.23	25.08	0.299	15.30
30	20	44.72	25.00	0.326	14.57
35	18	32.88	26.85	0.399	13.11
40	14	28.25	25.77	0.361	10.20
45	6	25.07	24.56	0.174	4.37

Figure 7.1 provides a graphical representation of the data in Table 7.3. As can be seen, 35 °C represents the optimum temperature in terms of specific productivity. Product-to-catalyst mass ratio decreases with temperature, the highest value corresponding to the lower temperature of the range under consideration. Indeed, more batches can be conducted at lower temperatures but this will be at the expense of a higher reaction time, meaning that the amount of product produced on a time basis will be lower, and this will have a negative impact on operating cost.

Answer: From the data in Table 7.1, the following temperature-explicit expressions are obtained for the kinetic and inactivation parameters of immobilized β-galactosidase.

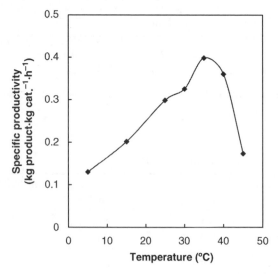

Figure 7.1 *Effect of temperature on reactor productivity in the hydrolysis of concentrated whey permeate with immobilized β-galactosidase.*

From Equation 2.32:

$$K_M = 5.05 \cdot 10^5 \cdot exp\left(-\frac{2947}{T}\right) \qquad r = 0.99$$

The value of K_{IC} is independent of temperature in the range considered.

From Equation 2.34:

$$k = 8.0 \cdot 10^6 \cdot exp\left(-\frac{4985}{T}\right) \qquad r = 0.99$$

From Equation 2.40:

$$k_D = 1.09 \cdot 10^4 \cdot exp\left(-\frac{3862}{T}\right) \qquad r = 0.99$$

The temperature-explicit expression for the equation describing reactor performance (Equation 7.3) is:

$$\frac{8 \cdot 10^6 \cdot exp\left(-\frac{4985}{T}\right) \cdot e_0 \cdot \left(1 - exp(-1.09 \cdot 10^4 \cdot exp\left(\frac{-3862}{T}\right) \cdot t\right)}{1.09 \cdot 10^4 \cdot exp\left(-\frac{3862}{T}\right) \cdot 5.05 \cdot 10^5 \cdot exp\left(-\frac{2947}{T}\right)} =$$

$$= s_i \cdot X \cdot \left[\frac{1}{5.05 \cdot 10^5 \cdot exp\left(-\frac{2947}{T}\right)} - \frac{1}{16.65}\right] - \left[1 + \frac{s_i}{16.65}\right] ln(1 - X)$$

7.2 A process for inverted sugar manufacturing uses invertase (β-fructofuranosidase, EC 3.2.1.26) from bakers' yeast immobilized in chitin. Invertase is uncompetitively inhibited by its substrate (sucrose) so the reaction is carried out with a CSTR followed by CPBR reducing the consumption of the catalyst (see Problem 4.11). The catalyst has a specific activity of $200\,IU \cdot g^{-1}$, a Michaelis constant of $48\,mM$, and an inhibition constant for the substrate of $250\,mM$. Reactors must process $1800\,L \cdot h^{-1}$ of concentrated cane juice ($200\,g \cdot L^{-1}$ of sucrose) with a substrate conversion at the end of the process (X_2) of 0.95. Assuming that diffusional restrictions are negligible, determine the optimum conversion in the first reactor (X_1) to minimize the catalyst mass used in the process.

Resolution: The total mass of catalyst (M_{cat}) in this case represents the objective function and is the sum of mass in the CSTR and the CPBR. From Equations 4.23 and 4.33, M_{cat} is given by:

$$M_{cat} = \frac{K_M \cdot F}{a_{sp}} \cdot \left(\frac{s_i \cdot X_1}{K_M} + \frac{X_1}{1 - X_1} + \frac{s_i^2 \cdot X_1 \cdot (1 - X_1)}{K_M K_S} \right)$$

$$+ \frac{K_M \cdot F}{a_{sp}} \cdot \left(\frac{s_i \cdot (X_2 - X_1)}{K_M} - \ln\left(\frac{1 - X_2}{1 - X_1} \right) + \frac{s_i^2 \cdot (X_2 - X_1)}{K_M K_S} - \frac{s_i^2 \cdot (X_2{}^2 - X_1{}^2)}{2 \cdot K_M K_S} \right)$$

Defining X_1^* as the optimum conversion in CSTR, $M_{cat}(X_1^*)$ must satisfy the necessary condition (see Equation 7.4):

$$\frac{dM_{cat}}{dX_1} = 0 = \frac{K_M \cdot F}{a_{sp}} \cdot \left(\left(\frac{s_i}{K_M} + \frac{1}{(1 - X_1^*)} + \frac{X_1^*}{(1 - X_1^*)^2} + \frac{s_i^2}{K_M K_S} - \frac{2 \cdot s_i^2 \cdot X_1^*}{K_M K_S} \right) \right.$$
$$\left. + \left(-\frac{s_i}{K_M} - \frac{1}{1 - X_1^*} - \frac{s_i^2 \cdot}{K_M K_S} + \frac{s_i^2 \cdot X_1^*}{K_M K_S} \right) \right)$$

$$\frac{dM_{cat}}{dX_1} = 0 = \frac{K_M \cdot F}{a_{sp}} \cdot \left(-\frac{s_i^2 \cdot X_1^*}{K_M K_S} + \frac{X_1^*}{(1 - X_1^*)^2} \right)$$

$$X_1^* = 1 \pm \sqrt{\frac{K_M K_S}{s_i^2}}$$

Note that only the minus sign in the square root term has physical meaning (by definition, $X \leq 1$). Replacing the values of K_M, K_S, and s_i in this equation gives:

$$X_1^* = 1 - \sqrt{0.03508} = 0.8126$$

Checking the sufficient condition (Equation 7.5) in $X_1^* = 0.8126$ gives:

$$\frac{d^2 M_{cat}}{dX_1^{*2}} = \left(-\frac{s_i^2}{K_M K_S} + \frac{2 \cdot X_1}{(1 - X_i^*)^3} + \frac{1}{(1 - X_i^*)^2} \right)$$

$$\frac{d^2 M_{cat}(X_1^*)}{dX_1^{*2}} = \left(-28.499 + \frac{2 \cdot 0.8126}{(1 - 0.8126)^3} + \frac{1}{(1 - 0.8126)^2} \right) = 246.92$$

Since the second derivative of M_{cat} evaluated in $X_1^* = 0.8126$ is positive, the value of X_1^* is the one that minimizes M_{cat}.

Answer: The conversion in the first reactor that minimizes catalyst consumption is 0.8126.

7.3 Consider the reaction of lactose hydrolysis into glucose and galactose catalyzed by a fungal β-galactosidase. A BSTR is loaded with $0.723\,\mathrm{IU} \cdot \mathrm{mL}^{-1}$ of soluble β-galactosidase with Michaelis constant 42 mM. The catalytic rate constant k is 1 at 40 °C and the energy of activation E_a for lactose hydrolysis is $29,000\,J\cdot mol^{-1}$. The first-order kinetic rate constant of inactivation $k_{D,0} = 1\cdot 10^{65}\,min^{-1}$ and the activation energy of inactivation $E_{ia} = 400,000\,J\cdot mol^{-1}$. The batch reaction has a 142 mM initial lactose concentration and a reaction time of 8 hours. Find the temperature that maximizes substrate conversion. Assume that the Michaelis constant is not dependent on temperature and that the hydrolysis reaction is described by simple Michaelis – Menten kinetics.

Resolution: In enzymatic reactions, temperature produces a trade-off between increasing enzyme activity, by increasing the catalytic rate constant, and reducing enzyme stability (preservation of activity through time), by increasing its inactivation rate constant. In this way, the choice of temperature during reactor operation determines a productive compromise for the reaction system.

The substrate to product conversion in a Michaelis – Menten-type reaction $S \rightarrow P$ is determined by the differential Equation 2.13, which can be conveniently rewritten as:

$$\frac{ds}{dt} = -\frac{k\cdot e\cdot s}{K_M + s} \tag{7.28}$$

The effect of temperature on the catalytic rate can be described using the Arrhenius equation:

$$k = k_0 \cdot exp\left(\frac{-E_a}{R\cdot T}\right) \tag{2.34}$$

On the other hand, assuming first-order enzyme inactivation kinetics:

$$\frac{e}{e_0} = exp(-k_D\cdot t) \tag{2.39}$$

where the effect of temperature on k_D can again be described by the Arrhenius equation:

$$k_D = k_{D,0} \cdot exp\left(\frac{-E_{ia}}{R\cdot T}\right) \tag{2.40}$$

Given Equation 4.2, the substrate concentration is given by $s = s_i\cdot(1-X)$ for any reaction time; for a BSTR (see Section 4.4.1), the following differential equation is obtained:

$$-s_i\cdot\frac{dX}{dt} = -\frac{k_0\cdot exp\left(\frac{-E_a}{R\cdot T}\right)\cdot e_0\cdot exp\left(-k_{D,0}\cdot exp\left(\frac{-E_{ia}}{R\cdot T}\right)\cdot t\right)\cdot s_i\cdot(1-X)}{K_M + s_i\cdot(1-X)} \tag{7.29}$$

Rearranging this equation, both sides can be integrated separately:

$$\int_0^X \frac{K_M + s_i \cdot (1-X)}{1-X} dX = k_0 \cdot exp\left(\frac{-E_a}{R \cdot T}\right) \cdot e_0 \cdot \int_0^t exp\left(-k_{D,0} \cdot exp\left(\frac{-E_{ia}}{R \cdot T}\right) \cdot t\right) dt$$

(7.30)

$$s_i \cdot X - K_M \, ln(1-X) = \frac{k_0}{k_{D,0}} e_0 \cdot exp\left(\frac{E_{ia} - E_a}{R \cdot T}\right) \cdot \left(1 - exp\left(-k_{D,0} \cdot exp\left(\frac{-E_{ia}}{R \cdot T}\right) \cdot t\right)\right)$$

(7.31)

Answer: The product concentration for a given batch duration t_f is determined as $p = s_i \cdot X|_{t_f}$; thus, in order to evaluate $X|_{t_f}$ at different reaction temperatures T_i, Equation 7.31 is solved numerically using Newton's method (see Section A.1). Product concentrations obtained at different reaction temperatures are shown in Figure 7.2. It can clearly be seen that a maximal product concentration is obtained by operating the BSTR at 33 °C.

7.4 Determine the reaction temperature profile in Problem 7.3 that maximizes the concentration of product at the final reaction time by allowing a variation in the operation temperature with lower and upper bounds of 0 and 40 °C.

Resolution: The case will be considered in which the temperature is changed along the reaction time. Assuming Michaelis – Menten kinetics, if changes in temperature are made with the purpose of favoring enzyme stability then an increase in substrate conversion might be observed. More temperature effects may be included, such as an effect on the

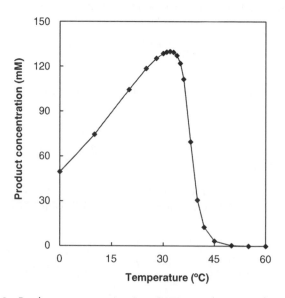

Figure 7.2   *Product concentration in a BSTR as a function of temperature.*

affinity constant K_M. Moreover, effects on complex kinetics involving inhibition by either substrate or product will be more difficult to predict and solve. Thus, the formulation results in the following optimal control problem, which can be solved recursively using the implicit Euler's method for a given Δt:

$$\max_{T,X,e,t} s_i \cdot X$$

subject to:

$$\frac{dX}{dt} = \frac{k_0 \cdot exp\left(\dfrac{-E_a}{R \cdot T}\right) \cdot e \cdot (1 - X)}{K_M + s_i \cdot (1 - X)} \tag{7.32}$$

$$\frac{de}{dt} = -k_{D,0} \cdot exp\left(\frac{-E_{ia}}{R \cdot T}\right) \cdot e \tag{7.33}$$

$$X|_{t_0} = 0$$

$$e|_{t_0} = e_0$$

$$T_{lo} \leq T \leq T_{up}$$

$$t_o \leq t \leq t_f$$

A different temperature might be chosen for every time interval Δt. Thus, using finite difference, discretized versions of Equations 7.32 and 7.33 can be written as:

$$X_{t+\Delta t} = X_t + \Delta t \cdot \frac{k_0 \cdot exp\left(\dfrac{-E_a}{R \cdot T_{t+\Delta t}}\right) \cdot e_{t+\Delta t} \cdot (1 - X_{t+\Delta t})}{K_M + s_i \cdot (1 - X_{t+\Delta t})} \tag{7.34}$$

$$e_{t+\Delta t} = e_t - \Delta t \cdot k_{D,0} \cdot exp\left(\frac{-E_{ia}}{R \cdot T_{t+\Delta t}}\right) \cdot e_{t+\Delta t} \tag{7.35}$$

Reducing the value of Δt to nearly zero will lead to an infinite number of equations and to instantaneous guess values for the temperature at every interval. Although the method is implicit, Michaelis – Menten rate expression is simple enough to obtain $X_{t+\Delta t}$ and $e_{t+\Delta t}$. The latter can be replaced in Equation 7.34 to yield a quadratic equation from which two roots for $X_{t+\Delta t}$ can be obtained, only one of which is a valid solution:

$$X_{t+\Delta t} = \frac{exp\left(\dfrac{-E_a}{R \cdot T_{t+\Delta t}}\right) \cdot \left(\Delta t \cdot exp\left(\dfrac{-E_{ia} - E_a}{R \cdot T_{t+\Delta t}}\right) \cdot e_o \cdot k_0 + A - \sqrt{B + C + D}\right)}{2 \cdot \left(exp\left(\dfrac{E_{ia}}{R \cdot T_{t+\Delta t}}\right) + \Delta t \cdot k_{D,0}\right) \cdot s_i} \tag{7.36}$$

where:

$$A = exp\left(\frac{-E_a}{R \cdot T_{t+\Delta t}}\right) \cdot \left(exp\left(\frac{2 \cdot E_{ia}}{R \cdot T_{t+\Delta t}}\right) + \Delta t \cdot k_0\right) \cdot (K_M + s_i \cdot (1 + X_t)) \quad (7.37)$$

$$B = \Delta t^2 \cdot e_0^2 \cdot k_0^2 \cdot exp\left(\frac{2 \cdot E_{ia}}{R \cdot T_{t+\Delta t}}\right) \quad (7.38)$$

$$C = 2 \cdot e_0 \cdot k_0 \cdot \Delta t \cdot exp\left(\frac{E_a + E_{ia}}{R \cdot T_{t+\Delta t}}\right) \cdot \left(exp\left(\frac{E_{ia}}{R \cdot T_{t+\Delta t}}\right) + \Delta t \cdot k_{D,0}\right) \\ \cdot (K_M + s_i \cdot (X_t - 1)) \quad (7.39)$$

$$D = exp\left(\frac{2 \cdot E_a}{R \cdot T_{t+\Delta t}}\right) \cdot \left(exp\left(\frac{E_{ia}}{R \cdot T_{t+\Delta t}}\right) + \Delta t \cdot k_{D,0}\right)^2 \cdot (K_M + s_i \cdot (1 - X_t))^2 \quad (7.40)$$

Answer: The discretized equations shown yield a system of nonlinear equations such that each $X_{t+\Delta t}$ and $T_{t+\Delta t}$ can be found. A fixed number of intervals is chosen to solve the problem corresponding to a Δt of 16 minutes, leading to 30 discretized versions of Equations 7.34 and 7.35 and 30 guess values of each temperature and conversion. This is a rather large system of equations, but using larger Δt values is not advisable as it could lead to a wrong solution. In this case, the problem can be solved by applying Newton's method for a multivariable system with a suitable software tool. Figure 7.3 shows the temperature profile at all times. It is interesting to note that in this case temperature starts at

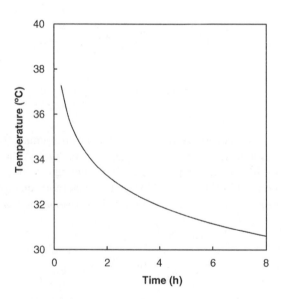

Figure 7.3 *Optimal temperature profile of a BSTR during operation.*

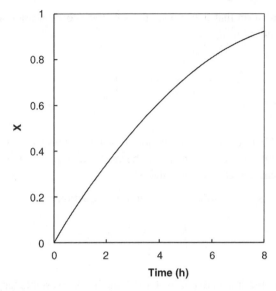

Figure 7.4 *Substrate conversion (X) for optimally controlled temperature of a BSTR.*

a high value at the beginning of the reaction but does not reach the upper bound and is then continuously reduced to compensate for deactivation. Figure 7.4 shows the product concentration values $p = s_i \cdot X$ at all reaction times, where final time concentration corresponds to a similar substrate conversion to that achieved when the reaction was carried out at constant optimal temperature (see Figure 7.2). Although there is no yield improvement in this case, the problem serves as an illustration of reactor operation at a variable temperature. This mode of operation might become relevant for more complex reaction kinetics, which may include substrate inhibition and/or a temperature effect on affinity parameters.

7.5 Consider the reaction of lactose hydrolysis in a CPBR with a diameter of 0.1 m and an effective length (H) of 0.5 m. The substrate concentration at the inflow s_i is 142 mM and the feed flow rate is $1 \, \text{L} \cdot \text{min}^{-1}$ ($1.667 \times 10^{-5} \, \text{m}^3 \cdot \text{s}^{-1}$). The reaction rate constant k is 1, the Michaelis constant K_M is 42 mM at 40 °C, and the energy of activation for reaction rate constant k is $E_a = 40{,}000 \, \text{J} \cdot \text{mol}^{-1}$. The pre-exponential term according to the Arrhenius equation $k_{D,0}$ (see Equation 2.40) is $10^{65} \, \text{min}^{-1}$ and the energy of activation of inactivation E_{ia} is $400{,}000 \, \text{J} \cdot \text{mol}^{-1}$. β-galactosidase is immobilized in particles of slab geometry with a width of 0.0015 m. The catalyst apparent density is $500 \, \text{kg} \cdot \text{m}^{-3}$ and its specific activity a_{sp} is $300 \, \text{UI} \cdot \text{g}^{-1}$. The void fraction in the packed bed ε is 0.55. The effective diffusion coefficient D_{eff} inside the catalyst particle is estimated as $1 \cdot 10^{-8} \, \text{m}^2 \cdot \text{s}^{-1}$. Based on this information, determine the optimal temperature profile by which to maximize lactose hydrolysis reaction conversion, for a total reactor operation time of 100 hours. Assume that external diffusional restrictions are negligible.

Resolution: The equation that determines the substrate concentration at any point in the reactor is (see Equation 5.8):

$$\frac{\varepsilon}{F}\int_{0}^{V_{bed}} dV = -\int_{s_i}^{s}\frac{ds}{\eta_G \cdot v}$$

(7.41)

where V_{bed} is the effective volume of reaction, F is the feed flow rate of substrate entering the reactor at substrate concentration s_i, ε is the void fraction in the packed bed, and η_G is the global effectiveness factor of the catalyst.

Equation 7.41 can also be written as a differential equation:

$$\frac{ds}{dh} = -\frac{\varepsilon \cdot A}{F}\cdot \eta_G \cdot v$$

(7.42)

where A is the cross-sectional area of the CPBR and h is the variable length of reactor bed.

The enzyme volumetric activity in the reactor and can be calculated as $e = a_{sp}\cdot \rho_{app}\cdot \varepsilon^{-1}$, where a_{sp} is the specific activity of the catalyst and ρ_{app} is the apparent density of the catalyst. Thus, Equation 7.42 can be expressed as:

$$\frac{ds}{dh} = -\frac{\varepsilon \cdot A}{F}\cdot \eta_G \cdot \frac{k \cdot e \cdot s}{K_M + s}$$

(7.43)

The effectiveness factor is determined assuming a slab geometry for the catalyst slab of width L, where $L_{eq} = 0.5 \cdot L$ is the distance from the surface of the slab to its center. Let z be the dimensionless distance from the surface, $z = x \cdot L_{eq}^{-1}$, x the depth from the surface, and β the normalized substrate concentration, $\beta = s \cdot s_S^{-1}$, where s_S is the substrate concentration at the surface of the catalyst at reactor length h. For a reaction under internal diffusional restrictions and Michaelis–Menten-type reaction kinetics (see Section 3.3.3):

$$\frac{\partial^2 \beta}{\partial z^2} = \Phi^2 \frac{\beta}{1+\beta}$$

(7.44)

where Φ is the Thièle modulus (see Equation 3.23), being:

$$\Phi^2 = \frac{V''_{max}\cdot L_{eq}^2}{K_M \cdot D_{eff}}.$$

where V''_{max} is the maximum reaction rate per unit volume of catalyst. By using the auxiliary variable $u = \frac{\partial \beta}{\partial z}$ and $\frac{\partial(\cdot)}{\partial z} = \frac{\partial(\cdot)}{\partial \beta}\cdot \frac{\partial \beta}{\partial z}$, Equation 7.44 can be modified to:

$$\frac{\partial^2 \beta}{\partial z^2} = \frac{\partial}{\partial z}\left(\frac{\partial \beta}{\partial z}\right) = \Phi^2 \frac{\beta}{1+\beta} \tag{7.45}$$

$$\frac{\partial}{\partial \beta}\left(\frac{\partial \beta}{\partial z}\right) \cdot \frac{\partial \beta}{\partial z} = \Phi^2 \frac{\beta}{1+\beta} \tag{7.46}$$

$$u \cdot \frac{\partial u}{\partial \beta} = \Phi^2 \frac{\beta}{1+\beta} \tag{7.47}$$

Equation 7.47 is separable and can be integrated to obtain:

$$\int_{u_c}^{u} u\,du = \Phi^2 \int_{\beta_c}^{\beta} \frac{\beta}{1+\beta}\,d\beta \tag{7.48}$$

The integration limits for u are the derivative $\dfrac{\partial \beta}{\partial z}$ at the center of the slab, which is zero, and any value u, and for β are the normalized concentration at the slab center β_c and any value β. Thus, integration yields:

$$\frac{u^2}{2} = \Phi^2\left(\beta - \beta_c - \ln\left(\frac{\beta+1}{\beta_c+1}\right)\right) \tag{7.49}$$

$$u = \Phi\sqrt{\left(\beta - \beta_c - \ln\left(\frac{\beta+1}{\beta_c+1}\right)\right)} \tag{7.50}$$

$$\frac{\partial \beta}{\partial z} = \Phi\sqrt{\left(\beta - \beta_c - \ln\left(\frac{\beta+1}{\beta_c+1}\right)\right)} \tag{7.51}$$

The total normalized concentration β variation at the surface at reactor length h is:

$$\left.\frac{\partial \beta}{\partial z}\right|_h = \Phi\sqrt{\left(\beta_h - \beta_{ch} - \ln\left(\frac{\beta_h+1}{\beta_{ch}+1}\right)\right)} \tag{7.52}$$

For strong diffusional effects, β_c may be considered zero. Thus:

$$\left.\frac{\partial \beta}{\partial z}\right|_h = \Phi\sqrt{(\beta_h - \ln(\beta_h + 1))} \tag{7.53}$$

The total normalized substrate reaction rate at the surface is:

$$v_a = \frac{D \cdot K_M}{L_{eq}^2} \cdot \frac{\partial \beta}{\partial z} \tag{7.54}$$

Now, by considering the normalized reaction rate in the liquid phase bulk we get:

$$v_a = \frac{V''_{max} \cdot \beta_h}{1 + \beta_h} \qquad (7.55)$$

Thus, the global effectiveness factor is determined as:

$$\eta_G = \frac{v_a}{v_b} \qquad (7.56)$$

Finally, replacing the non-normalized variables gives:

$$\eta_G = \frac{1}{\Phi} \cdot \frac{K_M + s}{s} \cdot \sqrt{\frac{2}{K_M} \cdot \left(s - K_M \cdot ln\left(\frac{s + K_M}{K_M}\right) \right)} \qquad (7.57)$$

By replacing Equation 7.44 into Equation 7.43, the following expression is obtained:

$$\frac{ds}{dh} = \frac{\varepsilon \cdot A \cdot k \cdot e}{\Phi \cdot F} \cdot \sqrt{\frac{2}{K_M} \cdot \left(s - K_M \cdot ln\left(\frac{s + K_M}{K_M}\right) \right)} \qquad (7.58)$$

where introducing substrate conversion as defined from Equation 4.2 gives:

$$\frac{dX}{dh} = \frac{\varepsilon \cdot A \cdot k \cdot e}{\Phi \cdot F \cdot s_i} \cdot \sqrt{\frac{2}{K_M} \cdot \left(s_i \cdot (1 - X) - K_M \cdot ln\left(\frac{s_i \cdot (1 - X) + K_M}{K_M}\right) \right)} \qquad (7.59)$$

The optimization objective in this problem is the maximization of the total amount of product produced in the reactor. The accumulated product over a fixed span of time can be determined as the integral J, where X(H,t) is the product conversion at length (H) and time (t):

$$J = \int_0^{t_f} s_i \cdot X(H, t) \cdot F \cdot dt \qquad (7.60)$$

Thus, the optimization problem is formulated as:

$$\max_{T,X,e,t} z$$

subject to Equations 7.33 and 7.59, giving:

$$X|_{t_0} = 0$$

$$e|_{t_0} = e_0$$

$$0 \leq h \leq H$$

$$T_{lo} \leq T \leq T_{up}$$

$$t_o \leq t \leq t_f$$

Answer: Discretization of Equations 7.59 and 7.33 is carried out for 10 time intervals and 10 length intervals, using the explicit Euler's method (shown in Equations 7.61 and 7.62). A Δh of 0.05 m and Δt of 10 hours are fixed. This yields 10 evaluations of conversion at length H for every time t considered, for a total of 100 discretized equations.

$$X_{h+\Delta h,t} = X_{h,t} + \frac{\varepsilon \cdot A \cdot k \cdot e_{h,t}}{\Phi \cdot F \cdot S_i} \cdot \Delta h \cdot$$

$$\cdot \sqrt{\frac{2}{K_M} \cdot \left(S_i \cdot (1 - X_{h,t}) - K_M \cdot ln\left(\frac{S_i \cdot (1 - X_{h,t}) + K_M}{K_M}\right)\right)} \tag{7.61}$$

$$e_{h,t+1} = e_{h,t} - \Delta t \cdot k_{D,0} \cdot exp\left(-\frac{Eia}{R \cdot T}\right) \cdot e_{h,t} \tag{7.62}$$

Objective J in Equation 7.60 is discretized and roughly approximated to the sum:

$$J = \sum_{t_o}^{t_f} x_{H,t} \cdot s_i \cdot F \cdot \Delta t \tag{7.63}$$

where $x_{H,t}$ is the conversion obtained at the exit of the reactor at discretized time t. The problem results in a system of 101 nonlinear equations, which can be solved using Newton's method. As this system is too large to be solved by hand, a suitable software tool can be used.

Figure 7.5 shows the optimal temperature profile that maximizes total product for a total time of 100 hours (J). Figure 7.6 shows the drop in conversion at the end point of the reactor with time as the enzyme inactivates.

7.6 Batch synthesis of galacto-oligosaccharides (GOS) with a β-galactosidase from a thermophilic microorganism is studied with the purpose of maximizing reaction yield, which represents the mass of GOS obtained per unit mass of lactose. Due to the complexity of the process, which involves several parallel reactions, the solution is approached using SRM, by performing a central composite experimental design with

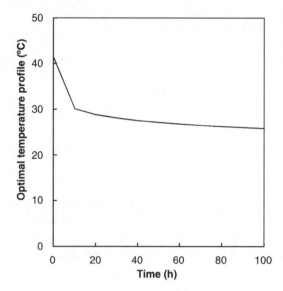

Figure 7.5   *Optimal temperature profile for the operation of CPBR with immobilized lactase.*

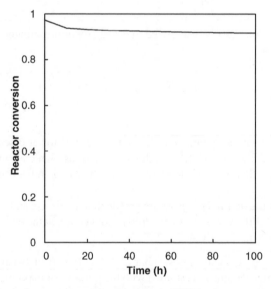

Figure 7.6   *Variation of conversion (X) with time in CPBR when operated under an optimal temperature profile.*

two blocks, assuming initial lactose concentration and temperature as variables. The following table summarizes the experimental results of this procedure:

Variables		Coded Variables		Response
Initial Lactose Concentration % (w/w)	Temperature (°C)	x_{Lac}	x_{Temp}	Yield · 100 (kg GOS · kg lactosa$^{-1)}$
40	50	−1	−1	31.00
60	50	1	−1	22.52
40	70	−1	1	28.03
50	60	0	0	31.13
50	60	0	0	30.93
60	70	1	1	30.30
50	60	0	0	31.20
50	60	0	0	30.99
50	74.14	0	1.41	30.75
50	60	0	0	31.01
50	60	0	0	31.02
35.86	60	−1.41	0	28.61
64.14	60	1.41	0	23.97
50	45.86	0	−1.41	29.87

x_{Lac} and x_{Temp} represent the coded variable for the initial lactose concentration and temperature, respectively.

Determine the best model by which to describe the effect of temperature and initial lactose concentration on yield, the values of the parameters, and their 95% CI. Determine also which variable has the strongest influence on yield and whether there are optimum values for temperature and initial lactose concentration (those maximizing yield).

Resolution: The simplest model that can be used to fit the experimental data is a first-order model: $\hat{y} = b_0 + b_1 \cdot x_{lac} + b_2 \cdot x_{temp}$. For this model, column vector Y and matrix X, assuming coded variables, correspond to:

$$
X = \begin{bmatrix}
1 & -1 & -1 \\
1 & 1 & -1 \\
1 & -1 & 1 \\
1 & 0 & 0 \\
1 & 0 & 0 \\
1 & 1 & 1 \\
1 & 0 & 0 \\
1 & 0 & 0 \\
1 & 0 & 1.41 \\
1 & 0 & 0 \\
1 & 0 & 0 \\
1 & -1.41 & 0 \\
1 & 1.41 & 0 \\
1 & 0 & -1.41
\end{bmatrix}
\qquad
Y = \begin{bmatrix}
31.00 \\
22.52 \\
28.03 \\
31.13 \\
30.93 \\
30.30 \\
31.20 \\
30.99 \\
30.75 \\
31.01 \\
31.02 \\
28.61 \\
23.97 \\
29.87
\end{bmatrix}
$$

$(X^T \cdot X)^{-1}$ and B^* can thus be easily calculated using any mathematical software, resulting in:

$$(X^T \cdot X)^{-1} = \begin{bmatrix} 0.07142 & 0 & 0 \\ 0 & 0.125 & 0 \\ 0 & 0 & 0.125 \end{bmatrix} \qquad B^* = (X^T \cdot X)^{-1} \cdot X^T \cdot Y = \begin{bmatrix} 28.3917 \\ -1.5977 \\ 0.7577 \end{bmatrix}$$

Note in this case that the covariance terms are zero (Equation A.21). This is because the experimental design is orthogonal, meaning that the estimated parameters are uncorrelated.

The same procedure can be employed to determine the parameters of the second-order model. The quality of fit of both models can be assessed by determining the CIs for each parameter (Equation A.24), R^2 (Equation A.25), and R^2_{adj} (Equation A.27) and whether there is a linear relationship between the response and regressor variables (Equation 7.4). The following table summarizes the results for both first- and second-order models:

Model	Parameter	95% CI	R^2	R^2_{adj}	SS_R	SS_E
First-order: $\hat{y} = b_0 + b_1 \cdot x_{lac} + b_2 \cdot x_{temp}$	$b_0 = 29.38$	± 1.54	0.25	0.11	25.01	76.08
	$b_1 = -1.6$	± 2.05				
	$b_2 = 0.76$	± 2.05				
Second-order: $\hat{y} = b_0 + b_1 \cdot x_{lac} + b_2 \cdot x_{temp}$ $b_3 \cdot x_{lac} \cdot x_{temp} + b_4 \cdot x_{lac}^2$ $+ b_5 \cdot x_{temp}^2$	$b_0 = 31.05$	± 0.46	0.98	0.97	99.21	1.88
	$b_1 = -1.6$	± 0.39				
	$b_2 = 0.76$	± 0.39				
	$b_3 = 2.69$	± 0.56				
	$b_4 = -2.46$	± 0.41				
	$b_5 = -0.45$	± 0.41				

For the first-order model, using Equation 7.15: $F_0 = (25.01/2)/(76.08/11) = 1.8$. From Montgomery and Runger [7], a value of $F_{0.05,2,11} = 3.98$ is obtained. Since $F_0 < F_{0.05,2,11}$, the model does not offer a statistically significant prediction of the behavior of reaction yield. Thus, first-order parameters (b_1 and b_2) should be considered equal to zero. Consistently, the CI for b_1 and b_2 exceeds the values estimated for these parameters, meaning that they make no contribution to explaining the response and should be removed from the model. Furthermore, the value of R^2 implies that only 25% of the variability of the data can be explained by the first-order model, which is a very low value. In the case of the second-order model, $F_0 = (99.21/5)/(1.88/8) = 84.43$ and $F_{0.05,5,8} = 3.69$ [6], meaning that the model is significant and at least one of the parameters (beside b_0) is nonzero. Analysis of the CI for the estimated parameters indicates that all of them are statistically significant. R^2 implies that 98% of the variability is explained by the second-order model. Also, R^2_{adj} increases significantly from the value obtained for the first-order model, showing that the incorporation of a second-order term contributes to better describing the behavior of the data. This is confirmed by the extra sum of squares test. The sum of squares due to the regression product of the incorporation of b_3, b_4, and b_5 into the model, given that b_0, b_1, and b_2 were already considerable, is $SS_R(b_3,b_4,b_5|b_0,b_1,b_2) = 99.21 - 25.01 = 74.2$. According to Equation 7.19, $F_0 = (74.2/3)/(1.88/8) = 105.25$, which is considerably higher than $F_{0.05,3,8} = 4.07$, implying that at least one of the added terms (b_3, b_4, or b_5) is nonzero and contributes significantly to the model. A more refined analysis can be made by the reader for each parameter, analyzing their contribution separately. On the other hand, from the magnitude of the first-order parameters it is clear that the initial lactose concentration exerts more influence on yield than does temperature, since $b_1 > b_2$. However, the signs of b_1 and b_2 indicate that, in the experimental range considered, the initial lactose concentration has a negative effect on yield. The opposite behavior is found for temperature.

Given that the second-order model offers a satisfactory prediction of the reaction yield, the best operational conditions (temperature and lactose concentration) can be determined using Equation 7.26:

$$x = -0.5 \cdot \begin{bmatrix} -2.46 & 1.345 \\ 1.345 & -0.45 \end{bmatrix}^{-1} \cdot \begin{bmatrix} -1.6 \\ 0.76 \end{bmatrix} = \begin{bmatrix} -0.215 \\ 0.201 \end{bmatrix}$$

Note that the previous values are coded; using Equation 7.12 transforms them to their respective uncoded values, which are 47.85% (w/w) initial lactose concentration and 62 °C. The predicted value of yield ($\hat{y}$) under this condition (x) and its CI are then calculated using Equations 7.20 and 7.21:

$$\hat{y} = xg \cdot B^* = 31.05 + 1.6 \cdot 0.215 + 0.76 \cdot 0.201 - 2.69 \cdot 0.043 - 2.46 \cdot 0.046 - 045 \cdot 0.040$$
$$= 31.29$$

The 95% CI of $\hat{y}$ at point x is given by:

$$CI = t(n - p, a/2) \cdot \sqrt{\frac{SS_E}{n - p} \cdot x_g \cdot (X^T \cdot X)^{-1} \cdot x_g^T} = 2.306 \cdot \sqrt{\frac{1.88}{14 - 6} \cdot 0.164} = 0.45$$

Thus, at an initial lactose concentration of 48.9% (w/w) and 62 °C, a yield of 0.313 ± 0.0045 (kg GOS · kg lactose^{-1}) is expected with 95% confidence. In order to verify whether this value corresponds to a maximum, minimum, or saddle point, the Hessian of $\hat{y}(x)$ must be analyzed. R^* is chosen to evaluate whether $\nabla^2\hat{y}(x)$ is a positive or negative definite. $v^T = [a\ b]$ with a and b ε. It is easy to prove that the Hessian is not a positive or negative definite because the sign $v^T\nabla^2\hat{y}(x) \cdot v = -4.92\,a^2 + 5.38 \cdot a \cdot b - 0.9 \cdot b^2$ is dependent on the values of a and b chosen. So the eigenvalues of Hessian must be determined by solving $|\nabla^2\hat{y}(x) - \lambda \cdot I| = 0$, where λ represents the eigenvalues, I is the identity matrix, and $|\ |$ denotes the determinant:

$$\left| \begin{bmatrix} -4.92 & 2.69 \\ 2.69 & -0.9 \end{bmatrix} - \lambda \cdot \begin{bmatrix} 1 & 0 \\ 0 & 1 \end{bmatrix} \right| = (-4.92 - \lambda) \cdot (-0.9 - \lambda) + 2.69^2 = 0$$

Solving this quadratic equation gives ($\lambda_1 = -6.26$ and $\lambda_2 = 0.45$. Since the eigenvalues have different signs, x is a saddle point.

Answer: The best statistic model by which to describe the behavior of the yield is the second-order one. The model and its parameters, using coded variables, are:

$$\hat{y} = 31,0 - 1.6 \cdot x_{lac} + 0.67 \cdot x_{temp} + 2.69 \cdot x_{lac} \cdot x_{temp} - 2.46 \cdot x_{lac}^2 - 0.45 \cdot x_{temp}^2$$

The 95% CIs for the model parameters from left to right are: $\pm0.46, \pm0.39, \pm0.39, \pm0.56, \pm0.41$, and ±0.41. The magnitude of the linear term suggests that lactose concentration has a greater effect than temperature. However, inside the experimental range an increase in the initial lactose concentration has a negative impact on yield. Regrettably, the model does not predict optimum values for the temperature and the initial lactose concentration by which to maximize yield.

Supplementary Problems

7.7 As in Problem 7.1, a glucose syrup is produced with immobilized β-galactosidase in a BSTR from a concentrated whey permeate with a lactose concentration of 285.71 mM. The enzyme is competitively inhibited by galactose. The BSTR, with 1000 L of working volume and loaded with $60\,g \cdot L^{-1}$ of biocatalyst with a specific activity of $368.5\,IU \cdot g^{-1}$, operates in repeated batch mode at 30 °C to a final conversion after each batch of 85%. The kinetic and inactivation parameters of the biocatalyst at 30 °C are: $K_M = 30.14\,mM$, $K_{IC} = 16.64\,mM$, and $k_D = 0031\,h^{-1}$.

Determine the residual activity (% of initial) after one cycle of reactor operation that maximizes the specific productivity of the reactor and the residual activity (% of initial) at the point of catalyst replacement (after one cycle of reactor operation) that maximizes the product-to-catalyst mass ratio (mass of product produced per unit mass of catalyst in one cycle of reactor operation). Discuss the results obtained. (One cycle of reactor operation considers all batches until catalyst replacement.)

Answer:

Residual Activity at Catalyst Replacement (% of initial)	Specific Productivity $(kg_{glucose} \cdot kg_{catalyst}^{-1} \cdot h^{-1})$	Product–Catalyst Ratio $(kg_{glucose} \cdot kg_{catalyst}^{-1})$
92.5	0.58	1.5
88.8	0.57	2.2
85.0	0.56	2.9
81.3	0.54	3.6
77.5	0.53	4.4
73.8	0.52	5.1
70.0	0.51	5.8
66.3	0.49	6.6
62.5	0.48	7.3
58.8	0.47	8.0
55.0	0.45	8.7
51.3	0.44	9.5
47.5	0.43	1.2
43.8	0.41	1.9

(*Continued*)

Residual Activity at Catalyst Replacement (% of initial)	Specific Productivity ($kg_{glucose} \cdot kg_{catalyst}^{-1} \cdot h^{-1}$)	Product–Catalyst Ratio ($kg_{glucose} \cdot kg_{catalyst}^{-1}$)
40.0	0.39	1.7
36.3	0.38	12.4
32.5	0.36	13.1
28.8	0.34	13.8
25.1	0.33	14.6
21.3	0.31	15.3
17.6	0.29	16.0
13.8	0.26	16.8
10.1	0.24	17.5

These results are depicted in Figure 7.7.

Specific productivity increases when the catalyst is replaced at high values of residual activity; however, operating under such conditions is inconvenient in terms of catalyst cost (due to frequent replacement). On the other hand, the product-to-catalyst mass ratio increases when the catalyst is replaced at low values of residual activity, meaning that extensive use is made of the catalyst potential of the enzyme and that more product mass is produced per unit mass of catalyst; however, operating time under such conditions is increased. As can be seen, there is a clear compromise between specific productivity and product-to-catalyst mass ratio. The intersection of the two curves, occurring in this case at 35% residual activity before catalyst replacement, may be a good operating condition by which to solve this compromise optimally. However, it is not necessarily the best one, since the impact on operating cost may be of a different magnitude. An expensive catalyst will

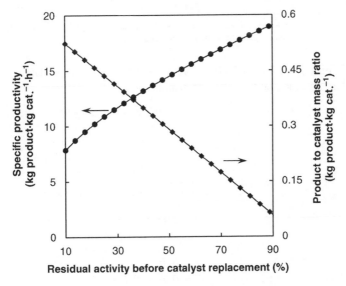

Figure 7.7 *Specific productivity and product-to-catalyst mass ratio as a function of catalyst replacement policy.*

favor an extensive use of the catalyst potential favoring catalyst replacement at lower residual activity; the opposite will hold when the catalyst is not expensive, and the cost of operating the reactor for prolonged time may have a stronger impact on production cost. In the end, the optimal policy of catalyst replacement will be based on economic considerations, taking into account both catalyst and reactor operating costs.

7.8 A process for the enzyme-aided extraction of phenolic antioxidant compounds from agroindustrial rejects is studied using a commercial enzyme preparation from *Aspergillus aculeatus* containing several carbohydrases and proteases. The temperature and enzyme-to-substrate ratio are selected for optimization by SRM. The following Box − Behnken experimental design is carried out for this purpose:

Temperature (°C)	Enzyme − Substrate Ratio $(mL \cdot kg^{-1})$	Total Phenolic Compounds $(g\,GA \cdot kg^{-1})^{*}$
x_1	x_2	y
40 (0)	0.004 (0)	80.33
40 (0)	0.004 (0)	80.27
40 (0)	0.004 (0)	81.30
40 (0)	0.004 (0)	79.87
25 (−1)	0.002 (−1)	57.77
25 (−1)	0.006 (1)	67.26
55 (1)	0.002 (1)	87.40
40 (0)	0.0068 (1.41)	85.95
40 (0)	0.004 (0)	80.36
18.78 (−1.41)	0.004 (0)	54.44
40 (0)	0.00117 (−1.41)	72.55
55 (1)	0.006 (1)	96.93
61.21 (1.41)	0.004 (0)	96.37

*Total phenolic compounds are expressed as equivalents of gallic acid (GA) per kilogram of rejects. In parenthesis are the values of the coded variables for each experimental point.

Find the best model by which to describe the experimental data, the CI of the parameters and the coefficient of determination.

Answer: The best model in terms of coded variables and the values of the parameters is:

$$y = b_o + b_1 \cdot x_1 + b_2 \cdot x_2 + b_{11} \cdot x_1^2 + b_{22} \cdot x_2^2$$

Parameter	Value	CI (95%)
b_o	80.43	±0.38
b_1	14.82	±0.30
b_2	4.75	±0.30
b_{11}	−2.51	±0.33
b_{22}	−0.58	±0.33
R^2	0.999	
R_{adj}	0.999	

References

1. Illanes, A., Wilson, L., and Tomasello, G. (2001) Effect of modulation of enzyme inactivation on temperature optimization for reactor operation with chitin-immobilized lactase. *Enzyme and Microbial Technology*, **11**, 531–540.
2. Rao, S.S. (2009) *Engineering Optimization Theory and Practice Fourth Edition*, John Wiley and Sons, New York.
3. Edgar, T.F., Himmelblau, D.M., and Lasdon, L.S. (2001) *Optimization of Chemical Process*, 2nd edn, McGraw Hill, New York.
4. Illanes, A. and Fajardo, A. (2001) Kinetically controlled synthesis of ampicillin with immobilized penicillin acylase in the presence of organic cosolvents. *Journal of Molecular Catalysis B: Enzymatic*, **11**, 587–595.
5. Montgomery, D.C. (2001) *Design and Analysis of Experiments*, 5th edn, John Wiley and Sons, New York.
6. Draper, N.R. and Smith, H. (1981) *Applied Regression Analysis*, 2nd edn, John Wiley and Sons, New York.
7. Montgomery, D.C. and Runger, G.C. (2010) *Applied Statistics and Probability for Engineers*, John Wiley and Sons, New York.

Appendix A

Mathematical Methods

A.1 Newton's Method

Newton's method is one of the most powerful tools for the numerical solution of nonlinear equations, or more precisely for finding their roots. It has several advantages, including fast convergence and easy numerical implementation. This section presents its derivation and use for a problem containing only one variable.

Consider the following equation:

$$g(x) = a \tag{A.1}$$

The aim of the Newton's method is to find the value of x for which the nonlinear function g(x) is equal to the constant a. For this purpose, Equation A.1 is rewritten as its equivalent:

$$f(x) = g(x) - a = 0 \tag{A.2}$$

To determine the value of x for which f(x) is equal to zero, the function f(x) is linearized around x_0 using Taylor series by taking only its first derivative ($f'(x)$) and neglecting the higher-order terms:

$$f(x) \approx f(x_0) + f'(x_0) \cdot (x - x_0) \tag{A.3}$$

Now, Equation A.3 is simply set to zero:

$$0 = f(x_0) + f'(x_0) \cdot (x - x_0) \tag{A.4}$$

Problem Solving in Enzyme Biocatalysis, First Edition. Andrés Illanes, Lorena Wilson and Carlos Vera.
© 2014 John Wiley & Sons, Ltd. Published 2014 by John Wiley & Sons, Ltd. Companion Website:
http://www.wiley.com/go/illanes-problem-solving

and x becomes:

$$x = x_0 - \frac{f(x_0)}{f'(x_0)} \qquad\qquad (A.5)$$

The problem is solved recursively, that is, the value of x becomes x_0 for the next evaluation of Equation A.5 and so on. The iteration is stopped when the approximate error is lower than a given δ; for instance, 10^{-6}. The value of δ depends on the user's goal, but it must not exceed computer precision. For common personal computers, precision is around 10^{-14}, so values of δ higher than 10^{-8} are advisable.

The relative error (E) obtained in the nth iteration is estimated as follows:

$$E = \left| \frac{x - x_0}{x} \right| \qquad\qquad (A.6)$$

In many cases, calculation of f'(x) may be difficult; if so, the first derivative can be approximated using finite differences:

$$f'(x) \approx \frac{f(x + \Delta x) - f(x)}{\Delta x} \qquad\qquad (A.7)$$

where values of Δx between 10^{-6} and 10^{-4} are suitable for most applications. Replacing Equation A.7 into Equation A.5 yields Equation A.8, which corresponds to the modified secant method:

$$x = x_0 - \frac{f(x_0) \cdot \Delta x}{f(x_0 + \Delta x) - f(x_0)} \qquad\qquad (A.8)$$

Figure A.1 illustrates the use of Newton's method to find the root of the function f(x). The search begins from a given initial value x_0, around which f(x) is linearized (dotted line)

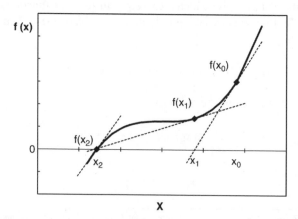

Figure A.1   *Newton's method. Continuous line: f(x); dashed line: linear approximation of f(x) around x_0, x_1, and x_2.*

using Taylor's series. This linearization is set to zero, obtaining x_1. Then $f(x)$ is linearized around x_1 and the resulting function is set to zero, to find x_2, and so on.

Although Newton's method is one of the best methods for solving nonlinear equations, it has some drawbacks. For example, it may fail if $f(x)$ has a saddle point ($f'(x_i) = 0$); in this scenario, according to Equation A.5, x tends to infinity. It may also fail if $f(x)$ is not continuously differentiable or if $f(x)$ has multiple roots, in which case it may fail to find the desired one. More details can be found in the book by Chapra and Canale [1]. In many cases these issues can be avoided with an adequate selection of δ and of the initial value of x.

Example A.1

Problem 2.7 is used to illustrate the application of Newton's method. As was shown in its resolution, Equation 2.55 describes the profile of sucrose concentration in time when invertase is added at a level of $1\,g \cdot L^{-1}$ to a 50 mM sucrose solution:

$$10 \cdot \ln\frac{50}{s} + (50 - s) = t \qquad (2.55)$$

If we want to know the concentration of sucrose at 3 minutes' reaction time, Equation 2.55 becomes:

$$10 \cdot \ln\frac{50}{s} + (50 - s) = 3$$

Given Equations A.1 and A.2:

$$f(s) = 10 \cdot \ln\frac{50}{s} + (50 - s) - 3$$

This equation is solved using Equations A.6 and A.8. The most critical factor is the choice of the initial value of s. By inspection of $f(s)$, s cannot be equal to zero, because the logarithm of 50/s with s close to zero tends to infinity. Also, for stoichiometric reasons, the concentration of s cannot be higher than 50 mM. An initial value of s of 30 mM seems reasonable. Taking conservative values for δ and Δs of 10^{-6} and 10^{-5}, respectively, the problem is solved as shown in the following table:

Iteration	s_0	f (s)	$s + \Delta s$	f (s + Δ s)	s	E
0	30	22.10826	30.00001	22.10824	46.58119	0.3559
1	46.58119	1.12707	46.58120	1.12706	47.50907	0.0195
2	47.50907	0.00196	47.50908	0.00195	47.51068	$3.404 \cdot 10^{-05}$
3	47.51068	$5.76 \cdot 10^{-9}$	47.51069	−0.00001	47.51068	$1.001 \cdot 10^{-10}$

In the third iteration, the criterion $E < \delta$ is accomplished and the routine is stopped. In this way, a value of s equal to 47.51068 mM is calculated as the solution of Equation 2.55 when t is equal to 3 minutes.

A.2 Curve Fitting by Least Squares

In this section, a matrix approach will be employed to show the fundamentals of linear and nonlinear regression by least squares. The formulation of the problem in matrix terms facilitates implementation of a computer routine to estimate the parameter and offer a clear framework of reference for both cases.

A.2.1 Linear Regression

It is important to emphasize that the concept of linear regression does not necessarily imply that the data set fits a straight line, but only to a linear model. A model is linear if its derivative with respect to each of the parameters is not dependent on any of them. Although the focus of this section is on the fit of the data to a straight line, the same procedure can be applied to others linear models, such as the polynomial models (e.g. $y = b_0 + b_1 \cdot x_1 + b_1 \cdot x_1^2$).

Given a set of n experimental observations $\{\{x_1,y_1\}, \{x_2, y_2\}, \ldots ,\{x_n,y_n\}\}$, where x_i represents the value of the independent variable that produces a value y_i in the dependent variable, and assuming that the relationship between the independent and dependent variables is represented by a straight line (Equation A.9), the problem can be written in matrix form as indicated by Equation A.10:

$$y_i = b_0 + b_1 \cdot x_1 + \varepsilon \tag{A.9}$$

$$\begin{bmatrix} y_i \\ \cdot \\ \cdot \\ \cdot \\ y_n \end{bmatrix} = \begin{bmatrix} 1 & x_1 \\ \cdot & \cdot \\ \cdot & \cdot \\ \cdot & \cdot \\ 1 & x_n \end{bmatrix} \cdot \begin{bmatrix} b_0 \\ b_1 \end{bmatrix} + \begin{bmatrix} \varepsilon_1 \\ \cdot \\ \cdot \\ \cdot \\ \varepsilon_n \end{bmatrix} \tag{A.10}$$

where b_0 and b_1 are the intercept and slope of such a straight line, respectively, and correspond to the parameters to be estimated, while ε_i represents the error or disturbance for the ith observation. At this point, it is convenient to define the vector of observations Y, the vector of parameters or regression coefficients B, the matrix of independent variables X, and the error vector ε as follows:

$$Y = \begin{bmatrix} y_i \\ \cdot \\ \cdot \\ \cdot \\ y_n \end{bmatrix} \tag{A.11}$$

$$X = \begin{bmatrix} 1 & x_1 \\ \cdot & \cdot \\ \cdot & \cdot \\ \cdot & \cdot \\ 1 & x_n \end{bmatrix} \tag{A.12}$$

$$B = \begin{bmatrix} b_0 \\ b_1 \end{bmatrix} \qquad (A.13)$$

$$\varepsilon = \begin{bmatrix} \varepsilon_1 \\ \cdot \\ \cdot \\ \cdot \\ \varepsilon_n \end{bmatrix} \qquad (A.14)$$

Given these definitions, Equation A.10 can be represented in its compact form:

$$Y = X \cdot B + \varepsilon \qquad (A.15)$$

The purpose of regression is to determine the value of the vector B that minimizes ε. In other words, the goal is to find the parameters that minimize the difference between the observed and the predicted values of the dependent variable. Vector ε can easily be written as function of B given Equation A.15:

$$\varepsilon = Y - X \cdot B \qquad (A.16)$$

Since there are n observations, it seems logical to set the sum of all ε_i ($S = \sum \varepsilon_i$) as the objective function to be minimized, and thereby minimize the global error. However, the value of ε_i can be positive or negative, so some values of ε_i might cancel others; hence, $S = \sum \varepsilon_i$ is not a sound objective function. To solve this problem, S is defined as the sum of all ε_i^2 (note that ε_i^2 is always positive). Thus, the problem of linear regression is:

$$\min_{b_0, b_1} S(B)$$

$$\qquad (A.17)$$

Subject to $\qquad S(B) = \sum_{i=1}^{n} \varepsilon_i^2 = \sum_{i=1}^{n} (y_i - b_0 - b_1 \cdot x_i)^2$

This methodology for estimating b_0 and b_1 is called "least-square regression." There are several procedures to calculate the parameters that minimize S. Here, the problem is solved using a matrix approach. Note that S can be written in terms of vectors[1] S(B) and expressed as:

$$S(B) = \sum_{i=1}^{n} (y_i - b_0 - \cdot b_1 \cdot x_i)^2 = \| Y - X \cdot B \|^2 = [Y - X \cdot B]^T \cdot [Y - X \cdot B]$$
$$= Y^T \cdot Y - Y^T \cdot X \cdot B - B^T \cdot X^T \cdot Y + B^T \cdot X^T \cdot X \cdot B$$
$$= Y^T \cdot Y - 2 \cdot Y^T \cdot X \cdot B + B^T \cdot X^T \cdot X \cdot B$$

$$\qquad (A.18)$$

---

[1] Useful properties of matrix and vector:

- For a vector $A \in R^n$, such that $A = \{a_1, a_2, \ldots, a_n\}$, the norm of A is defined as $\|A\| = \sqrt{a_1^2 + a_2^2 + \ldots + a_n^2}$.
- Transpose matrix properties: $(A \cdot C)^T = C^T \cdot A^T$; $(A + C)^T = A^T + C^T$; $(A^T)^T = A$; $(A^{-1})^T = (A^T)^{-1}$.
- For columns vectors such as Y and $X \cdot B$, the dot product has the following property: $u^T \cdot v = v^T \cdot u$.
- Matrix derivatives: $\frac{d}{dx}(A^T \cdot x) = A$ $\qquad \frac{d}{dx}(X^T \cdot A \cdot x) = 2 \cdot A \cdot X$

where $||A||$ represents the norm of A and A^T means "A transposed." In order to determine the value of B that minimizes S, the first derivative criterion must be applied. Taking the derivative of Equation A.18 with respect to B gives:

$$\frac{d}{dB}(S(B)) = -2 \cdot X^T \cdot Y + 2 \cdot X^T \cdot X \cdot B \tag{A.19}$$

Setting Equation A.19 to zero and solving for B, Equation A.20 is obtained:

$$B^* = \begin{bmatrix} b_o^* \\ b_1^* \end{bmatrix} = (X^T \cdot X)^{-1} \cdot X^T \cdot Y \tag{A.20}$$

where B^* represents the value of B that minimizes S.

Equation A.20 allows an easy computer estimation of the parameters. It is interesting to note that the variance–covariance matrix (VC) of vector B^* can be written as follows:

$$VC(B^*) = VC\begin{pmatrix} b_o^* \\ b_1^* \end{pmatrix} = \begin{bmatrix} V(b_o^*) & C(b_o^*, b_1^*) \\ C(b_o^*, b_1^*) & V(b_1^*) \end{bmatrix} = s^2 (X^T \cdot X)^{-1} \tag{A.21}$$

with:

$$s^2 = \frac{S(B^*)}{n - p} \tag{A.22}$$

where $V(b_i^*)$ is the variance of parameter b_i^*, $C(b_0^*, b_1^*)$ is the covariance of the parameters b_0^* and b_1^*, s^2 is the residual mean square or variance estimated for $n - p$ degrees of freedom and p is the total number of estimated parameters (two). The standard deviation of the estimated parameter i (σ_i) corresponds to the square root of $V(b_i^*)$; that is:

$$\sigma_i = \sqrt{V(b_i^*)} = s \cdot \sqrt{(X^T \cdot X)^{-1}}_{i,i} \tag{A.23}$$

In Equation A.23, the term $(X^T \cdot X)^{-1}{}_{i,i}$ represents the element i,i of the matrix $(X^T \cdot X)^{-1}$. The confidence interval (CI) of $100(1 - \alpha)\%$ for parameter b_i^* is given by:

$$b_i^* \pm \sigma_i \cdot t\left(n - p, \frac{\alpha}{2}\right) \quad \text{or} \quad b_i^* - \sigma_i \cdot t\left(n - p, \frac{\alpha}{2}\right) \leq b_i^* \leq b_i^* + \sigma_i \cdot t\left(n - p, \frac{\alpha}{2}\right) \tag{A.24}$$

where $t(n - p, \alpha/2)$ is the upper $\alpha/2$ quantile of Student's t distribution for $n - p$ degrees of freedom. Tables for $t(n - p, \alpha/2)$ values are available in any statistics textbook.

It is necessary to highlight that a standard deviation or CI larger than the value of the estimated parameter is indicative of a dummy parameter or a poor fit, meaning that the model and the experimental data should be checked. In this sense, the statistic parameters most frequently used to describe the quality of fit are the coefficient of determination (R^2) and the adjusted coefficient of determination (R_{adj}^2). The coefficient of determination is defined as the fraction of the data variability explained by the model and is calculated as follows:

$$R^2 = 1 - \frac{S(B^*)}{\sum\limits_{i=1}^{n}(y_i - \bar{Y})^2}$$ (A.25)

where $\bar{Y}$ corresponds to the mean of the observed values:

$$\bar{Y} = \frac{1}{n} \cdot \sum\limits_{i=1}^{n} y_i$$ (A.26)

R^2 can take values in the range from 0 to 1; usually, values of $R^2 > 0.95$ are indicative of a good fit. For models with more parameters, R^2_{adj} is preferred to R^2, as the latter increases its value by adding parameters to the model but these parameters do not contribute to explaining the behavior of the dependent variable. In contrast, R^2_{adj} decreases its value if additional unnecessary terms are added to the model. R^2_{adj} is defined as:

$$R^2_{adj} = 1 - \frac{(n-1) \cdot S(B^*)}{(n-p) \cdot \sum\limits_{i=1}^{n}(y_i - \bar{Y})^2}$$ (A.27)

Values of $R^2_{adj} > 0.7$ are considered acceptable for modeling purposes.

Example A.2

Problem 2.1 will be employed to illustrate the use of least-square linear regression. The experimental glucose release observed during lactose hydrolysis is given by:

t (minutes)	0	2	4	6	8	12	18	26	36	48
g (mM)	0	150	307	461	597	912	942	976	999	1015

Selecting the linear zone for g versus t (up to 12 minutes), the vector of observations, the matrix of independent variables and the vector of parameters are expressed according to Equations A.11 to A.14:

$$Y = \begin{bmatrix} 0 \\ 150 \\ 307 \\ 461 \\ 597 \\ 912 \end{bmatrix} \quad X = \begin{bmatrix} 1 & 0 \\ 1 & 2 \\ 1 & 4 \\ 1 & 6 \\ 1 & 8 \\ 1 & 12 \end{bmatrix} \quad B = \begin{bmatrix} b_0 \\ b_1 \end{bmatrix}$$

In order to find the parameters that minimize S(B), we must first calculate $X^T \cdot X$:

$$X^T \cdot X = \begin{bmatrix} 1 & 1 & 1 & 1 & 1 & 1 \\ 0 & 2 & 4 & 6 & 8 & 12 \end{bmatrix} \cdot \begin{bmatrix} 1 & 0 \\ 1 & 2 \\ 1 & 4 \\ 1 & 6 \\ 1 & 8 \\ 1 & 12 \end{bmatrix} = \begin{bmatrix} 6 & 32 \\ 32 & 264 \end{bmatrix}$$

The next step is to determine $(X^T \cdot X)^{-1}$:

$$(X^T \cdot X)^{-1} = \left(\begin{bmatrix} 6 & 32 \\ 32 & 264 \end{bmatrix} \right)^{-1} = \begin{bmatrix} \dfrac{33}{70} & -\dfrac{2}{25} \\ -\dfrac{2}{25} & \dfrac{3}{280} \end{bmatrix}$$

Finally, the parameters can be estimated as:

$$B^* = \begin{bmatrix} b_o^* \\ b_1^* \end{bmatrix} = (X^T \cdot X)^{-1} \cdot X^T \cdot Y$$

$$= \begin{bmatrix} \dfrac{33}{70} & -\dfrac{2}{25} \\ 1 - \dfrac{2}{25} & \dfrac{3}{280} \end{bmatrix} \cdot \begin{bmatrix} 1 & 1 & 1 & 1 & 1 & 1 \\ 0 & 2 & 4 & 6 & 8 & 12 \end{bmatrix} \cdot \begin{bmatrix} 0 \\ 150 \\ 307 \\ 461 \\ 597 \\ 912 \end{bmatrix} = \begin{bmatrix} 0.5 \\ 75.75 \end{bmatrix}$$

The intercept value (given by b_0^*) is thus 0.5 and the slope value (given by b_1^*) 75.75. Given these facts, the release of glucose is described by the following straight line: $g = 0.5 + 75.75 \cdot t$, where g is the mM glucose concentration, and t is the time in minutes. Equations A.16 and A.18 can then be used to calculate $S(B^*)$:

$$\varepsilon = Y - X \cdot B = \begin{bmatrix} 0 \\ 150 \\ 307 \\ 461 \\ 597 \\ 912 \end{bmatrix} - \begin{bmatrix} 1 & 0 \\ 1 & 2 \\ 1 & 4 \\ 1 & 6 \\ 1 & 8 \\ 1 & 12 \end{bmatrix} \cdot \begin{bmatrix} 0.5 \\ 75.75 \end{bmatrix} = \begin{bmatrix} -0.5 \\ -2 \\ 3.5 \\ 6 \\ -9.5 \\ 2.5 \end{bmatrix}$$

$$S(B^*) = \sum_{i=1}^{n} \varepsilon_i^2 = [Y - X \cdot B]^T \cdot [Y - X \cdot B]$$

$$= [-0.5 \quad -2 \quad 3.5 \quad 6 \quad -9.5 \quad 2.5] \cdot \begin{bmatrix} -0.5 \\ -2 \\ 3.5 \\ 6 \\ -9.5 \\ 2.5 \end{bmatrix} = 149$$

Equations A.21 and A.23 are employed to determine the VC and the parameter standard deviation:

$$VC(B^*) = s^2 (X^T \cdot X)^{-1} = \frac{149}{6-2} \begin{bmatrix} \dfrac{33}{70} & -\dfrac{2}{25} \\ -\dfrac{2}{25} & \dfrac{3}{280} \end{bmatrix} = \begin{bmatrix} 17.56 & -2.1285 \\ -2.1285 & 0.3991 \end{bmatrix}$$

$$\sigma_0 = \sqrt{17.56} = 4.1904$$

$$\sigma_1 = \sqrt{0.399} = 0.6317$$

The value of the t distribution for a 95% $(100 \cdot (1 - \alpha))$ confidence and four degrees of freedom can be found in any statistic book, such as Montgomery and Runger [2]:

$$t\left(n - p, \frac{\alpha}{2}\right) = t(4, 0.025) = 2.776$$

The CI for each parameter is calculated according to Equation A.24:

$$b_o^* = 0.5 \pm 4.1904 \cdot 2.776 \qquad \text{or} \qquad -11.13 \le b_o^* \le 12.13$$
$$b_1^* = 75.75 \pm 0.63174 \cdot 2.776 \qquad \text{or} \qquad 73.99 \le b_i^* \le 77.50$$

These results show that there is great uncertainty in the value determined for b_0, unlike with b_1, indicating that b_0 is a dummy parameter. Therefore, it would be advisable to use a simpler model for p versus t, such as: $g = b_1 \cdot t$. However, for the solution of Problem 2.1 this is not an issue and the fit to a straight line is good enough. The values of R^2 and R_{adj}^2 are calculated according to Equations A.25 and A.27. Taking:

$$Y = \begin{bmatrix} 0 \\ 150 \\ 307 \\ 461 \\ 597 \\ 912 \end{bmatrix}$$

the mean of the observed values is:

$$\bar{Y} = \frac{\sum_{i=1}^{6} y_i}{6} = \frac{(0 + 150 + 307 + 461 + 597 + 912)}{6} = 404.5$$

The total variability is given by:

$$\sum_{i=1}^{n}(y_i - \bar{Y})^2 = (0 - 404.5)^2(150 - 404.5)^2 + (307 - 404.5)^2 + (461 - 404.5)^2$$

$$= 535701.5 + (597 - 404.5)^2 + (912 - 404.5)^2$$

and the values of R^2 and R_{adj}^2 are:

$$R^2 = 1 - \frac{S(B^*)}{\sum_{i=1}^{n}(y_i - \bar{Y})^2} = 1 - \frac{149}{535701.5} = 0.99972$$

$$R_{adj}^2 = 1 - \frac{(n-1) \cdot S(B^*)}{(n-p) \cdot \sum_{i=1}^{n}(y_i - \bar{Y})^2} = 1 - \frac{5 \cdot 149}{4 \cdot 535701.5} = 0.99965$$

The latter result indicates a good fit of the data to the model (straight line), explaining more than 99.9% of the variability in the experimental data.

A.2.2 Nonlinear Regression

Consider a set of n experimental observations, which depend on m independent variables. For instance, a set of 30 (n) measurements of initial reaction rates for an enzyme in the presence of different concentrations of substrate and inhibitor (m=2). Assume that the relationship between the response and the independent variables is represented by the nonlinear function f(X,B). In this function, X denotes the m independent variables and B the p parameters that are to be estimated. The square error or disturbance for the ith observation is given by:

$$\varepsilon_i^2 = (y_i - f(x_i, B))^2 \tag{A.28}$$

So, as in linear regression, the problem is to find the parameters that minimize the sum of the squares; that is:

$$\min_{B} S(B)$$

Subject to $$S(B) = \sum_{i=1}^{n}\varepsilon_i^2 = \sum_{i=1}^{n}(y_i - f(x_i, B))^2 \tag{A.29}$$

Usually, in order to solve this problem the nonlinear function f(X,B) is approximated using Taylor's series and thus linearized around a known value for each parameter (b_i) by taking only the first derivative of f(X,B) with respect to each parameter B_i. In other words, it is assumed that f(X,B) is a linear function for a B' close to B:

$$f(X, B') \approx f(X, B) + \sum_{i=0}^{p-1} \left(\frac{\partial f(X, B)}{\partial B_i} \cdot (b'_i - b_i) \right) \tag{A.30}$$

with:

$$B = \begin{bmatrix} b_0 \\ \cdot \\ \cdot \\ \cdot \\ b_{p-1} \end{bmatrix} \qquad B' = \begin{bmatrix} b'_0 \\ \cdot \\ \cdot \\ \cdot \\ b'_{p-1} \end{bmatrix} \tag{A.31}$$

Given this, the error vector is approximately:

$$\varepsilon \approx \begin{bmatrix} y_1 \\ \cdot \\ \cdot \\ \cdot \\ y_n \end{bmatrix} - \left(\begin{bmatrix} f(x_1, B) \\ \cdot \\ \cdot \\ \cdot \\ f(x_n, B) \end{bmatrix} + \begin{bmatrix} \frac{\partial f(x_1, B)}{\partial b_0} & \cdot & \cdot & \frac{\partial f(x_1, B)}{\partial b_{p-1}} \\ \cdot & \cdot & \cdot & \cdot \\ \cdot & \cdot & \cdot & \cdot \\ \frac{\partial f(x_n, B)}{\partial b_0} & \cdot & \cdot & \frac{\partial f(x_n, B)}{db_{p-1}} \end{bmatrix} \cdot \begin{bmatrix} b'_0 - b_0 \\ \cdot \\ \cdot \\ \cdot \\ b'_{p-1} - b_{p-1} \end{bmatrix} \right) \tag{A.32}$$

Z, J, and δ are defined as:

$$Z = \begin{bmatrix} y_1 \\ \cdot \\ \cdot \\ \cdot \\ y_n \end{bmatrix} - \begin{bmatrix} f(x_1, B) \\ \cdot \\ \cdot \\ \cdot \\ f(x_n, B) \end{bmatrix} \tag{A.33}$$

$$J = \begin{bmatrix} \frac{\partial f(x_1, B)}{\partial b_0} & \cdot & \cdot & \cdot & \frac{\partial f(x_1, B)}{\partial b_{p-1}} \\ \cdot & \cdot & \cdot & \cdot \\ \cdot & \cdot & \cdot & \cdot \\ \frac{\partial f(x_n, B)}{\partial b_0} & \cdot & \cdot & \cdot & \frac{\partial f(x_n, B)}{\partial b_{p-1}} \end{bmatrix} \tag{A.34}$$

and:

$$\delta = \begin{bmatrix} b'_0 - b_0 \\ \cdot \\ \cdot \\ \cdot \\ b'_{p-1} - b_{p-1} \end{bmatrix} \tag{A.35}$$

where J is the Jacobian matrix. Equation A.32 is rewritten as:

$$\varepsilon \approx Z - J \cdot \delta \tag{A.36}$$

Replacing this into Equation A.29 gives:

$$\min_{B} S(B)$$

$$\text{Subject to} \qquad S(B) = \sum_{i=1}^{n} \varepsilon_i^2 = [Z - J \cdot \delta]^T \cdot [Z - J \cdot \delta] \tag{A.37}$$

Observe that Equation A.37 is analogous to Equation A.18. Applying the first derivative criterion to minimize S(B) with respect to δ gives:

$$\delta = (J^T \cdot J)^{-1} \cdot J^T \cdot Z \qquad \text{or} \qquad B' = B + (J^T \cdot J)^{-1} \cdot J^T \cdot Z \tag{A.38}$$

Hence, the nonlinear regression problem can be solved by employing linear regression in an iterative way, meaning that from the initial values of the parameters (B), better values (B') will be found in the next iteration, and so on. Usually, the iteration is stopped when the improvement in the value estimated for the parameter is insignificant; for example, $\max(|\delta_i|) < 10^{-6}$.

The CI for the estimated parameter can be calculated using a linear approximation. Thus, the VC for this case corresponds to:

$$VC(B^*) = VC \begin{pmatrix} b_o^* \\ \cdot \\ \cdot \\ \cdot \\ b_{p-1}^* \end{pmatrix} = \begin{bmatrix} V(b_o^*) & \cdot & \cdot & \cdot & C(b_o^*, b_{p-1}^*) \\ \cdot & & & & \cdot \\ \cdot & & & & \cdot \\ \cdot & & & & \cdot \\ C(b_o^*, b_{p-1}^*) & \cdot & \cdot & \cdot & V(b_{p-1}^*) \end{bmatrix} = s^2 \cdot (J^T \cdot J)^{-1} \tag{A.39}$$

with:

$$s^2 = \frac{S(B^*)}{n - p} \tag{A.40}$$

The standard deviation and CI are given by:

$$\sigma_i = \sqrt{V(b_i^*)} \tag{A.41}$$

$$b_i^* \pm \sigma_i \cdot t\left(n - p, \frac{\alpha}{2}\right) \tag{A.42}$$

The values of R^2 and R^2_{adj} can thus be determined by Equations A.25 and A.27.

Example A.3

Penicillin acylase (EC 3.5.1.11) catalyzes the deacylation of penicillin G. The effect of substrate concentration on the reaction rate of penicillin G hydrolysis is studied for the penicillin acylase of *Penicillium notatum*, with the follows results being obtained:

Pen G (mM)	0.1	0.2	0.4	0.8	1	2	10	50	100	150	200	300	400
v (μM/min)	0.2	0.38	0.74	1.37	1.65	2.78	5	2.77	1.65	1.17	0.91	0.62	0.46

where v is the initial rate of penicillin G hydrolysis.

It can be seen that the enzyme is subjected to inhibition at high concentrations of penicillin G. It therefore seems reasonable to assume uncompetitive inhibition by penicillin G, in mechanistic terms; that is:

$$E + Pen \underset{}{\overset{K_M}{\rightleftharpoons}} EPen \overset{k}{\longrightarrow} PAA + 6\text{-}APA + E$$

$$+$$
$$Pen$$

$$K_S \updownarrow$$

$$EPenPen$$

The constants of dissociation are defined as:

$$K_M = \frac{[E] \cdot [Pen]}{[EPen]} \qquad K_S = \frac{[EPen] \cdot [Pen]}{[EPenPen]}$$

where E is penicillin acylase, Pen is penicillin G, PAA is phenylacetic acid, and 6-APA is 6-aminopenicillanic acid (square brackets represent the molar concentration of each species). Hence, the reaction rate is given by:

$$v = \frac{V_{max} \cdot [Pen]}{K_M + [Pen] + \dfrac{[Pen]^2}{K_S}} \tag{A.43}$$

Inverting Equation A.43 gives:

$$\frac{1}{v} = \frac{1}{V_{max}} + \frac{K_M}{V_{max} \cdot [Pen]} + \frac{[Pen]}{V_{max} \cdot K_S} \tag{A.44}$$

For low [Pen] ([Pen] $\ll K_M$), Equation A.44 becomes:

$$\frac{1}{v} \approx \frac{1}{V_{max}} + \left(\frac{K_M}{V_{max}}\right) \cdot \frac{1}{[Pen]} \tag{A.45}$$

and for high [Pen] ([Pen] $\gg K_M$):

$$\frac{1}{v} \approx \frac{1}{V_{max}} + \left(\frac{1}{V_{max} \cdot K_S}\right) \cdot [Pen] \tag{A.46}$$

Plotting the reaction rate versus [Pen], two zones are clearly observed, as shown in Figure A.2. In the first, v increases with the substrate (closed symbols), while in the second it decreases (open symbols).

Given that the first zone in the graph, where [Pen] is low $(0.1 - 10.0\,\text{mM})$, is represented by Equation A.45, the values of V_{max} and K_M are estimated by fitting the data to this equation, as shown in Figure A.3. Equation A.45 indicates that the reciprocal of the intercept is equal to the V_{max} value; thus, $V_{max} = 0.1326^{-1} = 7.54\,\mu\text{M} \cdot \text{min}^{-1}$ and the value of K_M corresponds to the quotient between the slope and the intercept, so that $K_M = 0.489/0.1326 = 3.69\,\text{mM}$.

On the other hand, given that the behavior of the reaction rate is described by Equation A.46 for [Pen] $> 20\,\text{mM}$, the inhibition constant K_S is determined by fitting the data to this equation, as shown in Figure A.4. From Equation A.46, K_S is given by the multiplication of the inverse of the slope and the intercept. Thus $K_S = 0.1149 \cdot 0.005^{-1} = 22.98\,\text{mM}$ and the values of the kinetic constants estimated by linear regression are:

$$V_{max} = 7.54\,\mu\text{M} \cdot \text{min}^{-1}; \quad K_M = 3.69\,\mu\text{M}; \quad K_S = 22.98\,\mu\text{M}.$$

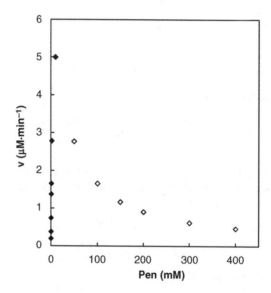

Figure A.2 *Initial rates of hydrolysis reaction at the different penicillin G concentrations experimentally determined.*

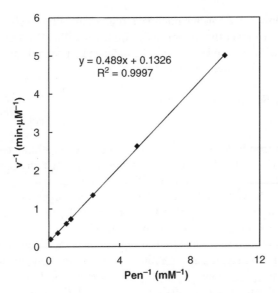

Figure A.3 Double-reciprocal plot at low penicillin concentrations. Symbol represents experimental data.

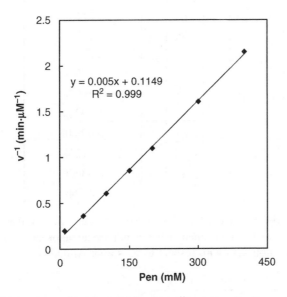

Figure A.4 Double-reciprocal plot at high penicillin concentrations. Symbol represents experimental data.

The problem is now solved using nonlinear regression. Previously, the kinetic model corresponding to uncompetitive inhibition at high penicillin G concentration had to be validated. In such a kinetic model, [Pen] is the independent variable and V_{max}, K_M, and K_S are the parameters to be fitted. Accordingly:

$$f(X, B) = f(Pen, V_{max}, K_M, K_S) = v = \frac{V_{max} \cdot [Pen]}{K_M + [Pen] + \dfrac{[Pen]^2}{K_S}} \qquad (A.47)$$

In nonlinear regression it is not necessary to separate the experimental observations into two groups ($s \ll K_M$ and $s \gg K_M$) as is required for linear regression, in which linear relationships between the reciprocals of initial reaction rate and substrate concentration occur only under such extreme conditions. The whole set of experimental data will now be used. This problem of nonlinear fitting thus has one independent variable (m), thirteen observations (n), and three parameters to be estimated (p).

One of the most critical steps in nonlinear regression is the election of the initial values for each parameter. An inappropriate election may cause the computational routine not to converge or to converge on values without physical meaning. In this exercise, the values previously obtained using linear regression ($V_{max} = 7.54 \ \mu M \cdot min^{-1}$, $K_M = 3.69 \ mM$, and $K_S = 22.98 \ mM$) will be employed as the initial values. Vector Z is calculated according to Equation A.33 for the initial values of the parameters:

$$
Z = \begin{bmatrix} y_1 \\ \cdot \\ \cdot \\ \cdot \\ y_n \end{bmatrix} - \begin{bmatrix} f(x_1, B) \\ \cdot \\ \cdot \\ \cdot \\ f(x_n, B) \end{bmatrix} = \begin{bmatrix} 0.2 \\ 0.38 \\ 0.74 \\ 1.37 \\ 1.65 \\ 2.78 \\ 5 \\ 2.77 \\ 1.65 \\ 1.17 \\ 0.91 \\ 0.62 \\ 0.46 \end{bmatrix} - \begin{bmatrix} \dfrac{7.54 \cdot 0.1}{3.69 + 0.1 + \dfrac{0.1^2}{22.98}} \\ 0.38748 \\ 0.73615 \\ 1.33515 \\ 1.5929 \\ 2.5716 \\ 4.17923 \\ 2.32028 \\ 1.39927 \\ 0.99840 \\ 0.77558 \\ 0.53600 \\ 0.40943 \end{bmatrix} = \begin{bmatrix} -0.00108 \\ -0.00748 \\ 0.00384 \\ 0.03485 \\ 0.05710 \\ 0.20840 \\ 0.82077 \\ 0.44971 \\ 0.25072 \\ 0.17159 \\ 0.13441 \\ 0.08399 \\ 0.05056 \end{bmatrix}
$$

The next step is to determine the Jacobian matrix. The derivative of f(xi,B) with respect to the parameters is easily determined using a numerical approach, given that the first derivative of a function g(x) with respect to X is approximately:

$$\frac{\partial g(x, y)}{\partial x} = \frac{g(x + \Delta x, y) - g(x, y)}{\Delta x}$$

Usually, values of ΔX between 10^{-6} and 10^{-4} are suitable. Using the approximation to the derivative given in this equation, J can be readily calculated. For example, setting $\Delta X = 10^{-5}$, the derivatives of $f(X,B)$ evaluated at the first concentration of penicillin G ($X_1 = 0.1$) with respect to each parameter are:

$$\frac{\partial f(x_1, B)}{\partial V_{max}} \approx \frac{\dfrac{(7.54 + 10^{-5}) \cdot 0.1}{3.69 + 0.1 + \dfrac{0.1^2}{22.98}} - \dfrac{7.54 \cdot 0.1}{3.69 + 0.1 + \dfrac{0.1^2}{22.98}}}{10^{-5}} = 0.02638$$

$$\frac{\partial f(x_1, B)}{\partial K_M} \approx \frac{\dfrac{7.54 \cdot 0.1}{(3.69 + 10^{-5}) + 0.1 + \dfrac{0.1^2}{22.98}} - \dfrac{7.54 \cdot 0.1}{3.69 + 0.1 + \dfrac{0.1^2}{22.98}}}{10^{-5}} = -0.05247$$

$$\frac{\partial f(x_1, B)}{\partial K_S} \approx \frac{\dfrac{7.54 \cdot 0.1}{3.69 + 0.1 + \dfrac{0.1^2}{(22.98 + 10^{-5})}} - \dfrac{7.54 \cdot 0.1}{3.69 + 0.1 + \dfrac{0.1^2}{(22.98)}}}{10^{-5}} = 9.93785 \cdot 10^{-07}$$

By applying the same procedure for the other initial concentrations of penicillin G, matrix J is determined. Thus, J, $J^T \cdot J$, and $(J^T \cdot J)^{-1}$ are:

$$J = \begin{bmatrix} \dfrac{\partial f(x_1, B)}{\partial V_{max}} & \dfrac{\partial f(x_1, B)}{\partial K_M} & \dfrac{\partial f(x_1, B)}{\partial K_S} \\ & \cdot & \\ \cdot & \cdot & \cdot \\ & \cdot & \\ \cdot & \cdot & \cdot \\ \dfrac{\partial f(x_n, B)}{\partial V_{max}} & \dfrac{\partial f(x_n, B)}{\partial K_M} & \dfrac{\partial f(x_n, B)}{\partial K_S} \end{bmatrix} = \begin{bmatrix} 0.02638 & -0.05248 & 9.94 \cdot 10^{-7} \\ 0.05139 & -0.09957 & 7.54 \cdot 10^{-6} \\ 0.09763 & -0.17968 & 5.44 \cdot 10^{-5} \\ 0.17708 & -0.29553 & 0.000358 \\ 0.21126 & -0.33651 & 0.000637 \\ 0.34106 & -0.43853 & 0.003322 \\ 0.55427 & -0.23164 & 0.043865 \\ 0.30773 & -0.01428 & 0.067605 \\ 0.18558 & -0.0026 & 0.049174 \\ 0.13242 & -0.00088 & 0.037552 \\ 0.10286 & -0.0004 & 0.030215 \\ 0.07109 & -0.00013 & 0.021646 \\ 0.05430 & -5.6 \cdot 10^{-7} & 0.016841 \end{bmatrix}$$

Note that V_{max}, K_M, and K_S have been set as b_0, b_1, and b_2, respectively.

$$J^T \cdot J = \begin{bmatrix} 0.67765 & -0.43047 & 0.066114 \\ -0.43047 & 0.491713 & -0.01309 \\ 0.066114 & -0.01309 & 0.012 \end{bmatrix}$$

and:

$$
\left(J^T \cdot J\right)^{-1} = \begin{bmatrix} 22.51224 & 16.8971 & -105.602 \\ 16.8971 & 14.7706 & -76.9769 \\ -105.602 & -76.9769 & 581.1924 \end{bmatrix}
$$

Finally, the values of δ and B' are calculated using Equation A.38:

$$
\delta = \left(J^T \cdot J\right)^{-1} \cdot J^T \cdot Z = \begin{bmatrix} 2.27161 \\ 1.24997 \\ -3.43095 \end{bmatrix} \qquad B' = B + \delta = \begin{bmatrix} 7.54 \\ 3.69 \\ 22.98 \end{bmatrix} + \begin{bmatrix} 2.27161 \\ 1.24997 \\ -3.43095 \end{bmatrix}
$$

$$
= \begin{bmatrix} 9.81161 \\ 4.93997 \\ 19.549 \end{bmatrix}
$$

Therefore, the estimated values for the parameters in the first iteration are: $V_{max} = 9.81 \ \mu M \cdot min^{-1}$, $K_M = 4.94 \ mM$, and $K_S = 19.55 \ mM$. The maximum absolute value of the elements in vector δ $(\max (|\delta_i|))$ is 3.43095. It is therefore advisable to keep on iterating until a lower value is obtained, such as $\max (|\delta|) < 10^{-6}$. The results of one such procedure are as follows:

Iteration	Parameter Value			max $(|\delta|)$	S(B)
	V_{max}	K_M	K_S		
Initial values	7.54	3.69	22.98	–	1.0438879
1	9.811613	4.939971	19.549047	3.430952	0.0277054
2	10.026808	5.019411	19.865989	0.3169421	0.0002993
3	10.029232	5.020131	19.854457	0.0115324	0.00029849
4	10.029233	5.020132	19.854458	$1.191 \cdot 10^{-6}$	0.00029849
5	10.029233	5.020132	19.854458	$1.555 \cdot 10^{-9}$	0.00029849

In the fifth iteration, the $\max (|\delta|) < 10^{-6}$ criterion is achieved and the routine is stopped. Therefore, the estimated values for V_{max}, K_M, and K_S using nonlinear regression are $10.03 \ \mu M \cdot min^{-1}$, $5.02 \ mM$, and $19.85 \ mM$, respectively. For the purpose of computer calculation, a mantissa of 8–12 decimal positions is suitable and advisable for avoiding the truncation error but not for expressing the final results. For the fifth iteration, the value of VC(B) is:

$$
VC(B^*) = s^2 \cdot \left(J^T \cdot J\right)^{-1} = \frac{0.00029849}{13 - 3} \cdot \begin{bmatrix} 38.40697 & 28.46253 & -111.14721 \\ 28.46253 & 23.95723 & -80.05889 \\ -111.14721 & -80.05889 & 365.30459 \end{bmatrix}
$$

$$
= \begin{bmatrix} 0.001146 & 0.00085 & -0.00332 \\ 0.00085 & 0.000715 & -0.00239 \\ -0.00332 & -0.00239 & 0.010904 \end{bmatrix}
$$

The value of the t distribution for a 95% CI and 10 $(n-p)$ degrees is obtained from Montgomery and Runger [2]:

$$t\left(n-p,\frac{\alpha}{2}\right) = t(10, 0.025) = 2.228$$

The 95% CI for each parameter is calculated accordingly:

$$V_{max}^* = 10.03 \pm \sqrt{0.001146} \cdot 2.228 = 10.03 \pm 0.07543$$
$$K_M^* = 5.02 \pm \sqrt{0.000715} \cdot 2.228 = 5.02 \pm 0.05958$$
$$K_S^* = 19.85 \pm \sqrt{0.010904} \cdot 2.228 = 19.85 \pm 0.23653$$

Finally, the values of R^2 and R_{adj}^2 are calculated using Equations A.23 and A.24:

$$R^2 = 1 - \frac{S(B^*)}{\sum_{i=1}^{n}(y_i - \bar{Y})^2} = 1 - \frac{0.00029849}{21.381012} = 0.999986$$

$$R_{adj}^2 = 1 - \frac{(n-1) \cdot S(B^*)}{(n-p) \cdot \sum_{i=1}^{n}(y_i - \bar{Y})^2} = 1 - \frac{(13-1) \cdot 0.00029849}{(13-3) \cdot 21.3810129} = 0.999983$$

The values of V_{max}, K_M, and K_S obtained by nonlinear regression in this case (10.03 $\mu M \cdot min^{-1}$, 5.02 mM, and 19.85 mM, respectively) differ significantly from those obtained by linear regression (7.54 $\mu M \cdot min^{-1}$, 3.69 mM, and 22.98 mM, respectively). Figure A.5 shows a comparison of the fits obtained by linear and nonlinear regression:

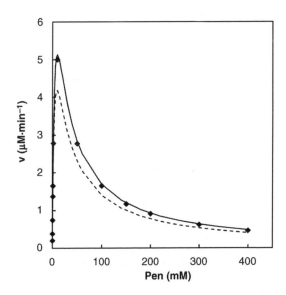

Figure A.5 *Comparison of the fit obtained using linear and nonlinear regression.* ◆ *: experimental data. Dashed line: linear regression; solid line: nonlinear regression.*

nonlinear regression allows in this case a better fit than linear regression. As explained before, the parameter estimation for uncompetitive inhibition produced by linear regression considers only the two extreme regions of the substrate concentration range (s $\ll K_M$ and s $\gg K_M$), which are precisely the more prone to error, so nonlinear regression should be used in this case.

A.3 Solving Ordinary Differential Equations

The solution of ordinary differential equations is a broad field in mathematics, impossible to cover in its full extension in this appendix. However, the most useful and simple techniques for the analytical and numerical solution of differential equations will be presented here. This appendix is focused on the solution of ordinary equations, meaning that only problems involving exact derivatives will be considered; this covers most of the differential equations that arise in the modeling of enzyme biocatalysis.

A.3.1 Solving First-Order Ordinary Differential Equations by the Separation of Variables

The order of a differential equation is given by its highest derivative, so a first-order equation is one involving only first derivatives.

Many first-order ordinary differential equations (FO-ODEs) that appear in bioprocesses can be solved using the method of separation of variables. Consider the general type:

$$\frac{dy}{dx} = f(x, y) \tag{A.48}$$

Sometimes it is possible to rearrange this as follows:

$$\frac{dy}{dx} = f(x, y) = g(x) \cdot h(y) \tag{A.49}$$

$$\frac{1}{h(y)} \cdot dy = \cdot g(x) \cdot dx \tag{A.50}$$

The separation of f(x,y) into two functions, g(x) and h(x), that depend on only one variable allows Equation A.48 to be solved by integration:

$$\int \frac{1}{h(y)} \cdot dy = \cdot \int g(x) \cdot dx + C \tag{A.51}$$

$$\int_{y_0}^{y_f} \frac{1}{h(y)} \cdot dy = \int_{x_0}^{x_f} g(x) \cdot dx \tag{A.52}$$

where C is a constant.

Example A.4

As shown in Section 2.7.2, enzyme inactivation is usually represented by a single-stage irreversible mechanism:

$$E \xrightarrow{k_D} E_{inactive}$$

The variation in the active form of the enzyme (E) with time is represented by the differential equation:

$$\frac{de}{dt} = -k_D \cdot e \tag{2.38}$$

Given that at time zero (t_0) the initial concentration of active enzyme is e_0, the concentration of active enzyme e at any given time t is obtained by integration of Equation A.53 under these limits:

$$\int_{e_0}^{e} \frac{1}{e} \cdot de = - \int_{t_0}^{t} k_D \cdot dt \tag{A.53}$$

$$\ln\left(\frac{e}{e_0}\right) = -k_D \cdot t \tag{A.54}$$

From Equations A.54 and 2.39 we obtain:

$$\frac{e}{e_0} = \exp(-k_D \cdot t) \tag{2.39}$$

The usefulness of the separation of variables is not limited to the solution of simple problems; for example, the mathematical expressions that describe the behavior of BSTR and CPBR in Chapter 4 were determined using this method.

A.3.2 Solving First-Order Ordinary Differential Equations Using an Integration Factor

This method is quite helpful in solving FO-ODEs that can be represented as:

$$\frac{dy}{dx} + p(x) \cdot y = q(x) \tag{A.55}$$

Multiplying Equation A.55 by the integration factor $e^{\int p(x)dx}$ gives:

$$e^{\int p(x)dx} \cdot \frac{dy}{dx} + e^{\int p(x)dx} \cdot p(x) \cdot y = e^{\int p(x)dx} \cdot q(x) \qquad (A.56)$$

Note that the right-hand side of Equation A.56 can be written in compact form using the chain rule for the product of two functions:

$$\frac{d}{dx}\left(y \cdot e^{\int p(x)dx}\right) = e^{\int p(x)dx} \cdot \frac{dy}{dx} + p(x) \cdot e^{\int p(x)dx} \cdot y \qquad (A.57)$$

Replacing Equation A.57 into Equation A.56 gives:

$$\frac{d}{dx}\left(y \cdot e^{\int p(x)dx}\right) = e^{\int p(x)dx} \cdot q(x) \qquad (A.58)$$

and integrating Equation A.58 gives:

$$y \cdot e^{\int p(x)dx} = \int e^{\int p(x)dx} \cdot q(x)dx + C \qquad (A.59)$$

Finally, by solving y in Equation A.59 the primitive form of Equation A.55 is obtained:

$$y = e^{-\int p(x)dx}\left(\int e^{\int p(x)dx} \cdot q(x)dx + C\right) \qquad (A.60)$$

The use of this method is exemplified in Section A.3.3 for the determination of the substrate concentration profile inside a catalytic slab.

A.3.3   Solving Second- and Higher-Order Linear Homogeneous Differential Equations with Constant Coefficients Using their Characteristic Equations

This technique can be used to solve differential equations with the following structure:

$$\frac{d^n y}{dx^n} + P_1 \cdot \frac{d^{n-1}y}{dx^{n-1}} + \ldots + P_{n-1} \cdot \frac{dy}{dx} + P_n \cdot y = 0 \qquad (A.61)$$

where P_i are constant. By defining operator D as d/dx, and by extension D^2 as d^2/dx^2 and so on, Equation A.61 becomes:

$$D^n \cdot y + P_1 \cdot D^{n-1} \cdot y + \ldots + P_{n-1} \cdot D \cdot y + P_n \cdot y = 0 \qquad (A.62)$$

Note that $D^n \cdot y$ means $d^n y/dx^n$. Equation A.62 can be treated as a polynomial equation with D as the independent variable and rearranged as indicated in Equation A.63:

$$(D - m_1) \cdot (D - m_2) \ldots (D - m_{n-1}) \cdot (D - m_n) \cdot y = 0 \qquad (A.63)$$

where the values of m_i correspond to the roots of the characteristic equation of (A.63); that is:

$$T^n + P_1 \cdot T^{n-1} + \ldots + P_{n-1} \cdot T + P_n = (T - m_1) \cdot (T - m_2) \ldots (T - m_{n-1}) \cdot (T - m_n)$$
$$= 0$$

(A.64)

The general solution of Equation A.61 has the following form:

$$y = C_1 \cdot e^{m_1 \cdot x} + C_2 \cdot e^{m_2 \cdot x} + \ldots + C_{n-1} \cdot e^{m_{n-1} \cdot x} + C_n \cdot e^{m_n \cdot x}$$

(A.65)

where C_i are constant and their value must be determined by evaluation of Equation A.65 according to the initial and boundary conditions of Equation A.61.

Example A.5

Equation 3.35 describes the profile of a substrate inside a biocatalyst slab, assuming first-order kinetics, as presented in Section 3.3.3:

$$\frac{d^2\beta}{dz^2} - \Phi^2 \cdot \beta = 0$$

(3.35)

Using the operator D, Equation 3.35 is expressed as:

$$(D^2 - \Phi^2) \cdot \beta = 0$$

(A.66)

This can be solved easily with its characteristic equation:

$$(T^2 - \Phi^2) = 0$$

(A.67)

Noting that Equation A.67 corresponds to the difference of two squares, it can be factorized as:

$$(T - \Phi) \cdot (T + \Phi) = 0$$

(A.68)

This allows Equation A.66 to be written in its factorized form:

$$(D - \Phi) \cdot (D + \Phi) \cdot \beta = 0$$

(A.69)

U is defined in such a way that:

$$(D + \Phi) \cdot \beta = U$$

(A.70)

Replacing Equation A.70 into Equation A.69 gives:

$$(D - \Phi) \cdot U = 0$$

(A.71)

$$\frac{dU}{dz} = \Phi \cdot U$$

(A.72)

Solving Equation A.72 by separation of variables gives:

$$U = e^C \cdot e^{\Phi \cdot Z} = C_1 \cdot e^{\Phi \cdot Z} \tag{A.73}$$

Replacing Equation A.73 into Equation A.70 gives:

$$(D + \Phi) \cdot \beta = \frac{d\beta}{dz} + \Phi \cdot \beta = C_1 \cdot e^{\Phi \cdot Z} \tag{A.74}$$

which is solved using the integration factor method.

Multiplying Equation A.74 by $e^{\Phi z}$ (corresponding to $e^{\int \Phi dz}$) gives:

$$\frac{d}{dz}(\beta \cdot e^{\Phi \cdot z}) = C_1 \cdot e^{2 \cdot \Phi \cdot z} \tag{A.75}$$

Integrating, and then solving for β, gives:

$$\beta(z) = \frac{C_1}{2 \cdot \Phi} \cdot e^{\Phi \cdot z} + C_2 \cdot e^{-\Phi \cdot z} \tag{A.76}$$

Finally, by evaluating Equation A.76 under the boundary conditions established in Equation 3.21, the following linear equation system is obtained:

$$\text{For } \beta(0) \qquad \beta_0 = \frac{C_1}{2 \cdot \Phi} + C_2 \tag{A.77}$$

$$\text{and for } \beta'(0.5) \qquad 0 = \frac{C_1}{2} \cdot e^{0.5 \cdot \Phi} - \Phi \cdot C_2 \cdot e^{-0.5 \cdot \Phi} \tag{A.78}$$

Solving Equations A.77 and A.78 gives:

$$C_1 = \frac{2 \cdot \Phi \cdot \beta_0 \cdot e^{-0.5 \cdot \Phi}}{e^{0.5 \cdot \Phi} + e^{-0.5 \cdot \Phi}} \tag{A.79}$$

$$C_2 = \frac{\beta_0 \cdot e^{0.5 \cdot \Phi}}{e^{0.5 \cdot \Phi} + e^{-0.5 \cdot \Phi}} \tag{A.80}$$

Replacing Equations A.79 and A.80 into Equation A.76 and reordering gives:

$$\beta(z) = \frac{2 \cdot \beta_0}{(e^{0.5 \cdot \Phi} + e^{-0.5 \cdot \Phi})} \cdot \frac{\left(e^{\Phi \cdot (z-0.5)} + e^{-\Phi \cdot (z-0.5)}\right)}{2} \tag{A.81}$$

Now, using the identity $\cosh(x) = 0.5 \cdot (e^x + e^{-x})$, we get:

$$\beta(z) = \beta_0 \frac{\cosh(\Phi \cdot (z - 0.5))}{\cosh(0.5 \cdot \Phi))} \tag{A.82}$$

Here the solution of a linear homogeneous differential equation is exemplified for a second-order equation, but the procedure can easily be extended for higher orders.

A.3.4 Solving Second- and Higher-Order Linear Homogeneous Differential Equations with Variable Coefficients

There is no general procedure available by which to solve this kind of ordinary differential equations (ODE). However, two main strategies can be proposed: the use of Taylor series and the change of a variable to transform the original ODE into one with constant coefficients. We will focus on the latter procedure, since it enables the solution of most problems in this field, including the one represented in Equation 3.36, whose resolution describes the substrate profile inside a spherical catalyst catalyzing a reaction following first-order kinetics:

$$\frac{d^2\beta}{d\rho^2} + \frac{2}{\rho}\frac{d\beta}{d\rho} - 9 \cdot \Phi^2 \cdot \beta = 0 \tag{3.36}$$

Observe that the term $2/\rho$ represents a variable coefficient in this equation with the following boundary conditions:

$$\beta|_{\rho=1} = \beta_0 \qquad \frac{d\beta}{d\rho}\bigg|_{\rho=0} = 0$$

Making $\beta = U \cdot \rho^{-1}$ and then applying the chain rule gives:

$$\frac{d\beta}{d\rho} = \left(\frac{dU}{d\rho}\right) \cdot \left(\frac{1}{\rho}\right) - U \cdot \left(\frac{1}{\rho^2}\right) \tag{A.83}$$

$$\frac{d^2\beta}{d\rho^2} = \left(\frac{d^2U}{d\rho^2}\right) \cdot \left(\frac{1}{\rho}\right) - \left(\frac{dU}{d\rho}\right) \cdot \left(\frac{1}{\rho^2}\right) - \left(\frac{dU}{d\rho}\right) \cdot \left(\frac{1}{\rho^2}\right) + 2 \cdot U \cdot \left(\frac{1}{\rho^3}\right) \tag{A.84}$$

By substituting Equations A.83 and A.84 into Equation 3.36 we get:

$$\left(\frac{1}{\rho}\right) \cdot \left(\frac{d^2U}{d\rho^2}\right) - \cdot \left(\frac{2}{\rho^2}\right) \cdot \left(\frac{dU}{d\rho}\right) + \left(\frac{2U}{\rho^3}\right) + \frac{2}{\rho} \cdot \left(\left(\frac{1}{\rho}\right) \cdot \left(\frac{dU}{d\rho}\right) - \left(\frac{U}{\rho^2}\right)\right) - 9\Phi^2 \cdot \beta$$
$$= 0$$

$$\tag{A.85}$$

Resulting in:

$$\frac{d^2U}{d\rho^2} - 9 \cdot \Phi^2 \cdot U = 0 \tag{A.86}$$

Equation A.86 can be solved easily by following the procedure given in Section A.3.3:

$$U = C_1 \cdot e^{-3 \cdot \Phi \cdot \rho} + C_2 \cdot e^{3 \cdot \Phi \cdot \rho} \tag{A.87}$$

By definition $U = \beta \cdot \rho$, this becomes:

$$\beta = \left(\frac{1}{\rho}\right) \cdot \left(C_1 \cdot e^{-3 \cdot \Phi \cdot \rho} + C_2 \cdot e^{3 \cdot \Phi \cdot \rho}\right) \tag{A.88}$$

By evaluating Equation A.88 under the boundary conditions and using the identity $\sinh(x) = 0.5 \cdot (e^x - e^{-x})$, we obtain:

$$\beta = \beta_0 \cdot \frac{\sinh[3\Phi \cdot \rho]}{\rho \cdot \sinh[3\Phi]} \tag{3.39}$$

A.4   Numerical Methods for Solving Differential Equations

In bioprocesses, differential equations are usually employed to describe the variation of a property with respect to an independent variable. However, interest is often focused on the value of the property at the end of the process, and for that purpose it is necessary to solve the differential equations, which is not always an easy task as they can be too complex or have no analytical solution. Numerical methods allow these problems to be overcome, offering various procedures by which to determine the value of the property for a given value of the independent variable. This appendix presents the simplest techniques for solving differential equations numerically. The reader who often deals with this kind of problem is recommended to consult a dedicated textbook, such as Chapra and Canale [1].

Two of the classic methods for solving ODEs with defined initial conditions—Euler and Runge–Kutta—will be presented first, followed by some techniques for solving ODEs defined by a boundary condition, including the finite-difference method in particular.

A.4.1   The Euler Method

Consider the following problem to be solved numerically:

$$\frac{dy}{dx} = f(x, y) \tag{A.48}$$

From Equation A.3, it is possible to linearize y around a small interval using Taylor's series $[x_0, x]$. Replacing Equation A.48 into Equation A.3 gives:

$$y(x) = y(x_0) + f(x_0, y_0) \cdot (x - x_0) \tag{A.89}$$

In this manner, the trajectory of y can be calculated in an iterative way using a small integration step (h) from an initial value.

$$y(x_{i+1}) = y(x_i) + f(x_i, y_i) \cdot h \tag{A.90}$$

and
$$x_{i+1} = x_i + h \tag{A.91}$$

The main advantage of this method is its simplicity. Nevertheless, it has important drawbacks: the global error in its prediction is highly dependent on h and the method is conditionally stable. The error can be decreased by reducing h. However, values of h lower than 10^{-8} are not recommended in most cases. In the same way, the method improves its stability at lower values of h.

A.4.2 The Fourth-Order Runge–Kutta Method

This method is one of the most popular numerical techniques for solving ODEs; in fact, in several software packages it is considered the default option, due to its low error and relative simplicity. There are many versions of this technique; the following is the most common or classic form:

$$y(x_{i+1}) = y(x_i) + \frac{1}{6} \cdot (k_1 + 2 \cdot k_2 + 2 \cdot k_3 + k_4) \cdot h \tag{A.92}$$

where:

$$k_1 = f(x_i, y_i) \tag{A.93}$$

$$k_2 = f\left(x_i + \frac{h}{2}, y_i + \frac{k_1 \cdot h}{2}\right) \tag{A.94}$$

$$k_3 = f\left(x_i + \frac{h}{2}, y_i + \frac{k_2 \cdot h}{2}\right) \tag{A.95}$$

$$k_4 = f(x_i + h, y_i + k_3 \cdot h) \tag{A.96}$$

A.4.3 The Finite-Difference Method

This method is based on the substitution of the derivatives by their corresponding finite-difference forms, transforming the problem into a set of algebraic equations to be solved. The finite-difference method is quite appropriate to solve differential equations subject to boundary conditions since they can be easily incorporated in the numerical procedures for the solution of the problem. The application of this technique is quite intuitive, and for this reason it will be explained using an example. The formulas for the finite differences of the first and second derivatives are listed in Table A.1. The centered finite-difference formulas offer a better estimation of the derivative value, so they should be preferred.

Table A.1 *Derivative estimation by finite difference.*

Derivative	Forward	Centered	Backward
$\dfrac{dy}{dx}$	$\dfrac{f(x_{i+1}) - f(x_i)}{h}$	$\dfrac{f(x_{i+1}) - f(x_{i-1})}{2h}$	$\dfrac{f(x_i) - f(x_{i-1})}{h}$
$\dfrac{d^2 y}{dx^2}$	$\dfrac{f(x_{i+2}) - 2 \cdot f(x_{i+1}) + f(x_i)}{h^2}$	$\dfrac{f(x_{i+1}) - 2 \cdot f(x_i) + f(x_{i-1})}{h^2}$	$\dfrac{f(x_i) - 2 \cdot f(x_{i-1}) + f(x_{i-2})}{h^2}$

Example A.6

Determine the substrate profile inside a biocatalytic slab. Assume that the enzyme follows Michaelis–Menten kinetics.

Resolution: The behavior of the dimensionless substrate concentration is given by Equation 3.21:

$$\frac{d^2\beta}{dz^2} - \Phi^2 \cdot \frac{\beta}{1+\beta} = 0 \tag{3.21}$$

where:

$$\beta|_{z=0} = \beta_0 \qquad \frac{d\beta}{dz}\bigg|_{z=0.5} = 0$$

The boundary condition at $z = 0.5$ is normally called the symmetry condition; it establishes that the substrate profile at one side of the slab is the mirror image of that at the other. It is thus not necessary to solve the problem for the whole interval of z but only to determine the behavior in one half of the slab. The first step in solving Equation 3.21 is to divide the dimensionless slab width into N equal sections. Taking 20 sections:

$$h = \frac{Interval}{N} = \frac{0.5 - 0}{20} = 0.025 \tag{A.97}$$

This procedure is represented in Scheme A.1, where h represents Δz. Lower values of h (a higher number of sections) reduce the error in the predicted result. However, values of h below 10^{-6} are not adequate for most personal computers, which have precision levels of around 10^{-14}. Note that the division of the width of the dimensionless slab into 20 sections generates 21 nodes (β_i).

The next step is to set the finite difference for each node by substituting the second derivative with the expression given in Table A.1. Using the formula for centered differences, according to the boundary condition, for node zero:

$$\beta_0 = \beta(0) \tag{A.98}$$

For node one:

$$\frac{\beta_2 - 2 \cdot \beta_1 + \beta_0}{h^2} - \Phi^2 \cdot \frac{\beta_1}{1+\beta_1} = 0 \tag{A.99}$$

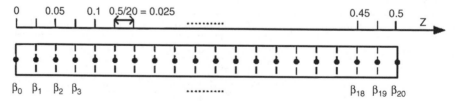

Scheme A.1 *Finite-difference approach for a catalytic slab.*

For any node:

$$\frac{\beta_{i+1} - 2 \cdot \beta_i + \beta_{i-1}}{h^2} - \Phi^2 \cdot \frac{\beta_i}{1 + \beta_i} = 0 \qquad \text{(A.100)}$$

For node 20, the symmetry condition implies:

$$\frac{d\beta(0.5)}{dz} = 0 = \frac{\beta_{21} - \beta_{19}}{2h} \qquad \text{(A.101)}$$

$$\beta_{21} = \beta_{19} \qquad \text{(A.102)}$$

So:

$$\frac{2 \cdot \beta_{19} - 2 \cdot \beta_{20}}{h^2} - \Phi^2 \cdot \frac{\beta_{20}}{1 + \beta_{20}} = 0 \qquad \text{(A.103)}$$

Reordering, the following set of nonlinear equations is obtained:

$$\beta_0 - \beta(0) = 0 \qquad \text{(A.104)}$$

For nodes 1–19:

$$\beta_{i+1} + \beta_{i+1} \cdot \beta_i - 2 \cdot \beta_i + \beta_{i-1} - 2 \cdot \beta_i^2 + \beta_i \cdot \beta_{i-1} - \Phi^2 \cdot \beta_i \cdot h^2 = 0 \qquad \text{(A.105)}$$

For node 20:

$$2 \cdot \beta_{19} + 2 \cdot \beta_{19} \cdot \beta_{20} - 2 \cdot \beta_{20} + 2 \cdot \beta_{20}^2 - \Phi^2 \cdot \beta_{20} \cdot h^2 = 0 \qquad \text{(A.106)}$$

The former system of nonlinear equations can be solved in an iterative manner using Newton's method. For a system of equations, Equation A.5 becomes:

$$J(X) \cdot \Delta X = -F(X) \qquad \text{(A.107)}$$

$$X' = X + \Delta X \qquad \text{(A.108)}$$

where:

$$F(X) = \begin{bmatrix} f_1(X) \\ \cdot \\ \cdot \\ \cdot \\ f_n(X) \end{bmatrix} \qquad \text{(A.109)}$$

$$J(X) = \begin{bmatrix} \dfrac{\partial(f_1(X))}{\partial x_1} & & & \dfrac{\partial(f_1(X))}{\partial x_n} \\[2mm] & \cdot & \cdot & \cdot \\ \cdot & & \cdot & \cdot \\ \cdot & & & \cdot \\[2mm] \dfrac{\partial(f_n(X))}{\partial x_1} & \cdot & & \dfrac{\partial(f_n(X))}{\partial x_n} \end{bmatrix} \tag{A.110}$$

$$X = \begin{bmatrix} x_1 \\ \cdot \\ \cdot \\ \cdot \\ x_n \end{bmatrix} \tag{A.111}$$

$$X' = \begin{bmatrix} x'_1 \\ , \\ \cdot \\ \cdot \\ x'_n \end{bmatrix} \tag{A.112}$$

where: $f_i(X)$ is equation i set to zero, as indicated in Equation A.2; J is the Jacobian matrix, which can be calculated numerically, as indicated in Section A.2.2; x_i are the initial value for X; x'_i are the new values X'.

Observe that Equation A.107 represents a linear system of equations, which can be solved easily using the Gauss or Gauss–Jordan method [1]. Just as in the case for a single variable (Section A.1), a nonlinear problem is linearized to solve it in an iterative way. The iteration is stopped when the maximum relative error E is lower than a predefined value δ; for example, a value between 10^{-8} and 10^{-5}. For highly nonlinear systems, very small values of δ are not recommended as sometimes such levels of error simply cannot be achieved and the criterion is not satisfied. This is especially important when a self-made computing routine is employed, because the implementation of an alternative criterion for stopping iteration is not generally considered.

$$E = \left| \frac{x'_i - x_i}{x'_i} \right| \tag{A.113}$$

$$max(E) < \delta \tag{A.114}$$

As in most numerical methods, the convergence of Newton's method is largely dependent on the initial value of the variables (X); appropriate initial values with which to start iteration can be obtained from Equation 3.37, which describes the substrate profile inside the slab for first-order kinetics. The following table shows the numerical solution of Equation 3.21 for $\beta_0 = 10$, $\Phi = 4$, and $\delta = 10^{-4}$. The rapid convergence of Newton's method can be appreciated, with only four iterations required to achieve an E lower than 0.0001.

Node	i	0	2[a]	4	6	8	10	12	14	16	18	20	max(δ)·100
Iteration	Z	0	0.05	0.1	0.15	0.2	0.25	0.3	0.35	0.4	0.45	0.5	
Initial values	β	10	8.2597	6.851	5.7171	4.8128	4.102	3.5549	3.1510	2.8735	2.7114	2.6580	
1	β	10	9.6212	9.279	8.9727	8.7037	8.472	8.2800	8.1280	8.0178	7.9510	7.9286	198
2	β	10	9.6591	9.354	9.0858	8.8533	8.657	8.4960	8.3710	8.2818	8.2283	8.2105	3.55
3	β	10	9.6591	9.354	9.0858	8.8533	8.657	8.4961	8.3711	8.282	8.2285	8.2106	0.001
4	β	10	9.6591	9.354	9.0858	8.8533	8.657	8.4961	8.3711	8.282	8.2285	8.2106	3.50·10⁻¹⁰

[a] The odd positions have been omitted to save space

Figure A.6 shows the substrate profile for different values of β_0 and Φ. It can be seen that at values of Φ lower than 0.1 the substrate concentration inside the catalyst is almost the same as that in the bulk, regardless of reaction kinetics. Equation 3.37 is often used to describe the substrate profile for a broad range of β values. However, as can be appreciated from Figure A.6b and c, this is not a realistic assumption for β greater than 1. Despite this, as Φ increases through values above 10, these differences are minimized as the effect of diffusional restrictions becomes predominant.

In the case of the spherical particle, the procedure is analogous. The substrate profiles obtained are presented in Figure A.7.

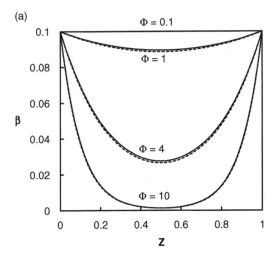

Figure A.6 *Substrate profile within a catalytic slab. (a) $\beta_0 = 0.1$. (b) $\beta_0 = 1$. (c) $\beta_0 = 10$. Solid lines: numerical solution of Equation 3.21; dashed lines: analytical solution for first-order kinetics (Equation 3.37).*

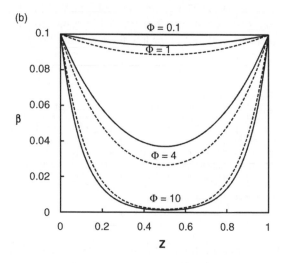

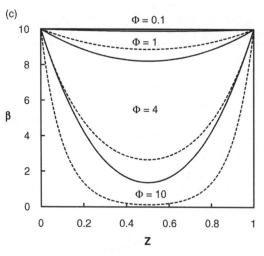

Figure A.6 (*Continued*)

(a)

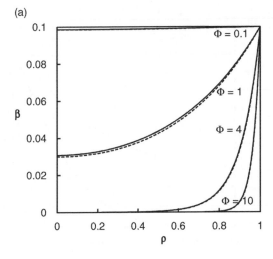

(b)

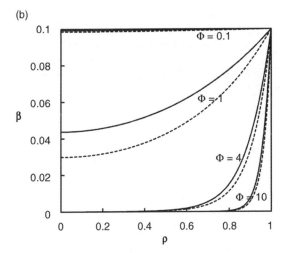

Figure A.7 *Substrate profile within a catalytic sphere. (a) $\beta_0 = 0.1$. (b) $\beta_0 = 1$. (c) $\beta_0 = 10$. Solid lines: numerical solution of Equation 3.22; dashed lines: analytical solution for first-order kinetics (Equation 3.39).*

(c)

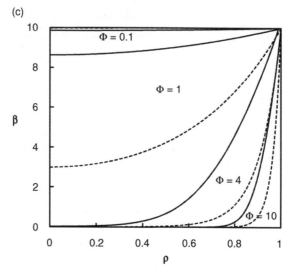

Figure A.7 *(Continued)*

References

1. Chapra, S.C. and Canale, R.P. (2010) *Numerical Methods for Engineers*, 6th edn, McGraw Hill, New York.
2. Montgomery, D.C. and Runger, G.C. (2010) *Applied Statistics and Probability for Engineers*, John Willey & Sons, New York.

Index

acrylamide   4, 63, 64, 66, 89, 175, 233
acrylonitrile   63, 66, 175, 233
activator   159, 160
active conformation   203
active site   1, 15, 17, 18, 25, 34, 55, 159
acyl donor   138, 172, 230
adsorption   89, 107, 245
agitation   113, 245
agitation rate   104, 113, 117, 130, 131, 182
alkaline phosphatase   79
alternative hypothesis   251, 252
ammonia   35, 81, 200, 237
α-amylase   3, 4, 142, 190, 199
α-galactosidase   74, 75, 137, 158, 175, 219
α-glucosidase   2, 76
amino acid   1, 15, 25, 88, 92, 200
amino acid ester hydrolase   51, 178
aminoacylase   4, 92
7-amino-3-deacetoxy cephalosporanic
    acid   47, 51, 174, 178, 230
6-amino penicillanic acid   80, 136, 161, 174,
    178, 289
amoxicillin   6, 80, 178
animal feed   3, 4, 7
annual cost   245
antibiotic   4, 5, 6, 80, 136, 137, 172, 178, 230
apparent density   135, 176, 177, 191, 200, 224,
    263, 264
apparent Michaelis constant   94
apparent kinetic parameters   18, 21, 26, 35, 54,
    78, 104, 115, 128, 135
aqueous media   5, 6, 17, 25, 141
arabinose isomerase   133
Arrhenius equation   27, 67, 82, 108, 185,
    259, 263
arylesterase   49
ascorbic acid   53, 54, 55, 238, 239
ascorbyl palmitate   238

asparaginase   2, 200
asparagine   200
aspartame   6, 169, 170
aspartic acid   4, 169, 172, 200
auto-aggregation   2

backmixing   142, 157
bakers' yeast   257
basket-type reactor   183
batch process   148, 213
batch reactor   41, 142, 147, 150, 158, 171, 172,
    174, 183, 206, 244
batch recirculating reactor   125, 126
bed volume   150, 162, 191, 224
Beer-Lambert   32
β-amylase   4, 83, 201, 234
β-fructofuranosidase   5, 36, 38, 119, 128, 135,
    163, 257
β-galactosidase   3, 4, 29, 30, 105, 114, 122,
    135, 166, 167, 174, 201, 206, 215, 224, 247,
    254, 256, 258, 263, 267, 272
β-glucosidase   3, 114, 236
β-lactam antibiotics   4, 6, 80, 136, 137, 172
β-lactamase   113
bifunctional reagent   89
bioethanol   3, 4, 7
biodiesel   6
biofuel   3, 7
biomedicine   183
bioremediation   2
biosensor   134, 183
Biot number   101, 119, 123, 193, 196
boundary condition   97, 98, 100–102, 120,
    121, 193, 195, 299–304
Box-Benhken   250, 274
Briggs-Haldane   15, 16
BSTR   44, 145, 146, 151, 157, 160, 170, 172,
    174, 183, 184, 197, 198, 207, 212, 213, 215,

Problem Solving in Enzyme Biocatalysis, First Edition. Andrés Illanes, Lorena Wilson and Carlos Vera.
© 2014 John Wiley & Sons, Ltd. Published 2014 by John Wiley & Sons, Ltd. Companion Website:
http://www.wiley.com/go/illanes-problem-solving